SINGLE-MODE FIBER OPTICS

OPTICAL ENGINEERING

Series Editor

Brian J. Thompson
Provost
University of Rochester
Rochester, New York

Laser Engineering Editor:
Peter K. Cheo
United Technologies Research Center
Hartford, Connecticut

Laser Advances Editor:
Leon J. Radziemski
Associate Dean, College of Arts and Sciences
New Mexico State University
Las Cruces, New Mexico

Optical Materials Editor:
Solomon Musikant
Paoli, Pennsylvania

1. Electron and Ion Microscopy and Microanalysis: Principles and Applications, *by Lawrence E. Murr*
2. Acousto-Optic Signal Processing: Theory and Implementation, *edited by Norman J. Berg and John N. Lee*
3. Electro-Optic and Acousto-Optic Scanning and Deflection, *by Milton Gottlieb, Clive L. M. Ireland, and John Martin Ley*
4. Single-Mode Fiber Optics: Principles and Applications, *by Luc B. Jeunhomme*
5. Pulse Code Formats for Fiber Optical Data Communication: Basic Principles and Applications, *by David J. Morris*
6. Optical Materials: An Introduction to Selection and Application, *by Solomon Musikant*
7. Infrared Methods for Gaseous Measurements: Theory and Practice, *edited by Joda Wormhoudt*
8. Laser Beam Scanning: Opto-Mechanical Devices, Systems, and Data Storage Optics, *edited by Gerald F. Marshall*
9. Opto-Mechanical Systems Design, *by Paul R. Yoder, Jr.*

10. Optical Fiber Splices and Connectors: Theory and Methods, *by Calvin M. Miller with Stephen C. Mettler and Ian A. White*
11. Laser Spectroscopy and Its Applications, *edited by Leon J. Radziemski, Richard W. Solarz, and Jeffrey A. Paisner*
12. Infrared Optoelectronics: Devices and Applications, *by William Nunley and J. Scott Bechtel*
13. Integrated Optical Circuits and Components: Design and Applications, *edited by Lynn D. Hutcheson*
14. Handbook of Molecular Lasers, *edited by Peter K. Cheo*
15. Handbook of Optical Fibers and Cables, *by Hiroshi Murata*
16. Acousto-Optics, *by Adrian Korpel*
17. Procedures in Applied Optics, *by John Strong*
18. Handbook of Solid-State Lasers, *edited by Peter K. Cheo*
19. Optical Computing: Digital and Symbolic, *edited by Raymond Arrathoon*
20. Laser Applications in Physical Chemistry, *edited by D. K. Evans*
21. Laser-Induced Plasmas: Physical, Chemical, and Biological Applications, *edited by Leon J. Radziemski and David A. Cremers*
22. Infrared Technology Fundamentals, *by Irving J. Spiro and Monroe Schlessinger*
23. Single-Mode Fiber Optics: Principles and Applications, Second Edition, Revised and Expanded, *by Luc B. Jeunhomme*

LASER HANDBOOKS—*Edited by Peter K. Cheo*

Handbook of Molecular Lasers

Handbook of Solid-State Lasers

Other Volumes in Preparation

Photoconductivity: Art, Science, and Technology, *by N. V. Joshi*

Image Analysis Applications, *edited by Rangachar Kasturi and Mohan M. Trivedi*

SINGLE-MODE FIBER OPTICS

PRINCIPLES AND APPLICATIONS

Second Edition, Revised and Expanded

Luc B. Jeunhomme
Photonetics
Marly le Roi, France

MARCEL DEKKER, INC. New York and Basel

Library of Congress Cataloging-in-Publication Data

Jeunhomme, Luc B.
Single-mode fiber optics : principles and applications / Luc B. Jeunhomme ; photonetics, Marly le Roi. -- 2nd ed., rev. and expanded.
p. cm -- (Optical engineering ; 4)
Includes bibliographical references and index.
ISBN 0-8247-8170-8 (alk. paper)
1. Fiber optics. I. Title. II. Series: Optical engineering (Marcel Dekker, Inc.) ; v. 4.
TA1800.J48 1989
621.36'92--dc20 89-23417
CIP

This book is printed on acid-free paper.

MARCEL DEKKER, INC.
270 Madison Avenue, New York, New York 10016

Current printing (last digit):
10 9 8 7 6 5 4 3 2 1

PRINTED IN THE UNITED STATES OF AMERICA

About the Series

Optical science, engineering, and technology have grown rapidly in the last decade so that today optical engineering has emerged as an important discipline in its own right. This series is devoted to discussing topics in optical engineering at a level that will be useful to those working in the field or attempting to design systems that are based on optical techniques or that have significant optical subsystems. The philosophy is not to provide detailed monographs on narrow subject areas but to deal with the material at a level that makes it immediately useful to the practicing scientist and engineer. These are not research monographs, although we expect that workers in optical research will find them extremely valuable.

Volumes in this series cover those topics that have been a part of the rapid expansion of optical engineering. The developments that have led to this expansion include the laser and its many commercial and industrial applications, the new optical materials, gradient index optics, electro- and acousto-optics, fiber optics and communications, optical computing and pattern recognition, optical data reading, recording and storage, biomedical instrumentation, industrial robotics, integrated optics, infrared and ultraviolet systems, etc. Since the optical industry is currently one of the major growth industries this list will surely become even more extensive.

Brian J. Thompson
University of Rochester
Rochester, New York

Preface

When the first edition of this book was published in mid-1983, single-mode fibers were just about to take a major share of the explosive growth of optical fiber based communication systems in the United States. Up to 1982, single-mode fibers had received increasing attention from many research laboratories around the world. Research studies and early experimental systems in the telecommunication area had shown that single-mode fibers were almost ready to serve increasing needs in long-distance communications. Then in 1982 deregulation of the U.S. telecommunication services that happened then opened up a large opportunity for installing new long-distance communication systems. Single-mode fibers thus became dominant as compared with multimode—the North-American market for single-mode fibers grew from about 10,000 km in 1982 to 100,000 km in 1983 and 1,200,000 km in 1985, while remaining at an almost constant level of 250,000 km per year for multimode fibers.

The first edition of this book was intended to provide to the scientific and technical community a basic text presenting in a single place the scientific and engineering principles of single-mode fibers. This second edition combines the scientific background of the first edition with an update on the technological developments spurred by the strong development of the single-mode fiber industry since 1983. Very few new developments in the theoretical analysis of single-mode fibers have occurred since the first edition; thus the main additions to the initial text concern practical developments such as high-birefringence fibers (Chap. 2), improvements in fiber attenuation control (Chap. 3), dispersion-shifted fibers (Chap. 4), standard measurement procedures (Chap. 5), connectors and splicing equipment (Chap. 6), long-distance terrestrial and undersea communication systems (Chap. 7), coherent transmission systems (Chap. 7), and the fiber optic gyro (Chap. 8).

The first chapter provides a simple analytical description of the properties and characteristics of single-mode fibers, most of them being approximated under a form workable with pocket calculators. The introduction to this chapter provides a simplified description of optical waveguides for the unspecialized reader, so this chapter is almost self-contained. Chapters 2 to 4 establish the light transmission characteristics of single-mode fibers, and Chapters 5 to 7 describe specific characterization methods, passive hardware, and telecommunication applications. In all these chapters, the specific features of single-mode fibers are extensively discussed, while problems common to multimode and single-mode fibers (e.g., description manufacturing techniques, terrestrial cables, etc.) are discussed only briefly, a complete discussion of these topics being found in previously published books. Chapter 8 is devoted to a description of single-mode-fiber sensors, which are finding important applications and attracting growing interest. Finally, Chapter 9 deals with the most commonly encountered nonlinear optical effects, which may find future practical applications in various broadband optical sources and also limit the power that can be handled by single-mode fibers.

Practicing engineers and graduate students should find in the theoretical chapters an accessible introduction to optical waveguides and all the necessary theoretical material on single-mode fibers, directly usable in the form of figures, diagrams, and approximations workable with pocket calculators, together with an underlying description of the physical phenomena. In the applications chapters, engineers involved in fiber systems design should find a convenient discussion of the specific hardware and applications associated with single-mode fibers. Some state-of-the-art experimental applications are analyzed, but the applications chapters are intended to provide a stable reference for extrapolating to future developments, because of the very quickly evolving applications. The emphasis is thus on basic features that will always be present in future applications, rather than on a detailed description of today's experimental systems.

Luc B. Jeunhomme

Contents

SINGLE-MODE FIBER OPTICS

1

Basic Theory

1.1 INTRODUCTION

This chapter is devoted to an extended view of the theory of light propagation in single-mode fibers. However, before treating the case of the circular core single-mode fiber, for which the physical meaning of the results is somewhat obscured by the mathematics involved, we discuss in this section the general guiding properties of the dielectric slab waveguide. This case involves much less complicated equations and allows a better physical discussion of the main features of the modes, which are qualitatively similar to those of the circular core fiber. This section can thus be seen as an introduction to the electromagnetic formalism used in the remainder of the chapter, for readers not familiar with the modes of dielectric circular waveguides.

Readers already familiar with the theory of dielectric waveguides may jump directly to Sec. 1.2, which is devoted to a detailed study of the first two modes of ideal step-index fibers. In Sec. 1.3 graded-index single-mode fibers are treated in terms of equivalent step-index fibers after exact theories have been reviewed. Finally, fibers with cladding refractive index variations are studied in Sec. 1.4. This chapter thus provides all the necessary material for the study of the transmission characteristics of various single-mode fibers in subsequent chapters.

1.1.1 Modes of the Symmetric Dielectric Slab Waveguide

A typical example of a symmetric dielectric slab waveguide is shown in Fig. 1.1. It consists of a sheet of material with refractive index

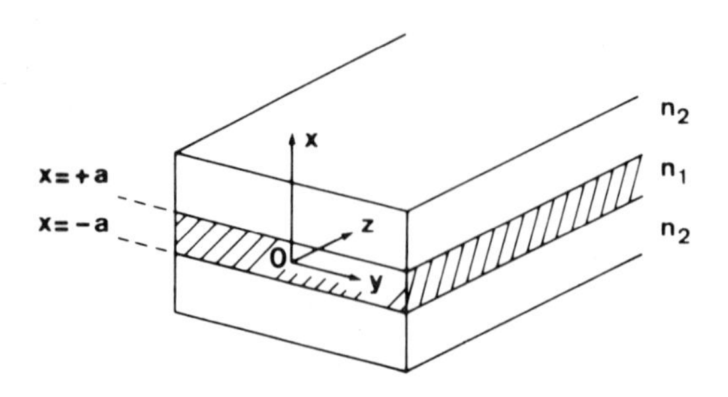

FIGURE 1.1 Example of a dielectric slab waveguide consisting of a sheet of material with refractive index n_1 and thickness 2a between two layers of material with refractive index n_2 lower than n_1.

n_1 and thickness 2a, sandwiched between two infinite layers of material with refractive index n_2 lower than n_1. In the following we assume that the structure is unlimited in the y and z directions, and that the materials are isotropic and without loss (permittivities are real and scalar). We write the electric (resp. magnetic) field of a purely monochromatic wave propagating in the z direction as

$$\bar{E}\ (\text{resp. } \bar{H})(x, y, z) = \bar{E}\ (\text{resp. } \bar{H})(x) \exp[i(\omega t - \beta z)] \quad (1.1)$$

where ω is the optical pulsation, β the propagation constant of the wave, and (x, y, z) the usual cartesian coordinates shown in Fig. 1.1. There is no y dependence of the fields on the right-hand side of Eq. (1.1) because of the assumed infinite extent along Oy.

The components of the electric and magnetic field vectors (E_x, E_y, E_z, . . .) should obey the well-known Maxwell equations, which are written here for the absence of charges or currents:

$$\beta E_y = -\omega\mu_0 H_x \quad (1.2a)$$

$$i\beta E_x + \frac{\partial E_z}{\partial x} = i\omega\mu_0 H_y \quad (1.2b)$$

$$\frac{\partial E_y}{\partial x} = -i\omega\mu_0 H_z \quad (1.2c)$$

$$\beta H_y = \omega\varepsilon_0 n_j^2 E_x \quad (1.2d)$$

$$i\beta H_x + \frac{\partial H_z}{\partial x} = -i\omega\varepsilon_0 n_j^2 E_y \tag{1.2e}$$

$$\frac{\partial H_y}{\partial x} = i\omega\varepsilon_0 n_j^2 E_z \tag{1.2f}$$

where μ_0 is the vacuum permeability, ε_0 is the vacuum dielectric constant, and n_j is equal to either n_1 or n_2 depending on the layer of the slab where the fields are considered. Looking in detail at Eqs. (1.2) reveals that we may obtain two self-consistent types of solutions. The first involves only E_y, H_x, and H_z and is called a transverse electric (TE) solution as the electric field is entirely contained in the plane transverse to the propagation direction ($E_z = 0$). Similarly, the second type of solution involves only H_y, E_x, and E_z and is referred to as a transverse magnetic (TM) solution. For each case, Maxwell's equations (1.2) reduce to:

TE: (1.2a) + (1.2c) and

$$\frac{\partial^2 E_y}{\partial x^2} = -(k^2 n_j^2 - \beta^2)E_y \tag{1.3}$$

TM: (1.2d) + (1.2f) and

$$\frac{\partial^2 H_y}{\partial x^2} = -(k^2 n_j^2 - \beta^2)H_y \tag{1.4}$$

where $k = \omega/c = 2\pi/\lambda$ is the plane-wave propagation constant in vacuum (c and λ are the light velocity and light wavelength in vacuum). It is thus apparent that the field variations along 0x will always exhibit sinusoidal behavior where $k^2 n_j^2$ is greater than β^2 (oscillating field), and exponential behavior where $k^2 n_j^2$ is smaller than β^2 (evanescent field). More generally, even in waveguides where the index profile is much more complicated, *the electromagnetic field is oscillating in regions where the longitudinal propagation constant is smaller than the plane-wave propagation constant in this region, and evanescent with an exponential-like behavior elsewhere.* This very important feature should be committed to memory, as it is the basis for the notion of mode cutoff and bending losses.

It can be deduced that modes having a physical significance (carrying a nonzero power) can exist only with propagation constants smaller than kn_1 (as otherwise they are evanescent everywhere). On the other hand, propagation constants smaller than kn_2 correspond to fields oscillating everywhere and thus nonvanishing at $x \to \pm\infty$. These are radiating modes (as they radiate their power laterally) which are not guided by the structure. Guided modes thus have propagation constants greater than kn_2 and smaller than kn_1:

$$kn_2 \leq \beta \leq kn_1 \tag{1.5}$$

In the following discussion we use a transverse propagation constant u/a and a transverse decay constant v/a defined as

$$\beta^2 = k^2 n_1^2 - \frac{u^2}{a^2} = k^2 n_2^2 + \frac{v^2}{a^2} \tag{1.6}$$

(u and v are chosen > 0 without loss of generality because of the symmetry of the guiding structure about the plane y0z).

From Eq. (1.6) we may define a dimensionless parameter V called the normalized frequency, as it is proportional to the light frequency

$$V^2 = u^2 + v^2 = a^2k^2(n_1^2 - n_2^2) \tag{1.7}$$

This normalized frequency depends only on the characteristics of the guide and the light wavelength (or frequency).

From Eqs. (1.3)–(1.7), it appears that the field (E_y for TE modes and H_y for TM modes) is a linear combination of cos (ux/a) and sin (ux/a) inside the central layer ($|x| \leq a$), and takes the exponentially decaying form outside: $\exp(-vx/a)$ for $x \geq a$ and $\exp(vx/a)$ for $x \leq -a$. In the following, we concentrate only on TE modes, as the characteristics of TM modes are very similar from the mathematical point of view.

We now have to take into account the continuity of the field components tangential at the dielectric discontinuities located at $x = \pm a$. On these surfaces E_y, H_y, E_z, and H_z will be continuous, and when this condition is combined with the general solutions noted above and Eqs. (1.2a), (1.2c), and (1.3) we obtain two kinds of solutions:

Even TE modes:

$$\left.\begin{aligned} E_y, H_x &\sim \cos\left(\frac{ux}{a}\right) \\ H_z &\sim \sin\left(\frac{ux}{a}\right) \end{aligned}\right\} \quad |x| \leq a \tag{1.8a}$$

$$E_y, H_x, H_z \sim \exp\left(-\frac{v|x|}{a}\right) \qquad |x| \geq a \tag{1.8b}$$

$$v = u \tan u \tag{1.8c}$$

Odd TE modes:

$$\left.\begin{aligned} E_y, H_x &\sim \sin\left(\frac{ux}{a}\right) \\ H_z &\sim \cos\left(\frac{ux}{a}\right) \end{aligned}\right\} \qquad |x| \leq a \tag{1.9a}$$

$$E_y, H_x, H_z \sim \exp\left(-\frac{v|x|}{a}\right) \qquad |x| \geq a \tag{1.9b}$$

$$v = -\frac{u}{\tan u} \tag{1.9c}$$

At this point, comparison of Eq. (1.8c) or (1.9c) with Eq. (1.7) shows that the guiding structure can support only discrete modes (i.e., u, v, and thus β can take only discrete values) whose characteristics can be found by a graphical method illustrated in Fig. 1.2. The solutions of u and v for the various modes are found from the intersections of circles of radius V [corresponding to Eq. (1.7) in cartesian coordinates] and of the curves representative of Eqs. (1.8c) and (1.9c).

1.1.2 Analysis of Cutoff Properties

If we start from a zero optical frequency (and thus $V = 0$) or an infinite wavelength, we observe that we always have at least one guided mode, the even TE_0 mode, which is thus the lowest-order or fundamental mode of the structure (in fact, we simultaneously have an even TM_0 mode). As long as $V < \pi/2$, Fig. 1.2 shows that the mode remains the only guided mode, with increasing u and v values. When V reaches $\pi/2$, the odd TE_0 mode begins to be guided and thus appears as the second-order mode. A third mode (the even TE_1 mode) appears at $V = \pi$, and so on. Each time V reaches a multiple integer of $\pi/2$, a new mode reaches its cutoff, which is seen to correspond to $v = 0$ and thus $\beta = kn_2$ for this mode. A mode at cutoff remains constant in the outer medium instead of decreasing exponentially.

1.1.3 Propagation Constant and Dispersion

Let us concentrate on the evolution of the even TE_0 mode propagation constant with optical frequency. At zero optical frequency, it

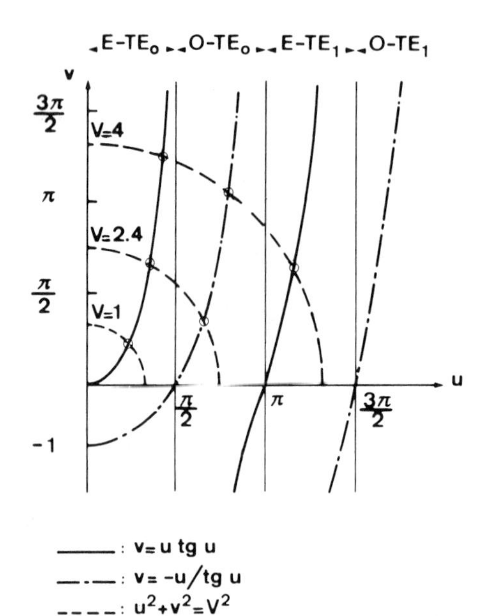

FIGURE 1.2 Graphical solution to Eqs. (1.7), (1.8c), and (1.9c) for determining the parameters u and v of TE modes in the dielectric slab waveguide of Fig. 1.1. The small circles represent various solutions for the corresponding value of the normalized frequency V.

is clear from Fig. 1.2 that we have $\beta = kn_2$. In the following we assume that there is no material dispersion, that is, that n_1 and n_2 are independent of the optical frequency. When the optical frequency increases, V increases proportionally and Fig. 1.2 shows that u tends toward an asymptotic value of $\pi/2$. Compared with Eq. (1.6), this shows that β tends toward kn_1. We can thus draw a diagram $\beta(\omega)$ as shown in Fig. 1.3.

Now, in many applications, the waveguide is not illuminated with a purely monochromatic light but rather with a time-dependent signal whose field can be written in the form of a Fourier expansion:

$$E_i(t) = \int_{-\infty}^{+\infty} e(\omega)\, \exp(i\omega t)\, d\omega$$

After propagation through the waveguide of length L, each component will have its phase changed by a quantity $\beta(\omega)L$, and if we assume that the spectral width of the signal is small compared

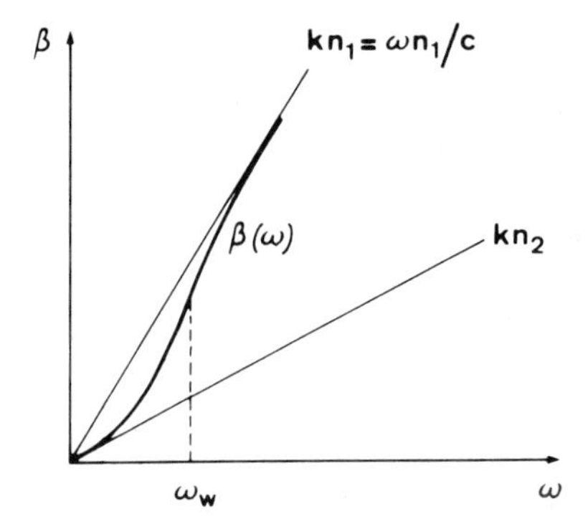

FIGURE 1.3 Qualitative display of the dispersion curve $\beta(\omega)$ in a dielectric waveguide with two media of refractive index n_1 and n_2 (assumed to be independent of the optical frequency $\omega/2\pi$). ω_W represents the value of ω, where $d^2\beta/d\omega^2 = 0$, and thus corresponds to the frequency of zero waveguide dispersion.

to the central optical carrier frequency $\omega_0/2\pi$, we can write the output field as

$$E_0(t) = \int_{-\infty}^{+\infty} e(\omega) \exp\left\{ i \left[\omega t - \beta_0 L - (\omega - \omega_0)\beta_0{}'L - \frac{(\omega - \omega_0)^2}{2} \beta_0{}''L \right] \right\} d\omega$$

where

$$\beta_0 = \beta(\omega_0)$$

$$\beta_0{}' = \frac{d\beta}{d\omega} \qquad \text{at } \omega = \omega_0$$

$$\beta_0{}'' = \frac{d^2\beta}{d\omega^2} \qquad \text{at } \omega = \omega_0$$

The equation above shows that the modulated signal has been delayed by a time per unit length equal to β_0' (group delay time) and that its shape will be affected by an amount controlled by the magnitude of $-\beta_0''$ which is called the chromatic dispersion.

As we assumed here that the refractive indices are independent of the light frequency (no material dispersion), we are left with the curvature of the curve $\beta(\omega)$ due to the waveguide itself. This contribution is called the waveguide dispersion and can be evaluated from Fig. 1.3. The waveguide dispersion is zero at zero optical frequency, becomes negative when the frequency increases up to $\omega_W/2\pi$, where the waveguide dispersion cancels and changes its sign, and tends toward zero again at infinite optical frequencies. It should be remembered that in practical situations, a contribution of material dispersion (variation of n with λ) has to be taken into account.

1.1.4 Field Shape

We will again concentrate almost entirely on the fundamental even TE_0 and even TM_0 modes. Equations (1.8) and Fig. 1.2 show that at zero optical frequency ($V = 0$), the field is uniformly constant throughout the structure. In this case the central layer with refractive index n_1 is so small (compared to the infinite optical wavelength and to the infinite field extent) that it is not unexpected that the propagation constant is equal to kn_2. Similarly, at infinite optical frequency, u tends toward $\pi/2$ and v toward infinity, as shown by Fig. 1.2, and we thus deduce from Eqs. (1.8) that the field is entirely contained inside the central layer, explaining why the propagation constant tends toward kn_1, as shown by Fig. 1.3. At intermediate frequencies, the field is distributed throughout the entire structure and becomes more and more tightly confined in the central layer as the frequency increases.

Practically, at a given optical frequency, a wave incident at the input of such a waveguide will excite the various TE and TM modes (together with radiation modes). Only TE (resp. TM) modes will be excited if the wave is polarized accordingly ($E_x = 0$ for TE or $E_y = 0$ for TM), and only even (resp. odd) modes will be excited if the incident field exhibits the same symmetry properties.

1.1.5 Summary

The example of the symmetric dielectric slab waveguide allowed us to illustrate simply some important features of the electromagnetic field in dielectric waveguides, which we summarize briefly here.

The guided electromagnetic field takes the form of *discrete modes* with propagation constants bounded by the plane-wave propagation constants in the guiding and outer regions. This quantification arises from a transverse resonance condition imposed by the dielectric discontinuities. The field follows an oscillating function inside the guiding region and an exponential-like decay in the outer region. A mode is said to be at cutoff when the optical frequency is such that its propagation constant becomes equal to the plane-wave propagation constant in the outer medium. We generally get an optical frequency range between zero and some second-order mode cutoff frequency where the guide supports a single mode with two possible states of polarization (even TE_0 and even TM_0, here). Contrary to (closed) metallic waveguides, dielectric waveguides (open) let some part of the guided power flow into the cladding, and this specific feature opens up the possibilities of coupling between guides through evanescent waves, which is the basic principle of most "integrated optics" components.

In circular core fibers, the general behavior of the electromagnetic field and of the modes is qualitatively similar to that described above. However, the presence of a dielectric discontinuity on a surface involving both the x and y variables has two consequences:

1. The mode's label will contain two indices instead of one, the first index being related to the "radial propagation constant" and the second index to the azimuthal periodicity of the field (in a fiber with a rectangular core, the first index would be related to the transverse propagation constant along 0x, while the second would hold for the transverse propagation constant along 0y).

2. It is no longer possible to assume that there is no variation of fields along 0y, and we will thus find that some eigenmodes (especially the fundamental one) are not purely transversely polarized (like TE or TM), but rather have a small longitudinal component for both the electric and magnetic fields. These modes are called HE and EH modes.

1.2 IDEAL STEP-INDEX FIBER

The structure of the ideal step-index fiber is shown in Fig. 1.4 together with the classical cylindrical coordinates that will be used hereafter. The index of refraction n(r) at a radial distance r is given by

$$
\begin{aligned}
n(r) &= n_1, \qquad 0 \leq r \leq a \\
n(r) &= n_2, \qquad r > a
\end{aligned}
\tag{1.10}
$$

where n_1 (resp. n_2) is the core (resp. cladding) refractive index ($n_2 < n_1$), a is the core radius, and the cladding is assumed to

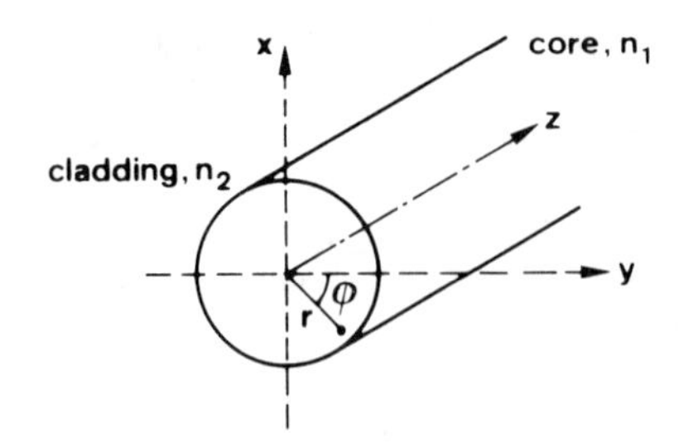

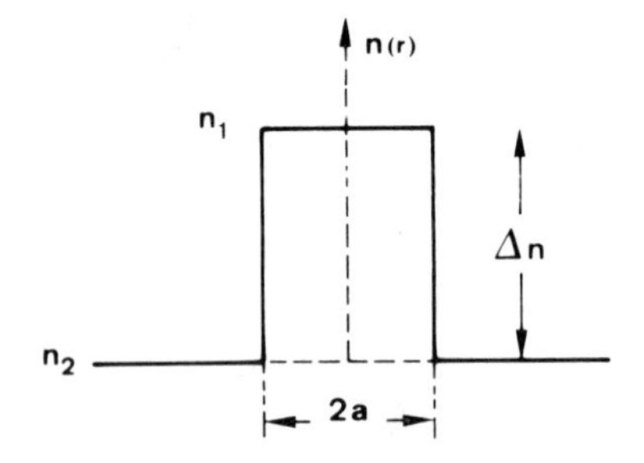

FIGURE 1.4 Coordinate system, structure, and refractive index profile of the ideal step-index fiber.

extend to infinity. The index difference Δn and the relative index difference Δ are given by

$$\Delta n = n_1 - n_2, \qquad \Delta = \frac{\Delta n}{n_2} << 1 \tag{1.11}$$

In practice, Δ is smaller than 1%.

1.2.1 Exact Field Expressions

Complete theoretical treatments are found in Refs. 1 and 2, and we will recall here the important results for the low-order modes of ideal step-index fibers. Most of them are obtained through the use of the scalar wave approximation, which is justified by the fact that $\Delta < 1\%$, as it has been shown that in this limit the error imposed on all mode characteristics by this approximation remains below 0.1% [3]. Let us write the electric (resp. magnetic) field as

$$\bar{E}\ (\text{resp. } \bar{H})(r, \varphi, z, t) = \bar{E}\ (\text{resp. } \bar{H})(r, \varphi) \exp[i(\omega t - \beta z)] \tag{1.12}$$

where ω is the optical pulsation, β the propagation constant of the wave, and (r, φ, z) the usual cylindrical coordinates shown in Fig. 1.4.

The scalar wave approximation allows to transform Maxwell's equations into the scalar wave equation governing the longitudinal field components, from which the other field components can be deduced through the standard Maxwell's equations

$$\left[\frac{\partial^2}{\partial r^2} + \frac{1}{r}\frac{\partial}{\partial r} + \frac{1}{r^2}\frac{\partial^2}{\partial \varphi^2} + (k^2 n_j^2 - \beta^2)\right]\begin{bmatrix} E_z \\ H_z \end{bmatrix} = 0 \qquad (1.13)$$

where E_z and H_z are the electric and magnetic field components along 0z and n_j is equal to n_1 in the core and n_2 in the cladding.

Apart from the use of this approximation, the formal mathematical treatment is equivalent to that described for the dielectric slab waveguide in Sec. 1.1.1 and will not be discussed here. (This approximation was not necessary in the dielectric slab waveguide because the structure is uniform along 0y, reducing the transverse field vector to a scalar as a direct solution of the scalar wave equation.)

Classically, we define the normalized frequency V as

$$V = ak(n_1^2 - n_2^2)^{\frac{1}{2}} \simeq ak(2n_2\Delta n)^{\frac{1}{2}} = akn_2(2\Delta)^{\frac{1}{2}} \qquad (1.14)$$

where $k = 2\pi/\lambda$ is the free-space propagation constant of the light at wavelength λ.

Fundamental Mode. As long as we have

$$0 \leq V < 2.405 \qquad (1.15)$$

only the two HE_{11} modes can propagate (one polarized along 0x and one polarized along 0y, for the transverse electric field). As we consider in this chapter perfectly circular core fibers, these two modes are perfectly degenerated (they have the same propagation constant), and Eq. (1.15) thus defines the so-called "single-mode" regime. Condition (1.15) can be understood physically to mean that the maximum guiding angle obtained from Snell's law must be smaller than some diffraction angle proportional to λ/a. Figure 1.5 shows the relationship between 2a and Δn for various wavelengths, deduced from Eqs. (1.14) and (1.15) with $n_2 = 1.46$ (refractive index of silica). Using Eqs. (1.14) and (1.15), we can define the cutoff wavelength λ_c as the wavelength above which a given fiber becomes single moded:

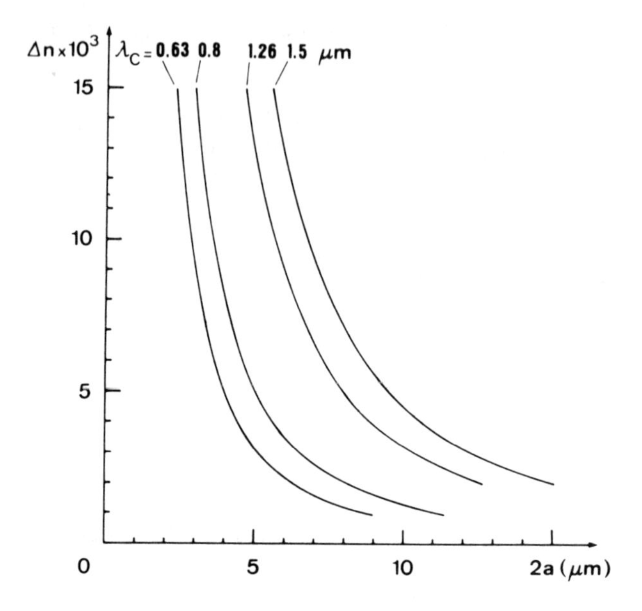

FIGURE 1.5 Curves illustrating the relationship between the core diameter 2a and the index difference Δn for obtaining various cutoff wavelengths λ_c. For each cutoff wavelength, the area below the corresponding curve corresponds to the single-mode (two-polarization) regime.

$$V = 2.405 \frac{\lambda_c}{\lambda}, \qquad \lambda_c = \frac{V\lambda}{2.405} \tag{1.16}$$

The fields of the HE_{11} mode are given by (either E_x or E_y can be taken as 0)

$$E_{y,x} = \mp \frac{Z_0}{n_2} H_{x,y} = E_0 \times \begin{cases} \dfrac{J_0(ur/a)}{J_0(u)}, & 0 \le r \le a \\[2ex] \dfrac{K_0(vr/a)}{K_0(v)}, & r \ge a \end{cases} \tag{1.17a}$$

$$E_z = -i \frac{E_0}{kan_2} (\sin\varphi, \cos\varphi) \begin{cases} \dfrac{uJ_1(ur/a)}{J_0(u)}, & 0 \le r \le a \\[2ex] \dfrac{vK_1(vr/a)}{K_0(v)}, & r \ge a \end{cases} \tag{1.17b}$$

$$H_z = -i\frac{E_0}{kaZ_0}(\cos\varphi,\ \sin\varphi)\begin{cases}\dfrac{uJ_1(ur/a)}{J_0(u)}, & 0 \le r \le a\\[2ex] \dfrac{vK_1(vr/a)}{K_0(v)}, & r \ge a\end{cases} \tag{1.17c}$$

where the first of $(\sin\varphi,\ \cos\varphi)$ for E_z and of $(\cos\varphi,\ \sin\varphi)$ for H_z holds if $E_x = 0$, and the second holds if $E_y = 0$. Z_0 is the vacuum impedance, J_0 and J_1 are the Bessel functions of order 0 and 1, and K_0 and K_1 are the modified Bessel functions. The terms u and v should simultaneously satisfy the two equations

$$u^2 + v^2 = V^2 \tag{1.18}$$

$$u\frac{J_1(u)}{J_0(u)} = v\frac{K_1(v)}{K_0(v)} \tag{1.19}$$

Equation (1.19) corresponds to the continuity of the field components tangential to the core-cladding interface at $r = a$. The total power carried by the mode along 0z is given by

$$P_t = \frac{1}{2}\int_0^{\infty}\int_0^{2\pi} \mathrm{Re}(\underline{E} \times \underline{H}^*) \cdot \underline{z}\ r\ dr\ d\varphi \tag{1.20}$$

where Re indicates the real part, * denotes the complex conjugate, × is the vector product, and z is the unitary vector along 0z. Normalization of P_t to 1 requires that

$$E_0 = \frac{u}{V}\frac{K_0(v)}{K_1(v)}\left(\frac{2Z_0}{\pi a^2 n_2}\right)^{\frac{1}{2}} = \frac{v}{V}\frac{J_0(u)}{J_1(u)}\left(\frac{2Z_0}{\pi a^2 n_2}\right)^{\frac{1}{2}} \tag{1.21}$$

Equations (1.17) show that the longitudinal components of the fields are on the order of u/akn with respect to the transverse components. Using Eqs. (1.18) and (1.14) and the fact that $\Delta < 1\%$ shows that the longitudinal components can be almost neglected and that we can consider the mode as transversely polarized with a linear polarization. This leads to the denomination of the LP_{01} mode [1], for which the field distribution is shown schematically in Fig. 1.6.

Second-Order Mode. When we have $2.405 \le V < 3.832$, another set of modes (TE_{01}, TM_{01}, HE_{21}) becomes guided by the fiber. As the longitudinal field components are again small compared to the

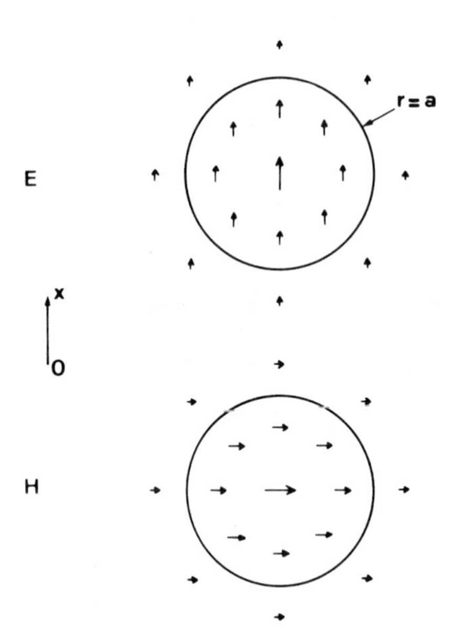

FIGURE 1.6 Electric and magnetic field distribution for an LP_{01} mode polarized along 0x. The length of the arrows is representative of the field amplitude at the corresponding point, and their direction corresponds to the field direction at a given time.

transverse field components, this set is called the LP_{11} mode, for which we have

$$E_{y,x} = E_1 \, (\cos\varphi, \, \sin\varphi) \begin{cases} \dfrac{J_1(u_1 r/a)}{J_1(u_1)}, & 0 \leq r \leq a \\[2ex] \dfrac{K_1(v_1 r/a)}{K_1(v_1)}, & r \geq a \end{cases} \tag{1.22}$$

There are four possible combinations of (x, y) and ($\cos\varphi$, $\sin\varphi$). u_1 and v_1 should satisfy

$$u_1^{\,2} + v_1^{\,2} = V^2 \tag{1.23}$$

$$u_1 \frac{J_2(u_1)}{J_1(u_1)} = v_1 \frac{K_2(v_1)}{K_1(v_1)} \tag{1.24}$$

To get a total power equal to 1, we need

$$E_1 = \left(\frac{2Z_0}{\pi a^2 n_2}\right)^{\frac{1}{2}} \frac{u_1}{V} \frac{K_1(v_1)}{[K_0(v_1)K_2(v_1)]^{\frac{1}{2}}} \tag{1.25}$$

Figure 1.7 illustrates the field distribution of the LP_{11} mode. Above $V = 3.832$, other modes, namely the LP_{02} (HE_{12}) and the LP_{21} ($HE_{31} + EH_{11}$) modes, become guided.

Field Shape. Table 1.1 gives (u, v) and (u_1, v_1) as solutions of Eqs. (1.18) and (1.19) for (u, v) and of Eqs. (1.23) and (1.24) for (u_1, v_1). Figure 1.8 shows u(V) and u_1(V) together with a useful approximation for u(V) [4]:

$$\left.\begin{aligned} v &\simeq 1.1428V - 0.9960 \\ &\simeq 2.7484\frac{\lambda_c}{\lambda} - 0.9960 \end{aligned}\right\} \quad u = (V^2 - v^2)^{\frac{1}{2}} \tag{1.26}$$

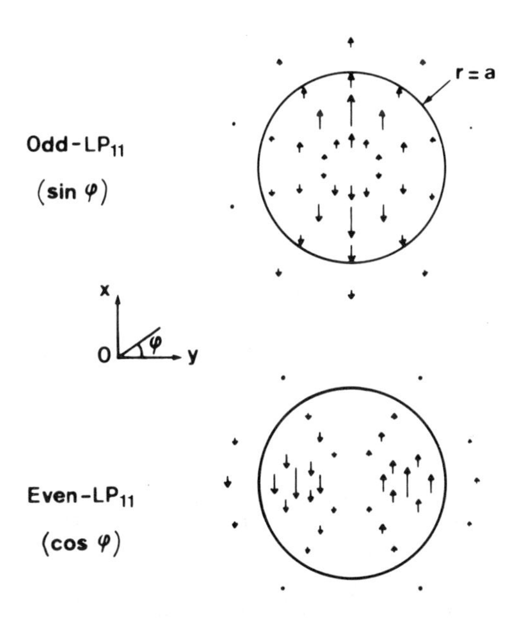

FIGURE 1.7 Electric field distribution for odd and even LP_{11} modes polarized along 0x. The arrows have the same significance as in Fig. 1.6.

TABLE 1.1 "Transverse propagation constants" u and u_1 in the core and exponential decay constants v and v_1 in the cladding for the fields of the LP_{01} mode (u, v) and the LP_{11} mode (u_1, v_1) as a function of the normalized frequency V or equivalently the light wavelength λ normalized to the LP_{11}-mode cutoff wavelength λ_c

V	λ/λ_c	u	v	u_1	v_1
0.6	4.01	0.59997	0.0056		
0.8	3.01	0.7974	0.0640		
1.0	2.41	0.9793	0.2024		
1.2	2.00	1.1341	0.3921		
1.4	1.72	1.2618	0.6065		
1.6	1.50	1.3670	0.8315		
1.8	1.34	1.4545	1.0604		
2.0	1.20	1.5282	1.2902		
2.2	1.09	1.5911	1.5194		
2.4	1.00	1.6453	1.7473		
2.6	0.93	1.6926	1.9736	2.53	0.60
2.8	0.86	1.7342	2.1983	2.63	0.95
3.0	0.80	1.7711	2.4214	2.72	1.27
∞	0	2.405	∞	3.832	∞

The relative error in u compared to the exact solution is less than 0.1% for $1.5 \leq V \leq 2.5$ ($1 \leq \lambda/\lambda_c \leq 1.6$) and increases to 1% for $1 \leq V \leq 3$ ($0.8 \leq \lambda/\lambda_c \leq 2.4$). Although this approximation should not be used for V smaller than 0.9, it is the simplest and the most accurate result published thus far for cases of practical interest [see Ref. 5, which compares the accuracy of several approximations for u(V)].

Once u and v are obtained, Eq. (1.17a) makes it possible to compute the field distribution of the LP_{01} mode. Observation of Table 1.1 and Fig. 1.8 shows that at zero optical frequency (infinite wavelength) the field is uniform across the whole structure, whereas at infinite frequency (zero wavelength) the field is entirely confined inside the core [as $J_0(2.405) = 0$]. At intermediate optical frequencies, the field is at a maximum on the fiber axis and

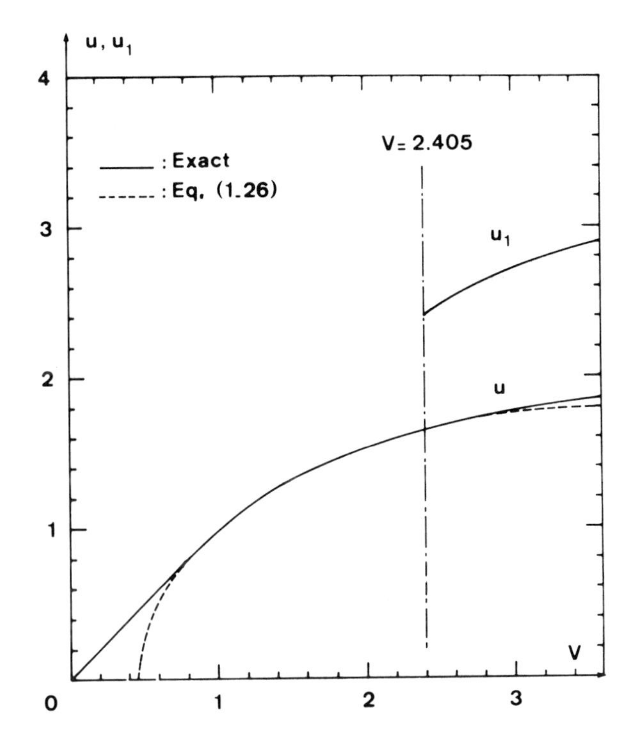

FIGURE 1.8 "Transverse propagation constants" of the LP_{01} mode (u) and of the LP_{11} mode (u_1) as a function of normalized frequency V. The dashed line represents a simple analytical approximation for u(V) which is very useful and accurate in most cases of practical interest.

decreases smoothly when the radial distance increases. Appendix 1.1 gives simple approximations for $J_0(x)$ and $J_1(x)$ in the range of interest, allowing simpler calculations of the field shape inside the core. Figure 1.9 shows the total guided intensity for various values of the normalized frequency V, assuming that both the LP_{01} and LP_{11} modes carry the same power and an incoherent illumination (no interferences).

1.2.2 Gaussian Approximation

When looking at Fig. 1.9, it appears that the shape of the fundamental LP_{01} mode is similar to a Gaussian shape, and this leads to an approximation of the exact field distribution by a Gaussian

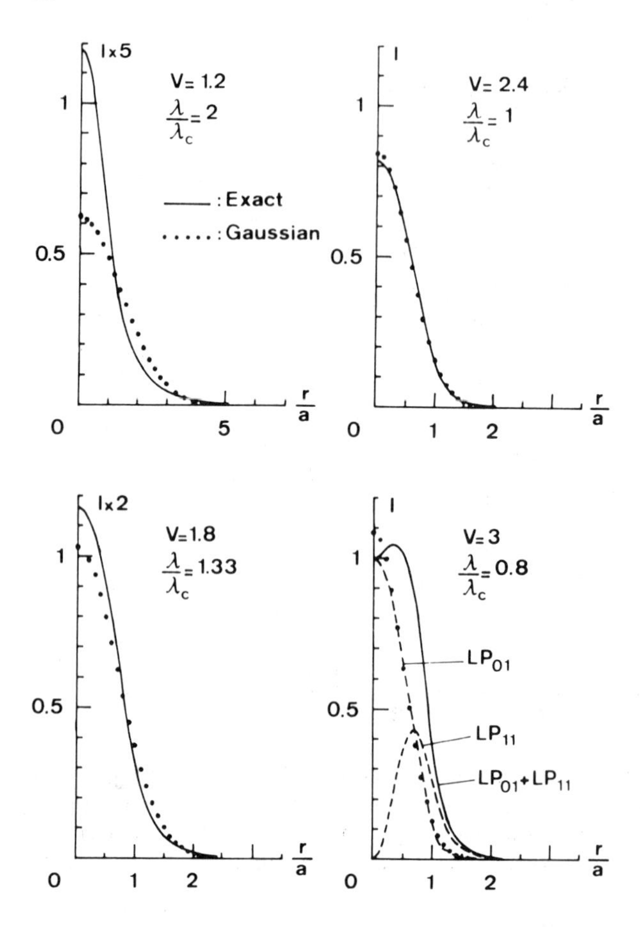

FIGURE 1.9 Shape of the guided intensity I(r) for various values of the normalized frequency V. In all cases the total power guided by each mode is constant. The solid and dashed lines are obtained from the Bessel functions, whereas the dotted lines correspond to the appropriate Gaussian approximation, with the same total power. Note the changes in horizontal and vertical scales.

function, which is the exact field shape of the LP_{01} mode in a non-truncated parabolic index profile. To find the best Gaussian approximation to the exact LP_{01} field, one could use different criteria. Equating the 1/e field widths usually leads to a very poor approximation, and it is generally preferred to choose the Gaussian field that leads to the maximum launching efficiency into the LP_{01} mode [6]. For the LP_{01} mode we use E_y as given by Eqs. (1.17) and (1.21), whereas the Gaussian field is given by

$$H_x = -\frac{2}{w}\left(\frac{n_2}{Z_0\pi}\right)^{\frac{1}{2}} \exp\left[-\left(\frac{r}{w}\right)^2\right] \tag{1.27}$$

The total power carried by the Gaussian field is thus normalized to 1 [see Eq. (1.20)] and the power launching efficiency is obtained as

$$\rho = \left(\frac{1}{2}\int_0^{2\pi}\int_0^{\infty} E_y H_x r \, dr \, d\varphi\right)^2 \tag{1.28}$$

By varying the parameter w, we can find the value w_0 that maximizes ρ as a function of V. w_0/a is shown in Fig. 1.10 as a function of both λ/λ_c and V, together with the corresponding values of ρ. It can be seen that within the usual range of λ/λ_c (0.8 to 1.8), ρ is greater than 96%, which indicates that the Gaussian approximation is good. For $0.8 \leq \lambda/\lambda_c \leq 2$, w_0/a can be approximated to better than 1% accuracy by [6]

$$\begin{aligned}\frac{w_0}{a} &= 0.65 + 0.434\left(\frac{\lambda}{\lambda_c}\right)^{3/2} + 0.0149\left(\frac{\lambda}{\lambda_c}\right)^{6} \\ &= 0.65 + 1.619V^{-3/2} + 2.879V^{-6}\end{aligned} \tag{1.29}$$

This approximation is also shown in Fig. 1.10 as a dashed line. The Gaussian field shape deduced from Eqs. (1.27) and (1.29) is also shown in Fig. 1.9 for comparison with the exact field. It is seen that the shape looks surprisingly different for extreme V values, but nevertheless the launching efficiency is good. The most restrictive part of the Gaussian approximation is the calculation of the evanescent field far from the core–cladding interface. When vr/a is greater than 2, $K_0(vr/a)$ can be approximated to better than 5% by

$$K_0\left(\frac{vr}{a}\right) \simeq \left(\frac{\pi}{2}\right)^{\frac{1}{2}}\left(\frac{a}{vr}\right)^{\frac{1}{2}} \exp\left[-\left(\frac{vr}{a}\right)\right] \tag{1.30}$$

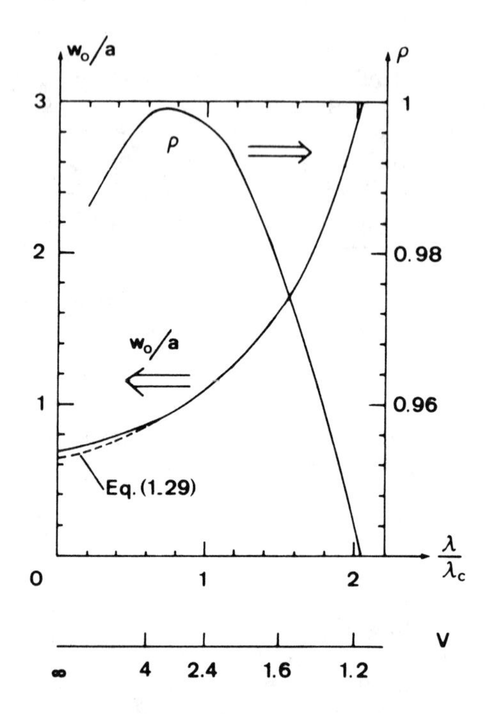

FIGURE 1.10 Normalized field radius w_0/a and launching efficiency of the Gaussian field distribution which gives the best launching efficiency into the LP_{01} mode, as a function of the normalized wavelength λ/λ_c or of the normalized frequency V. [After D. Marcuse, Loss—Splices, *Bell System Technical J.* *56*(5), 1977. Copyright 1977, American Telephone and Telegraph Company. Reprinted with permission.]

This indicates that the evanescent field actually decreases much more slowly than indicated by the Gaussian approximation. Hereafter, w_0 will be called the *mode field radius*.

1.2.3 Power Distribution

Besides the field shape itself, many problems require having a detailed knowledge of the LP_{01}-mode power distribution. For example, to determine the manufacturing conditions, it is necessary to know the power carried by the cladding (see Sec. 3.2.1).

Power Proportion Inside the Core. Using Eq. (1.20), the power proportion inside the core will be given by

$$\frac{P_c}{P_t} = \frac{\int_0^a \int_0^{2\pi} \mathrm{Re}\,(\underline{E} \times \underline{H}^*) \cdot \underline{z}\; r\, dr\, d\varphi}{\int_0^\infty \int_0^{2\pi} \mathrm{Re}\,(\underline{E} \times \underline{H}^*) \cdot \underline{z}\; r\, dr\, d\varphi} \tag{1.31}$$

where the components of $\bar{E}$ and $\bar{H}$ are given by Eqs. (1.17) for the LP_{01} mode. Using recurrence relations between J_0, J_1, K_0, and K_1 and Eq. (1.19), we get [1]

$$\frac{P_c}{P_t} = 1 - \left(\frac{u}{V}\right)^2 \left\{1 - \left[\frac{K_0(v)}{K_1(v)}\right]^2\right\} \tag{1.32}$$

If we replace the exact field expressions by the Gaussian approximation [eqs. (1.27) and (1.29)], we obtain

$$\frac{P_c}{P_t} \simeq 1 - \exp\left[-2\left(\frac{a}{w_0}\right)^2\right] \tag{1.33}$$

Figure 1.11a shows both expressions as a function of λ/λ_c. As seen in this figure, the Gaussian approximation leads to an acceptable accuracy except for high values of λ/λ_c. This figure also illustrates how the power carried by the mode spreads into the cladding when the wavelength increases.

Power Proportion Inside a Given Radius. The power proportion P_0/P_t inside a cylinder of radius r_0 is calculated by replacing a by r_0 in Eq. (1.31) and has been evaluated numerically, as there is no simple form for it. The Gaussian approximation leads to

$$\frac{P_0}{P_t} \simeq 1 - \exp\left[-2\left(\frac{r_0}{w_0}\right)^2\right] \tag{1.34}$$

Figure 1.11b shows the exact result and the result of the Gaussian approximation as a function of r_0/a for λ/λ_c = 1, 1.2, 1.5, and 2. We again observe that for large r_0/a values, the Gaussian approximation underestimates the evanescent field. In many cases one is also interested in knowing the residual power proportion above a given radius with good accuracy, and some results are given below.

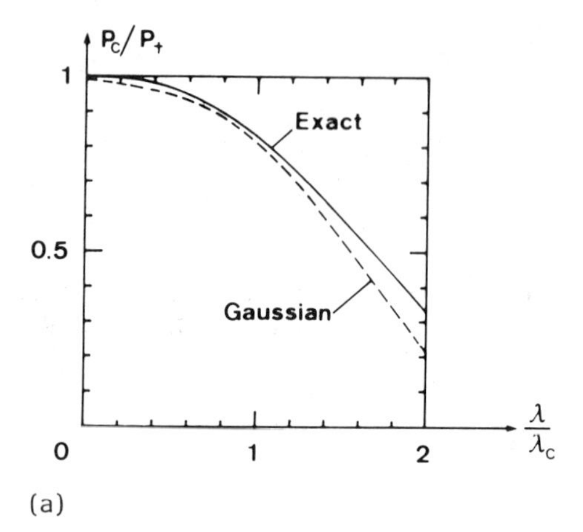

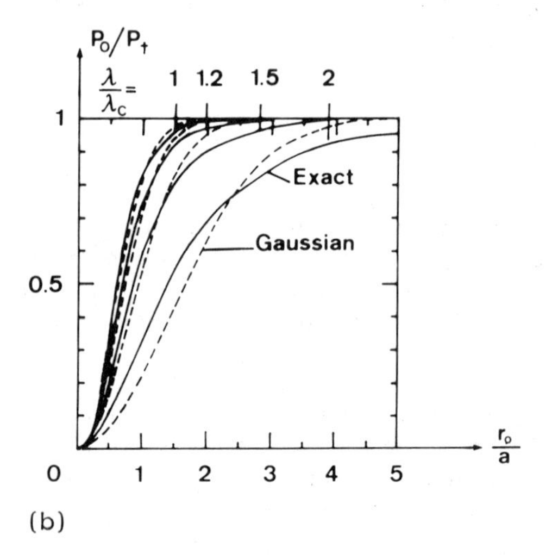

FIGURE 1.11 LP_{01}-mode power contained within given regions, normalized to the total guided LP_{01}-mode power P_t. (a) Power proportion in the core as a function of the normalized wavelength λ/λ_c, with the exact calculation and the result obtained when the LP_{01} field is approximated by a Gaussian function. (After Ref. 1.) (b) Power proportion within a radius r_0 for several values of the normalized wavelength λ/λ_c with the exact calculation and the Gaussian approximation.

Residual Power Proportion Above a Given Radius. The power proportion P_1/P_t above the radius r_1 is obtained from Eq. (1.31) by replacing 0 and a by r_1 and ∞, respectively, in the upper integration. The exact calculation should again be carried out numerically using the exact fields of Eqs. (1.17). We know that the Gaussian approximation cannot give correct results for $r_1 > a$, but we can use the approximation of the evanescent field given by Eq. (1.30) together with Eqs. (1.17) and (1.21):

$$\frac{P_1}{P_t} \underset{-}{\sim} \frac{\pi}{2}\left[\frac{u}{vVK_1(v)}\right]^2 \exp\left[-2\left(\frac{vr_1}{a}\right)\right] \tag{1.35}$$

This approximate formula gives an error of less than 10% for $vr_1/a > 1.5$. Plots of P_1/P_t as a function of r_1/a for $\lambda/\lambda_c = 0.8$, 1, 1.2, 1.5, and 2 are shown in Fig. 1.12.

1.2.4 Propagation Constant, Group Delay, and Dispersion

Propagation Constant. Classically, as in the case of the slab waveguide (see Sec. 1.1.1), the propagation constant β is related to u, v, and V by

$$\beta^2 = k^2 n_1^2 - \frac{u^2}{a^2} = k^2 n_2^2 + \frac{v^2}{a^2} \tag{1.36}$$

When $V \to 0$, $u \underset{-}{\sim} V$ and $v \underset{-}{\sim} V^2$, as shown by Eq. (1.8c) for the slab waveguide and Eq. (1.19) for the circular core fiber. It can thus be deduced that the variation of β with the optical frequency (or wavelength) will be similar in both types of guides. At infinite optical wavelength, the propagation constant of the LP_{01} mode is equal to the plane-wave propagation constant in the cladding, whereas at zero optical wavelength, it tends toward the propagation constant in the core. Instead of considering the absolute variations of β, it is more useful to consider a normalized propagation constant b, varying between 0 and 1.

$$b(V) = \frac{\beta^2 - k^2 n_2^2}{k^2 n_1^2 - k^2 n_2^2} = \left(\frac{v}{V}\right)^2 = 1 - \left(\frac{u}{V}\right)^2 \tag{1.37}$$

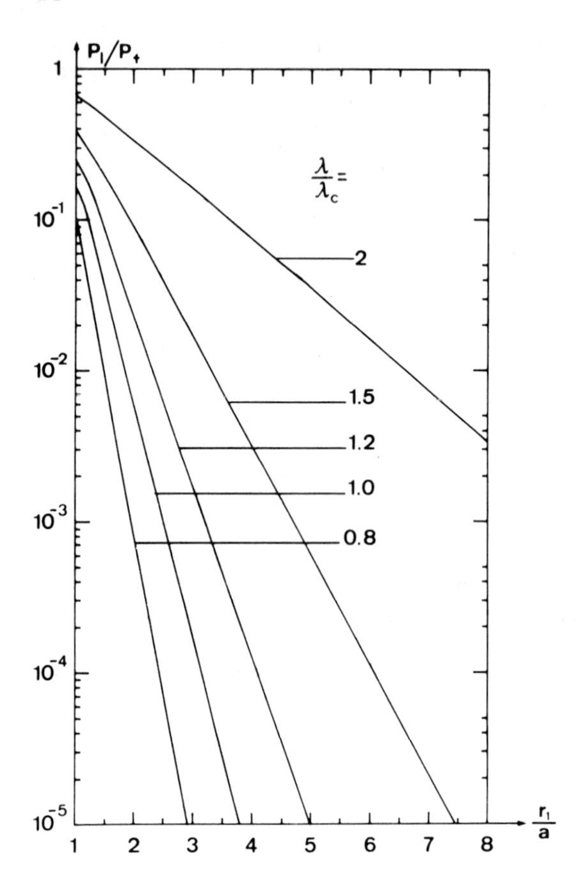

FIGURE 1.12 LP_{01}-mode power P_1 propagating outside a cylinder of radius r_1, normalized to the total LP_{01}-mode power P_t for various values of the normalized wavelength.

When the mode is at cutoff (infinite wavelength for the LP_{01} mode), its normalized propagation constant is zero, whereas it is equal to 1 at zero wavelength. Using the fact that the index difference is small, we can write

$$\beta \simeq kn_2(1 + b\Delta) \tag{1.38}$$

b is shown in Fig. 1.13 together with the approximation deduced from Eqs. (1.26) and (1.37).

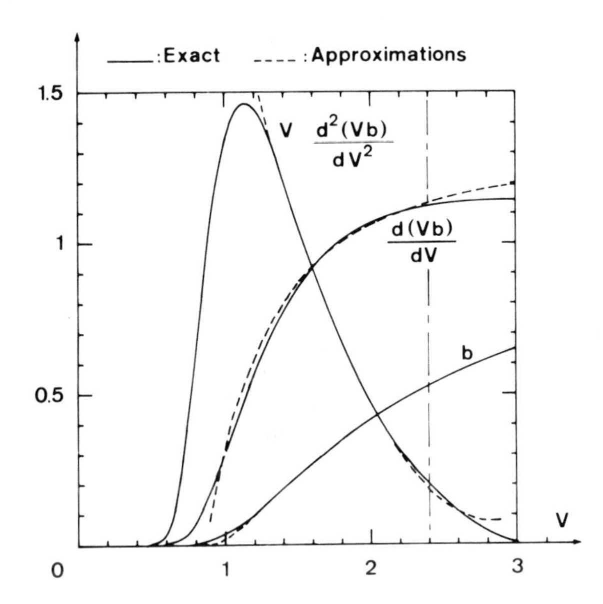

FIGURE 1.13 LP_{01}-mode normalized propagation constant b, normalized group delay d(Vb)/dV, and waveguide dispersion parameter $V[d^2(Vb)/dV^2]$ as a function of the normalized frequency V. Solid lines are exact results, dashed lines are the corresponding analytical approximations, and the dashed-dotted line recalls the single-mode limit.

$$b(V) \simeq \left(1.1428 - \frac{0.9960}{V}\right)^2 = \left(1.1428 - 0.4141\,\frac{\lambda}{\lambda_c}\right)^2 \qquad (1.39)$$

The relative error is less than 0.2% for $1.5 \leq V \leq 2.5$ ($1 \leq \lambda/\lambda_c \leq 1.6$) and less than 2% for $1 \leq V \leq 3$ ($0.8 \leq \lambda/\lambda_c \leq 2$). Obviously, the curve b(V) is equivalent to the curve $\beta(\omega)$ (shown in Fig. 1.3), but the use of the normalized propagation constant allows us to visualize more accurately the variations of the propagation constant, and b(V) has the advantage of being a universal curve (independent of the other fiber parameters).

Group Delay. As recalled in Sec. 1.1.3, the group delay τ characterizes the propagation delay time per unit length of a modulated signal transmitted by the optical wave. It is obtained as

$$\tau = \frac{d\beta}{d\omega} = \frac{1}{c}\frac{d\beta}{dk} \tag{1.40}$$

where c is the light velocity in vacuum.

Let us introduce the group index of refraction N_j in medium j of refractive index n_j:

$$N_j = \frac{d(kn_j)}{dk} \tag{1.41}$$

For silica in the usual wavelength range, the relative difference between N_j and n_j is less than 1.5% [7]. We use the same kind of calculations as detailed in Ref. 8. By starting from Eq. (1.37) and using Eqs. (1.14), (1.40), and (1.41), we obtain to first order in Δ,

$$\tau = \frac{N_2}{c}\,[1 + \Delta(Vb)'] - \frac{n_2\Delta}{c}\,P\left[\frac{b + (Vb)'}{2}\right] \tag{1.42}$$

where $(Vb)' = d(Vb)/dV$ and P expresses the difference in the dispersive properties between the core and the cladding [9]:

$$P = \frac{\lambda}{\Delta}\frac{d\Delta}{d\lambda} \tag{1.43}$$

When one neglects this difference in dispersive properties, Eq. (1.42) appears similar to Eq. (1.38), and this leads to the definition of (Vb)' as the normalized group delay. Within this approximation ($P \simeq 0$), a normalized group delay equal to zero means that the signal propagates at the group velocity of a plane wave in the cladding, whereas a normalized group delay equal to 1 leads to the plane-wave group velocity in the core. As $|P| \leq 0.1$ for the most usual dopant (GeO_2; see Ref. 9) in the wavelength range of interest, it is seen that the practical calculation of τ can be limited to its first term in Eq. (1.42).

For calculating (Vb)', we use a relationship giving du/dV [8]:

$$\frac{du}{dV} = \frac{u}{V}\,[1 - K(v)]; \qquad K(v) = \left[\frac{K_0(v)}{K_1(v)}\right]^2 \tag{1.44}$$

Equation (1.37) then gives

$$(Vb)' = 1 - \left(\frac{u}{V}\right)^2 [1 - 2K(v)] \tag{1.45}$$

The normalized group delay (Vb)' is plotted in Fig. 1.13. As expected from the behavior of b, when the frequency tends toward zero, the group delay tends toward that of a plane wave in the cladding [b and (Vb)' → 0]. Similarly (not shown in Fig. 1.13), when the frequency tends toward infinity, the group delay tends toward that of a plane wave in the core [(b and (Vb)' → 1) ⇒ ($\tau \to N_1/c$)]. However, there is a range of V values where the group delay of the guided mode exceeds that of plane waves in either the core or the cladding, as (Vb)' is greater than 1.

For getting an approximation, we can use Eq. (1.26) for calculating u and v and insert these values into Eq. (1.45). The accuracy is excellent for $1 \leq V \leq 3$, but this is not within the capability of a pocket calculator. We can also start from approximation equation (1.39) and obtain

$$(Vb)' \simeq 1.3060 - \left(\frac{0.9960}{V}\right)^2$$

$$\simeq 1.3060 - \left(0.4141 \frac{\lambda}{\lambda_c}\right)^2 \tag{1.46}$$

This approximation gives an error of less than 1% for $1.6 \leq V \leq 2.4$ ($1 \leq \lambda/\lambda_c \leq 1.5$) and less than 4% for $1 \leq V \leq 3$ ($0.8 \leq \lambda/\lambda_c \leq 2$), and is also shown in Fig. 1.13.

Dispersion. When the fiber is used with a light source emitting several different wavelengths, the variation of the group delay with wavelength around the center wavelength of emission is an important parameter (see Chap. 4). The variation of τ with λ will be due to the variations of N_2, Δ, and P with λ (material dispersion), and to the variations of b and (Vb)' with λ (or V) (waveguide dispersion). The resulting dispersion is called the chromatic dispersion. For computing $d\tau/d\lambda$, let us define the material dispersion parameters:

$$M_j = \frac{1}{c}\frac{dN_j}{d\lambda} = -\frac{\lambda}{c}\frac{d^2 n_j}{d\lambda^2} \tag{1.47}$$

From Eqs. (1.14) and (1.42), we get to first order in Δ and its derivatives:

$$\frac{d\tau}{d\lambda} = M_2 - \frac{\Delta n}{c\lambda} V(Vb)'' + (M_1 - M_2)\frac{b + (Vb)'}{2}$$

$$+ \frac{1}{cn_2}\frac{d(n_2 \Delta n)}{d\lambda}[V(Vb)'' + (Vb)' - b] \tag{1.48}$$

with

$$(Vb)'' = \frac{d^2(Vb)}{dV^2}$$

The first term represents a pure material dispersion contribution, while the second term corresponds to a pure waveguide dispersion effect. The last two terms are mixed contributions. From the behavior of b, (Vb)' and V(Vb)'' shown in Fig. 1.13, it is clear that the chromatic dispersion is equal to M_2 at zero frequency (V = 0) and M_1 at infinite frequency. The exact calculation of b + (Vb)' yields an approximation providing less than 2% error for $1.6 \leq V \leq 2.6$ $(0.9 \leq \lambda/\lambda_c \leq 1.5)$.

$$\frac{b + (Vb)'}{2} \simeq 1.306 - \frac{1.138}{V}$$

$$\simeq 1.306 - 0.4732 \frac{\lambda}{\lambda_c} \qquad (1.49)$$

For calculating V(Vb)'', we can use Eqs. (1.44) and (1.45) and classical recurrence relations between K_0, K_1, K_0', and K_1', to get

$$V(Vb)'' = 2\left(\frac{u}{V}\right)^2 \left\{K(v)[1 - 2K(v)] + \frac{2}{v}[v^2 + u^2K(v)]K^{\frac{1}{2}}(v)\right.$$

$$\left.\left[K(v) + \frac{1}{v}K^{\frac{1}{2}}(v) - 1\right]\right\} \qquad (1.50)$$

V(Vb)'' is shown in Fig. 1.13. It has been shown that all the known approximations for u(V) do not lead to acceptable approximations of V(Vb)'' if they are used directly [5]. We can use the approximation equation (1.26) for calculating u and v simply, and insert these values into Eq. (1.50). This leads to an error smaller than 4% for $1.2 \leq V \leq 3$ [10], but again this is not within the capability of a pocket calculator. However, we can get an empirical approximation which is useful in the usual range of operation:

$$V(Vb)'' \simeq 0.080 + 0.549(2.834 - V)^2$$

$$\simeq 0.080 + 0.549\left(2.834 - 2.405\frac{\lambda_c}{\lambda}\right)^2 \qquad (1.51)$$

The relative error is less than 5% for $1.3 \leq V \leq 2.6$ ($0.9 \leq \lambda/\lambda_c \leq 1.9$), and this approximation is also shown in Fig. 1.13. Finally, it should be noted that the present calculation is an approximation to first order in Δ, but it has been shown that the error is negligible compared to the result of an exact direct calculation [11].

1.2.5 Summary

We recall here the most important features of the ideal step-index fiber. When the normalized frequency is smaller than 2.405 (first zero of the zeroth-order Bessel function), there is only one guided mode (with two orthogonal polarizations), which is almost linearly polarized transversely to the propagation direction and is called the LP_{01} mode (or HE_{11}). The normalized frequency of 2.405 corresponds to the cutoff wavelength of the second-order mode (called LP_{11}), which is also transversely polarized, and no other mode appears up to a normalized frequency of 3.832 (second zero of the first-order Bessel function). The field distribution of the fundamental LP_{01} mode is almost Gaussian, with a diameter increasing with the light wavelength; usually, about 30 to 50% of the fundamental mode power is carried by the cladding. Finally, the variation of the group delay with wavelength (chromatic dispersion) is composed primarily of a pure material dispersion term proportional to the curvature of the curve $n_2(\lambda)$, and of a pure waveguide dispersion term proportional to the index difference and decreasing when the frequency increases (above $V = 1$).

1.3 GRADED-CORE FIBERS

In the early days of single-mode fiber technology, manufacturing imperfections often introduced some grading of the core—cladding interface and/or some index depression at the center of the core. Such index gradients may still be encountered from time to time, or even routinely in some specific manufacturing processes. Evaluating the transmission characteristics of such graded-core single-mode fibers is usually done by the use of *equivalent step-index fiber* methods (ESI). On the other hand, intentionally introduced index gradients in the core of single-mode fibers make it possible to obtain specific transmission characteristics such as longer zero chromatic dispersion wavelength around 1550 nm ("dispersion shifting"). In this case, one normally has to accurately solve the scalar wave equation for the index profile considered, but ESI methods still can provide rapid evaluations of the transmission parameters with an acceptable accuracy.

We will thus review accurate methods first and next give a simple analytical, yet accurate, ESI method.

We will consider a general index profile of the form

$$n^2(r) = n_2^2\left[1 + 2\Delta\ h\left(\frac{r}{a}\right)\right] = n_2^2 + 2n_2\ \Delta n(r) \tag{1.52}$$

$$\max h\left(\frac{r}{a}\right) = 1, \qquad 0 \leq r \leq a \tag{1.53a}$$

$$h\left(\frac{r}{a}\right) = 0, \qquad r \geq a \tag{1.53b}$$

$$\Delta << 1 \tag{1.53c}$$

We will use throughout a V parameter obtained by replacing n_1 by $n_2(1 + \Delta)$ in Eq. (1.14). It is common practice to consider more specifically two kinds of index gradients [12,13]:

$$h(x) = 1 - x^g, \qquad 0 \leq x \leq 1 \tag{1.54a}$$

$$h(x) = 1 - \gamma(1 - x)^g, \qquad 0 \leq x \leq 1 \tag{1.54b}$$

The first corresponds to a rounding of the core–cladding interface as seen in Fig. 1.14a, whereas the second corresponds to a central dip of relative depth γ as shown in Fig. 1.14b. Values of g greater than 8 can be considered as typical defects of both kinds in the modified chemical vapor deposition (MCVD) manufacturing process.

1.3.1 Exact Methods

All the published methods start from the scalar wave equation and solve it directly [12–14]. The most complete treatment is probably that of Refs. 12 and 13, where the fields are described by a power series expansion in r/a inside the core and by the modified Bessel functions inside the cladding. This general method is applied to index profiles described by Eqs. (1.54) and allows us to compute the cutoff wavelength λ_c, the field shape, bending losses, the propagation constant, and w_0/a for a Gaussian approximation based on the criterion of the maximum launching efficiency (care should be taken that w_0 as used here is a factor of $2^{\frac{1}{2}}$ larger than in Refs. 12 and 13). Among the interesting outcomes, it is notable that even for index profiles with a large central dip, the Gaussian approximation still leads to a good launching efficiency, and that a central

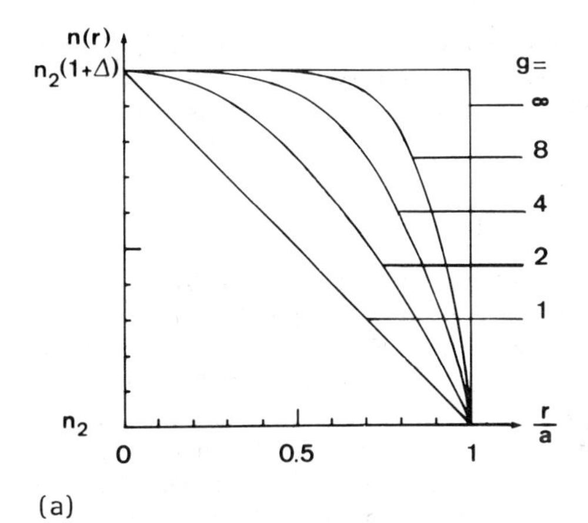

(a)

n(r)
$n_2(1+\Delta)$
$n_2[1+\Delta(1-\gamma)]$
$\gamma = 0.7$
n_2
0 0.5 1
r/a
g=
8
4
2
1

(b)

FIGURE 1.14 Core index profiles of graded-core fibers. (a) Rounding of the core–cladding interface as described by Eq. (1.54a). (b) Central index dip as described by Eq. (1.54b).

depression appears in the field shape only for $\gamma = 1$ and $g < 6$ in Eq. (1.54b). The results obtained by this method will serve as references hereafter, although we do not describe them in detail here.

1.3.2 Equivalent Step-Index-Fiber Methods

Various methods of this kind have been proposed in the literature [15–20], but either they are too complicated to be handled with a pocket calculator or even a small desktop computer, or the accuracy is not always good. We will see later that it is still possible to get both ease of handling and acceptable accuracy. All these methods start from two points: first, it has been observed that the fields of graded-core fibers usually look like the fields of a step-index fiber, and second, the step-index-fiber characteristics being well known, it is very convenient to replace the exact methods described in Refs. 12 and 13 by approximate methods that make reference to a step-index fiber. This should be done only if one is not interested in accurate knowledge of a specific characteristic that the index gradient is intended to modify in a given way.

We would thus like to find an equivalent step-index fiber (ESI) characterized by parameters a_e, Δ_e, and V_e which should have an LP_{11} mode cutoff wavelength λ_c, LP_{01} propagation constant β_e, evanescent field, and mode spot radius as close as possible to the corresponding parameters of the actual fiber, at least for $0.8 \leq \lambda/\lambda_c \leq 1.5$. If this is the case, the most important characteristics (bending losses, microbending losses, splicing losses, dispersion) of the actual fiber will be well approximated by the corresponding characteristics of the ESI.

General Method. To do this for an ESI as described above, we start from a perturbation theory for expressing β in the actual fiber with a stationary expression [15]

$$\beta^2 = \beta_e^2 + k^2 \frac{\int_0^\infty [n^2(r) - n_e^2(r)] E_e^2(r) r \, dr}{\int_0^\infty E_e^2(r) r \, dr} \tag{1.55}$$

where the subscript e denotes parameters of the equivalent fiber, $E_e(r)$ describing the electric field distribution in this equivalent fiber. The rigorous approach described in Ref. 15 consists of using for $E_e(r)$ the exact field shape of the ESI as given by Eq. (1.17) or (1.22), depending on whether we are interested in matching the propagation constant of the LP_{01} mode or the LP_{11} mode (for calculating λ_c). Once $E_e(r)$ has been replaced by the appropriate exact field shape, we have to find the set of independent parameters (V_e and a_e) (from which Δ_e can be deduced) that minimizes $|\beta^2 - \beta_e^2|$ in Eq. (1.55). Generally, each of these parameters is a function of both V and a, and it is thus impossible to get one ESI applicable throughout an extended wavelength range (at each wavelength there is one ESI). It is shown in Refs. 15, 19, and 20 that either for the LP_{01}-mode characteristics or for predicting the cutoff of the LP_{11} mode, this method is very accurate, although it requires complicated numerical calculations.

Fortunately, it is osbervable in Ref. 15 that for a limited range of wavelengths (e.g., $0.8 \leq \lambda/\lambda_c \leq 1.5$), a_e/a and Δ_e/Δ remain almost constant. If we assume, then, that a_e/a and Δ_e/Δ are totally independent of V in a limited but useful wavelength range, we are left with two independent adjustable parameters instead of one set. We can then start from Eq. (1.55) and use Eqs. (1.52) and (1.53) for obtaining

$$\Delta \int_0^\infty h\left(\frac{r}{a}\right) E_0^2(r) r\, dr = \Delta_e \int_0^{a_e} E_0^2(r) r\, dr \tag{1.56a}$$

$$\Delta \int_0^\infty h\left(\frac{r}{a}\right) E_1^2(r) r\, dr = \Delta_e \int_0^{a_e} E_1^2(r) r\, dr \tag{1.56b}$$

where $E_0(r)$ and $E_1(r)$ should describe modal field shapes related to the mode(s) and wavelength range of interest. A detailed discussion of the correspondence between this approach and other ESI approaches [16,18–20] is given in Appendix 1.2, and we concentrate here on the results.

Results and Accuracy. Our goal here is to obtain an acceptable trade-off between the accuracy and the simplicity of the results. We will thus use very crude approximations for $E_0(r)$ and $E_1(r)$ which should describe the LP_{01}-mode and LP_{11}-mode fields in the wavelength range of interest here: $E_0(r) = 1$ and $E_1(r) = (r/a)^{\frac{1}{2}}$ for $0 \leq r \leq \max(a, a_e)$, and $E_0(r) = E_1(r) = 0$ everywhere else. Equations (1.56) then lead to

$$\frac{V_e}{V} = \left[2\int_0^1 h(x)x\, dx\right]^{\frac{1}{2}} \tag{1.57a}$$

$$\frac{a_e}{a} = \frac{3}{2} \frac{\int_0^1 h(x)x^2\,dx}{\int_0^1 h(x)x\,dx}; \qquad a_e = \frac{3}{2} \frac{\int_0^a \Delta n(r)r^2\,dr}{\int_0^a \Delta n(r)r\,dr} \tag{1.57b}$$

$$\frac{\Delta_e}{\Delta} = \left(\frac{V_e}{V}\frac{a}{a_e}\right)^2; \qquad \Delta n_e = 2\,\frac{\int_0^a \Delta n(r)r\,dr}{a_e^2} \tag{1.57c}$$

It is noteworthy that V_e as given by Eq. (1.57a) is equivalent to the mode volume defined in Ref. 18, or V_e/V corresponds to the degree of guidance in Ref. 13. Physically, the significance of V_e can be compared to the average density of a disk with a local density equal to h(x). For the restricted class of profiles defined by Eqs. (1.54a) and (1.54b), we get

$h(x) = 1 - x^g$:

$$\frac{V_e}{V} = \left(\frac{g}{g+2}\right)^{\frac{1}{2}} \tag{1.58a}$$

$$\frac{a_e}{a} = \frac{g+2}{g+3} \tag{1.58b}$$

$h(x) = 1 - \gamma(1 - x)^g$:

$$\frac{V_e}{V} = \left[1 - \frac{2\gamma}{(g+1)(g+2)}\right]^{\frac{1}{2}} \tag{1.59a}$$

$$\frac{a_e}{a} = \frac{(g+1)(g+2)(g+3) - 6\gamma}{(g+3)[(g+1)(g+2) - 2\gamma]} \tag{1.59b}$$

Figures 1.15a and 1.16a show two graded-core fibers with the corresponding ESI.

For computing the LP_{11}-mode cutoff wavelength λ_c, or equivalently the cutoff normalized frequency V_c [both are related through Eq. (1.16)], we simply set $V_e = 2.405$ in Eq. (1.57a), (1.58a), or (1.59a). A comparison with exact values and other approximate results is shown in Table 1.2 for $h(x) = 1 - x^g$ only, as it has been shown that a central dip has little influence on this parameter [13]; physically, this limited influence corresponds to the fact

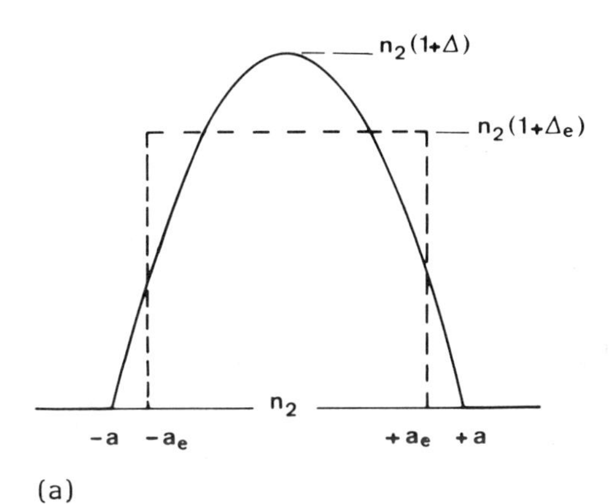

(a)

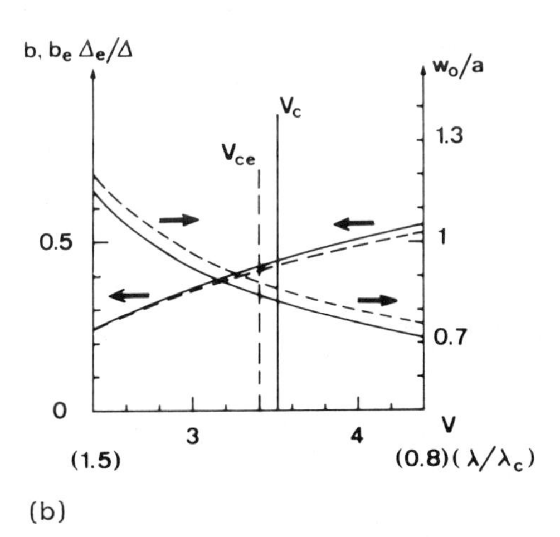

(b)

FIGURE 1.15 Characteristics of the equivalent step-index fiber corresponding to a parabolic index fiber [Eq. (1.52) with $h(x) = 1 - x^2$]. (a) Index profile of the actual fiber (solid line) and of the ESI (dashed line). (b) Exact normalized propagation constant b (Ref. 12, solid line) and equivalent normalized propagation constant $b_e\Delta_e/\Delta$ (dashed line). Exact normalized mode field radius w_0/a (Ref. 12, solid line) and equivalent normalized field radius [$= (w_0/a_e) \times (a_e/a)$, dashed line].

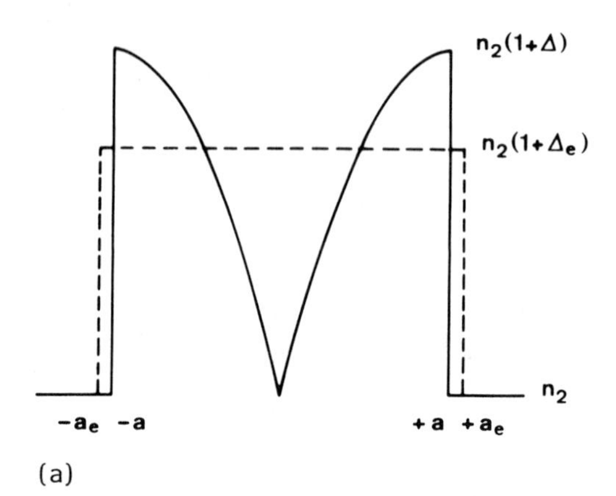

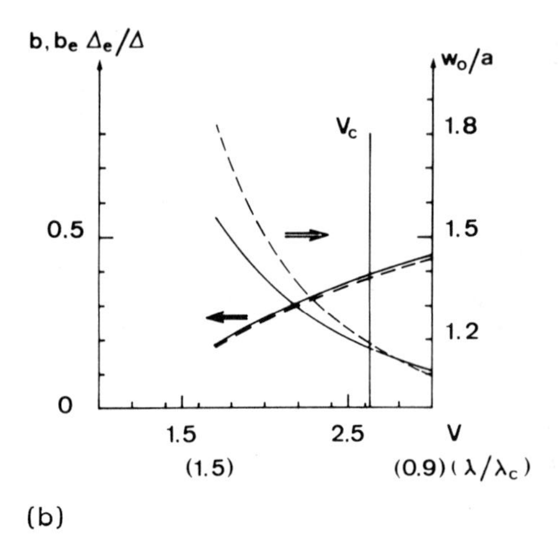

FIGURE 1.16 Characteristics of the equivalent step-index fiber corresponding to a parabolic index dip [Eq. (1.52) with $h(x) = 1 - (1 - x)^2$]. (a) Index profile of the actual fiber (solid line) and of the ESI (dashed line). (b) Exact normalized propagation constant b (Ref. 13, solid line) and equivalent normalized propagation constant $b_e\Delta_e/\Delta$ (dashed line). Exact normalized mode field radius w_0/a (Ref. 13, solid line) and equivalent normalized field radius [$= (w_0/a_e) \times (a_e/a)$, dashed line].

TABLE 1.2 Comparison of predicted LP_{11}-mode normalized cutoff frequencies V_c obtained by various methods for index profiles $h(x) = 1 - x^g$ (Fig. 1.14a)

g	Exact V_c: Ref. 12	Approximate V_c and relative error: Ref. 16	Ref. 18 and Eq. (1.58a)
1	4.381	3.674 (−16%)	4.166 (−5%)
2	3.518	3.227 (−8%)	3.401 (−3.3%)
3	3.181	3.015	3.105
4	3.000	2.890	2.946
5	2.886	2.808	2.846
8	2.710	2.673 (−1.4%)	2.689 (−0.8%)
10	2.649	2.624	2.635
20	2.527	2.520	2.522
∞	2.405	2.405	2.405

that the LP_{11} mode vanishes on the fiber axis and is thus little affected by a central perturbation. For computing the propagation constant of the LP_{01} mode through the ESI parameters, we simple deduce the ESI normalized propagation constant b_e from V_e, assuming a step-index profile [e.g., by using Eq. (1.39)], and then compute $b_e\Delta_e$ to obtain the propagation constant β_e through Eq. (1.38) (where $b\Delta$ is replaced by $b_e\Delta_e$). Figures 1.15b and 1.16b show the equivalent normalized propagation constants $b_e\Delta_e/\Delta$ compared to the exact normalized propagation constants b computed by the methods described in Refs. 12 and 13 as a function of V for two graded-core fibers. It can be seen that the accuracy is quite good, and more detailed data and comparisons are given in Appendix 1.2.

For the field shape of the LP_{01} mode, we notice first that the accuracy for the evanescent field in the cladding will be just as good as the accuracy over $b\Delta$, because Eqs. (1.14) and (1.37) show that $b\Delta$ is proportional to $(v/a)^2$. For the overall shape, we use the normalized mode field radius w_0/a; its exact value can be obtained by computing the exact field shape as in Refs. 12 and 13 and then using the definitions of Sec. 1.2.2. Its ESI value is obtained by deducing w_0/a_e from V_e through Eq. (1.29) (where V and a are replaced by V_e and a_e, respectively) and then multiplying the result by a_e/a as given by Eq. (1.57b). Figures 1.15b and 1.16b show both the exact and the ESI values of w_0/a for two graded-core fibers, and it is shown that the accuracy is not as good as for b, although it remains acceptable for the usual applications. More data and comparisons with other approximations are given in Appendix 1.2.

For the waveguide dispersion, Figs. 1.15b and 1.16b show that the curvature of the b(V) curve, and hence the waveguide dispersion, will be very poorly approached by the ESI method. Tests carried out and compared to exact results for the waveguide dispersion [21] show that the error becomes less than 10% when $g \geq 8$ only. A better approximation to the waveguide dispersion, even for very chaotic profiles, has been presented in Ref. 22, which uses V_e as given by Eq. (1.57a) and higher-order profile moments.

1.3.3 Summary

The waveguide characteristics of graded-core fibers can be obtained very accurately by exact methods applied to the resolution of the scalar wave equation with graded-index profiles. However, for most practical purposes, and when we are not interested in very accurate knowledge of a specific parameter, a graded-core fiber provides almost the same waveguide characteristics (usually within a few percent) as an equivalent step-index fiber determined by simple equations. This approximate method is significantly simpler than exact calculations, as it uses the well-documented characteristics of step-index fibers, although the accuracy is generally good at least over the wavelength range $0.8 \leq \lambda/\lambda_c \leq 1.5$ (except for the waveguide dispersion, when the index profile is too far from a step). Finally, it should be noted that central index dips have little influence on the LP_{11}-mode cutoff, but strongly increase the mode field radius compared to a step-index fiber. In Chap. 3 this will be seen to increase the microbending loss sensitivity of the fiber.

1.4 FIBERS WITH CLADDING INDEX VARIATIONS

Up to now we assumed that the cladding had a constant refractive index (matched-cladding fiber structure). Either for technological manufacturing reasons [23,24] or for enhancing some propagation characteristics [25], it may happen that a nominally step-index single-mode fiber presents index variations in the cladding. Usually, this index variation appears as either a positive or a negative step, and three typical cases are shown in Fig. 1.17. We thus have a general structure with a core (refractive index n_1, diameter 2a), an inner cladding (refractive index n_2, diameter 2a'), and an outer cladding (refractive index n_3, unlimited) (in the case of Fig. 1.17b, the inner cladding is separated from the core by an intermediate layer with refractive index n_3). We will consider only $n_1 > (n_2, n_3)$, as otherwise the propagation constant of any mode is always higher than the plane-wave propagation constant in the core, and thus all the modes are evanescent in the core. We will use reduced parameters:

$$\Delta n = n_1 - n_2, \qquad \Delta = \frac{\Delta n}{n_3} << 1 \tag{1.60a}$$

$$\Delta' n = n_2 - n_3, \qquad \Delta' = \frac{\Delta' n}{n_3} << 1 \tag{1.60b}$$

$$\delta = \frac{\Delta' n}{\Delta n} > -1 \tag{1.60c}$$

$\Delta' n$ and δ are positive in the case of a raised inner cladding (Fig. 1.17a and b), whereas they are negative in the case of the depressed inner cladding (Fig. 1.17c). It is also common to call fibers having a narrow depressed cladding ($a' < 2a$) "W-type fibers," but here we use only the term "depressed inner cladding."

We define two different normalized frequencies:

$$V_{12} = ak(n_1{}^2 - n_2{}^2)^{\frac{1}{2}} \simeq akn_3(2\Delta)^{\frac{1}{2}} \tag{1.61a}$$

$$V_{13} = ak(n_1{}^2 - n_3{}^2)^{\frac{1}{2}} \simeq V_{12}(1 + \delta)^{\frac{1}{2}} \tag{1.61b}$$

In the past these fibers have received much less attention than graded-core fibers, but recently there has been growing interest in structures with depressed inner cladding. The reasons for this are that these fibers can provide a very large and efficient control of their dispersion properties without simultaneously affecting their loss characteristics. Detailed treatments can be found in Refs. 25–28 for depressed inner cladding and in Ref. 28 for raised inner cladding.

1.4.1 Cutoff Conditions

Let us first recall that the exact definition of the cutoff for a mode is that the cutoff occurs when the mode propagation constant becomes equal to the plane-wave propagation constant in the outer cladding. In the case of Fig. 1.17, this condition can be written

$$\beta_{mode}(\lambda_c) = kn_3 \tag{1.62}$$

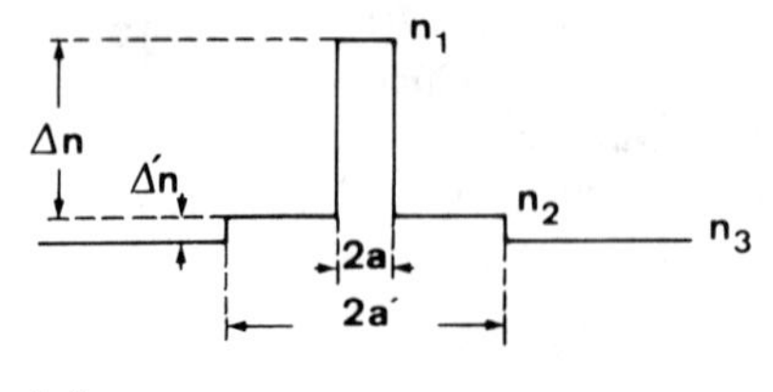

(a)

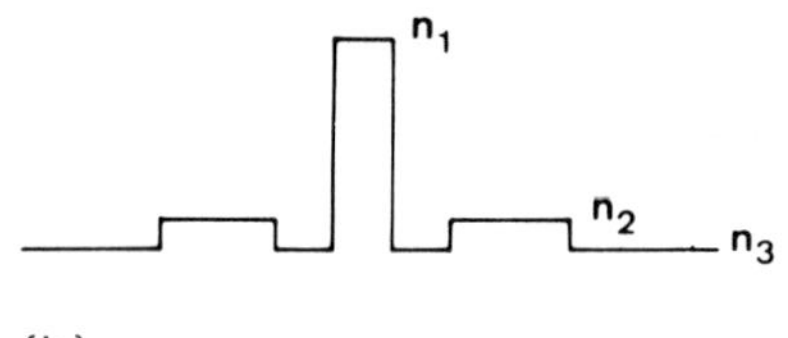

(b)

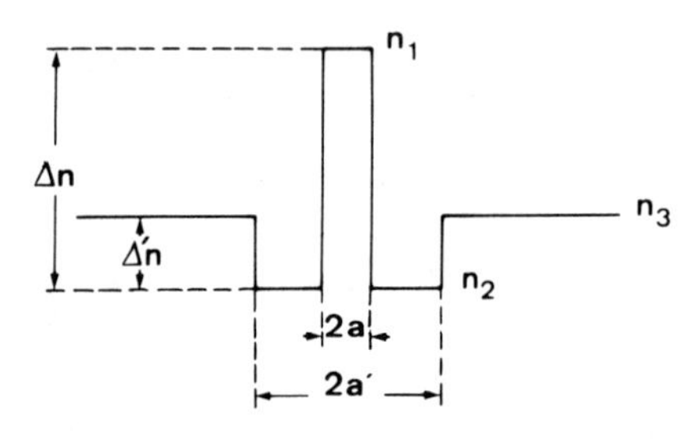

(c)

FIGURE 1.17 Typical index profiles of fibers with cladding index variations. (a) Raised inner cladding. (b) Raised inner cladding with barrier layer. (c) Depressed inner cladding.

where λ_c is the cutoff wavelength of the mode under consideration. When dealing with the cutoff conditions in these fibers, it is interesting to compare them to what is obtained with simple well-known step-index models in the two limiting cases $a'/a = 1$ and $a'/a \to \infty$. The definition of the mode cutoff recalled above clearly shows that the only possible comparison concerns the case $a'/a = 1$, as it would not make sense to suppress the outer cladding by using $a' \to \infty$. When $a'/a = 1$, we have a step-index fiber with core radius a, relative index difference $\Delta(1 + \delta)$, and normalized frequency V_{13}.

Raised Inner Cladding. This case corresponds to Fig. 1.17a and b, but we present results for the structure of Fig. 1.17a only. Starting from $a'/a = 1$ (or $\delta = 0$), it is clear that increasing a'/a and/or δ has the effect of increasing the propagation constant of all modes. This means that the cutoff wavelength as defined by Eq. (1.62) will increase, or equivalently V_{13c} (V_{13} at cutoff) will decrease, when a'/a or γ increase. This arises from the fact that the normalized propagation constant increases with frequency. Figure 1.18 shows V_{13c} for the LP_{11} and LP_{02} modes as a function of δ for $a'/a = 5$ and $a'/a = 2$, and Fig. 1.19 shows V_{13c} for the LP_{11} mode as a function of a'/a for $\delta = 0.5$ and $\delta = 0.25$.

As expected from the physical arguments above, V_{13c} decreases when a'/a or δ increase, that is, when the inner cladding increases its "weight." Noticeable in Fig. 1.18 is the fact that for $a'/a \geq 5$, there exists a range of δ values in which the LP_{02} mode has a smaller V_{13c} (longer cutoff wavelength) than the LP_{11} mode. It thus becomes the second-order mode (instead of the third-order mode in the usual step-index fibers) in this region. This is of practical interest as the values of δ in this range and $a'/a \geq 5$ are values often encountered in practice [23,24]. Except for this peculiar behavior, we can find an approximation for V_{13c} of the LP_{11} mode, by considering the region between $r = 0$ and $r = a'$ (core and inner gladding) as the core of a graded-core fiber, and thus use Eq. (1.57a) with $V_e = 2.405$. We get

$$V_{13c} \simeq 2.405 \left[\frac{1 + \delta}{1 + \delta (a'/a)^2}\right]^{\frac{1}{2}} \tag{1.63}$$

For $a'/a = 1$ (or $\delta = 0$) we obviously find the exact result, but the worst case occurs when $\delta \ll 1$ and $a'/a \gg 1$, as in this case the actual index profile is far from a step. Nevertheless, the worst-case accuracy is better than 10% on V_{13c}. Practically, however, it is expected that when a'/a is large, the LP_{11} mode becomes very sensitive to bends and microbends (because of its large radial extent; see Chap. 3), and that its effective cutoff wavelength is likely to be much shorter than the one predicted here.

Depressed Inner Cladding. Starting again from the limiting case $a'/a = 1$, we observe that increasing a'/a or $|\delta|$ has the effect of decreasing the propagation constant of all the modes. As a consequence, the cutoff wavelength defined by Eq. (1.62) will be decreased, or equivalently the value of V_{13c} will increase. This is especially true for the LP_{01} mode, which has an infinite cutoff

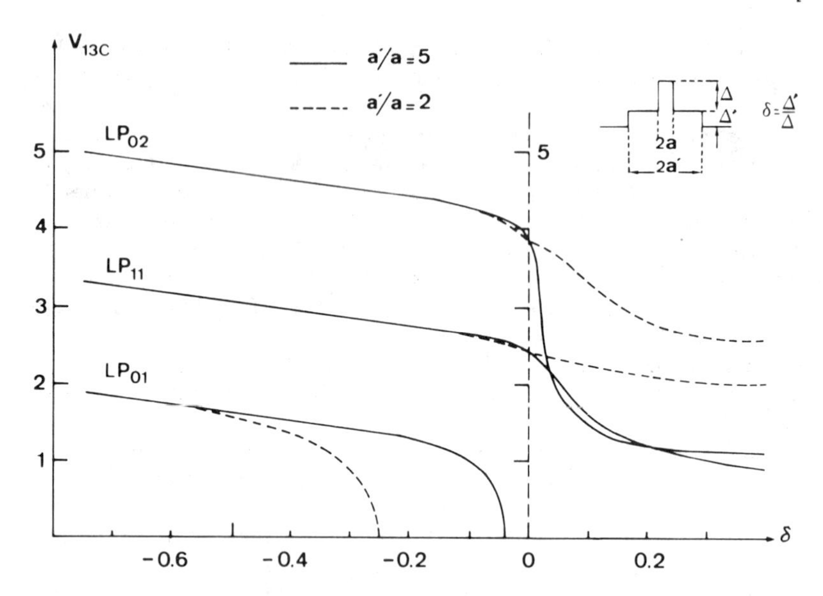

FIGURE 1.18 Cutoff value of V_{13} (V_{13c}) as a function of the inner cladding relative height δ for several values of the inner cladding relative radius a'/a, and for LP_{01}, LP_{11}, and LP_{02} modes. (From M. Monerie, Propagation in doubly-clad single-mode fibers, *IEEE Quantum. Electron.*, Apr. 1982. Reprinted with permission.)

wavelength (or zero cutoff frequency) in step-index fibers. In depressed inner cladding fibers, if the inner cladding has a sufficiently "negative" weight, the LP_{01} mode may exhibit a finite cutoff wavelength (or nonzero cutoff frequency). Using the mode volume, or the degree of guidance defined by Eq. (1.57a) when considering the core and the inner cladding as the core of a graded-index fiber, we observe that the mode volume becomes negative when

$$\frac{a'}{a} > (-\delta)^{-\frac{1}{2}} \tag{1.64}$$

This is exactly the relationship which indicates that the LP_{01} mode has a nonzero cutoff frequency, although rigorous demonstration of this point is more complicated than the arguments used here [28].

Figures 1.18 and 1.19 show the values of V_{13c} for the LP_{01}, LP_{11}, and LP_{02} modes for various a'/a and δ values. Again these figures illustrate quantitatively the qualitative discussion above, based on physical arguments. In is observable in Fig. 1.19 that, when condition (1.64) becomes fulfilled, the cutoff frequency V_{13c}

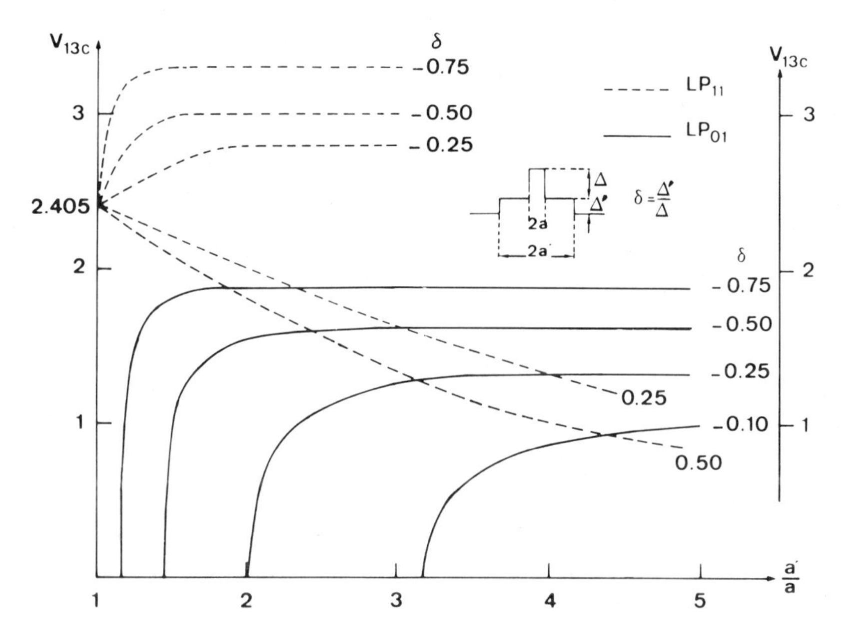

FIGURE 1.19 Cutoff value of V_{13} (V_{13c}) as a function of the inner cladding relative radius a'/a for several values of the inner cladding relative height δ and for LP_{01} and LP_{11} modes. (From M. Monerie, Propagation in doubly-clad single-mode fibers, *IEEE Quantum. Electron.*, Apr. 1982. Reprinted with permission.)

of the LP_{01} mode rises abruptly from zero to attain very quickly some limit value depending almost entirely on δ.

The cutoff illustrated here is based on the definition given by Eq. (1.62), which indicates that at cutoff the mode field remains constant in the outer cladding. When the frequency decreases below cutoff (wavelength increases), the field of the mode becomes oscillating (or radially traveling) in the outer cladding (as in the core), although it is still evanescent in the inner (depressed) cladding, as long as $\mathrm{Re}(\beta) > kn_2$. The mode then suffers from leakage losses which are analogous to a tunnel effect between the power trapped inside the core and the outer cladding region, through the inner cladding. Intuitively, one can say that these leakage losses will be decreased by a thick inner cladding and that the effective cutoff wavelength in fibers with very thick inner claddings will be almost entirely unaffected by the presence of the outer cladding. As an illustration, the leakage losses computed for the LP_{01} mode of a particular depressed inner cladding fiber (Δ = 0.5%, 2a = 7.5 μm)

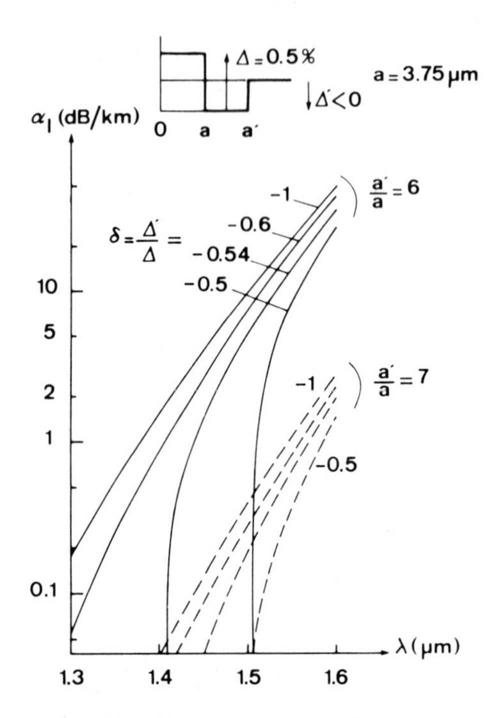

FIGURE 1.20 Leakage loss α_ℓ of the LP_{01} mode in a depressed inner cladding fiber as a function of wavelength. The fiber is characterized by the index profile shown in the insert, two values of a'/a are considered, and the variation of the parameter δ is obtained by varying Δ'. (From L. G. Cohen, D. Marcuse, and W. L. Mammel, in *Optical Fiber Communication 1982, Digest of Technical Papers*, Optical Society of America, Washington, D.C., Apr. 1982, paper THCC1.)

are shown in Fig. 1.20 for several values of a'/a and δ as a function of wavelength. For a'/a = 6, the leakage loss increases very abruptly for wavelengths above the LP_{01}-mode cutoff wavelength, the loss increase being much less dramatic for a'/a = 7.

1.4.2 Equivalent Step-Index Fiber

Because the cutoff properties of either raised inner cladding or depressed inner cladding fibers differ so widely from the cutoff properties of step-index fibers, it is obvious that it will not be possible to find step-index fibers as equivalent to fibers with cladding index variations as they were to graded-core fibers. However, the LP_{01}-mode field (when not close to its cutoff) again looks like the LP_{01}-mode field of some step-index fibers and it is thus desirable to find step-index fibers exhibiting almost the same field shape as that of the actual fibers. This can be done only for V_{13} values well above V_{13c} for the LP_{01} mode anr over a limited range of V_{13}. We will deal only with the LP_{01}-mode propagation constant and field shape, and no attention will be paid to either LP_{01} or LP_{11} cutoff wavelength. Moreover, we restrict the analysis to

$$-1 < \delta \leq 0.2 \tag{1.65}$$

which corresponds to most cases of practical interest. A numerical approach to this problem is presented in Ref. 17, but we shall obtain analytical results workable with a pocket calculator. The principle of the method consists of taking the core refractive index n_1 as a reference, and looking for a step-index fiber having a cladding refractive index which should be some average value of the actual indices n_2 and n_3. We thus write the actual and equivalent index profiles:

$$n^2(r) = n_1^2\left[1 - 2\Delta q\left(\frac{r}{a}\right)\right] \tag{1.66a}$$

$$q(x) = \begin{cases} 0, & 0 \leq x < 1 \\ 1, & 1 \leq x < \dfrac{a'}{a} \\ 1 + \delta, & \dfrac{a'}{a} \leq x \end{cases} \tag{1.66b}$$

$$n_e^2(r) = n_1^2\left[1 - 2\Delta_e q_e\left(\frac{r}{a_e}\right)\right] \tag{1.67a}$$

$$q_e(x) = \begin{cases} 0, & 0 \leq x < 1 \\ 1, & 1 \leq x \end{cases} \tag{1.67b}$$

We then start from Eq. (1.55) and use the same method as in Sec. 1.3.2. For the field shapes needed in Eqs. (1.56), we notice that here we need only LP_{01} fields (as we care only for this mode), and that the field shape is required only for $r > a$. As we exclude the case where the LP_{01} mode is close to its cutoff, we know that the required functions will be proportional to $K_0(v_e r/a_e)$. Limiting the analysis to a useful practical range of operation determined by

$1.5 \leq V_e \leq 3$, we determine from Table 1.1 that $1 \leq v_e \leq$ and replace $K_0(x)$ by its exponential approximation [Eq. (1.30)]

$$E_0(r) \sim \left(\frac{a}{r}\right)^{\frac{1}{2}} \exp\left[-\left(\frac{r}{a} - 1\right)\right]$$

$$E_1(r) \sim \left(\frac{a}{r}\right)^{\frac{1}{2}} \exp\left[-2\left(\frac{r}{a} - 1\right)\right]$$

When these fields are injected into Eq. (1.55) with Eqs. (1.66) and (1.67), we obtain

$$\frac{a_e}{a} = 1 + \frac{1}{2} \ln \left\{ \frac{1 + \delta \exp[-2(a'/a - 1)]}{1 + \delta \exp[-4(a'/a - 1)]} \right\} \tag{1.68a}$$

$$\frac{\Delta_e}{\Delta} = \frac{\{1 + \delta \exp[-2(a'/a - 1)]\}^2}{1 + \delta \exp[-4(a'/a - 1)]} \tag{1.68b}$$

$$V_e = \frac{a_e}{a}\left(\frac{\Delta_e}{\Delta}\right)^{\frac{1}{2}} V_{12} = \frac{a_e}{a}\left(\frac{\Delta_e}{\Delta}\right)^{\frac{1}{2}} \frac{V_{13}}{(1 + \delta)^{\frac{1}{2}}} \tag{1.68c}$$

It can immediately be seen that for $a'/a = 1$ and $a'/a \to \infty$, these ESI parameters tend toward the characteristics of the step-index fibers obtained from Fig. 1.17 in these limits. For testing the accuracy of these relationships, we compared the actual values of $kn_1 - \beta$ [28] to values of $kn_1 - \beta_e = (1 - b_e)\Delta_e$, where b_e is the normalized propagation constant of the LP_{01} mode in the ESI and is computed from Eqs. (1.68) and (1.39). As an example, the relative error in this parameter is less than 1% through the range $1.5 \leq V_e \leq 3$ for $\delta = -0.5$ and $a'/a = 1.3$. Figure 1.21 shows two examples of actual fibers with their ESI, and Fig. 1.22 shows some curves a_e/a, Δ_e/Δ, and V_e/V_{13} as functions of a'/a and δ. It can be seen that the more negative δ is, the more the fiber is tightly confining the LP_{01} mode (a_e/a decreases and V_e increases, thus w_0 should strongly decrease), and the opposite holds if δ is positive.

1.4.3 Dispersion Properties

Because of their special cutoff properties, these fibers also exhibit special dispersion properties, especially for the depressed inner cladding. In such fibers, at short wavelength (high V values), the LP_{01} mode does not "see" the outer cladding and thus behaves as if the inner cladding extended to infinity. When the wavelength increases (V decreases), the LP_{01} mode will quickly discover that

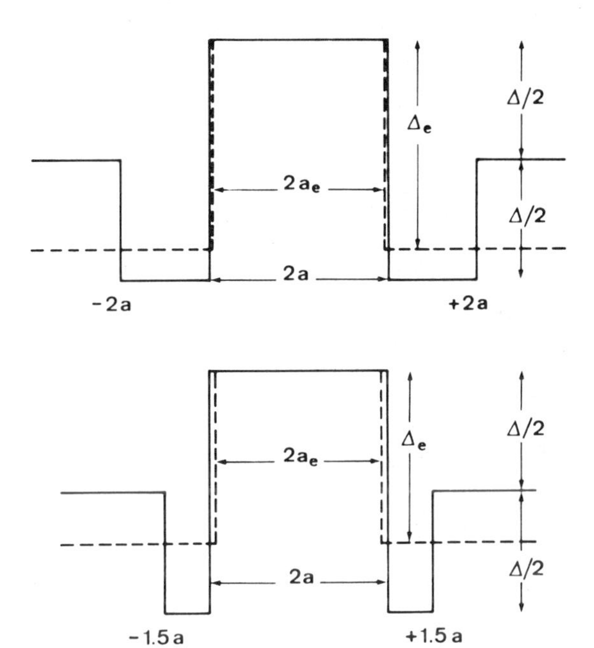

FIGURE 1.21 Examples of depressed inner cladding fibers (solid lines), together with the index profiles of the corresponding equivalent step-index fibers (dashed lines).

there is an outer cladding with a refractive index higher than that of the inner cladding. The LP_{01} mode then "realizes" that its propagation constant has to stop its decrease at $\beta = kn_3$ and not at $\beta = kn_2$ as the mode first "thought." Obviously, this will give a sharp curvature to the b(V) curve, thus inducing a strong waveguide dispersion. This situation is illustrated in Fig. 1.23.

It has been shown in Ref. 28 that the chromatic dispersion $d\tau/d\lambda$ obeys the same equation as Eq. (1.49) provided that M_2 is replaced by M_3 [from Eq. (1.47) with n_2 replaced by n_3], Δ is replaced by $\Delta(1 + \delta)$, N_2 is replaced by N_3, and V is understood as V_{13}, with P and b being given by

$$P = \frac{(\lambda/\Delta)(d\Delta/d\lambda) + \delta(\lambda/\Delta')(d\Delta'/d\lambda)}{1 + \delta} \tag{1.69}$$

$$b = \frac{(\beta/k)^2 - n_3^2}{n_1^2 - n_3^2} \tag{1.70}$$

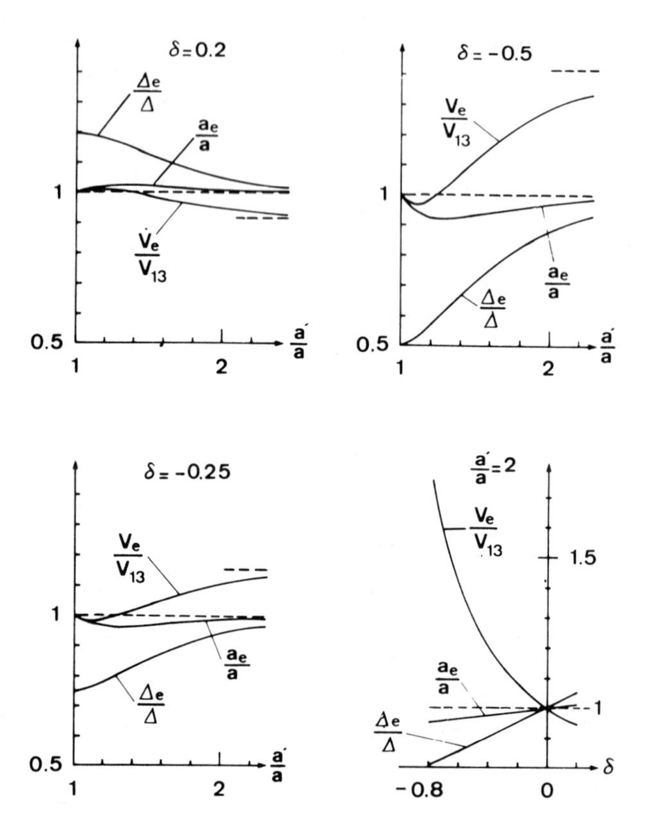

FIGURE 1.22 Equivalent step-index-fiber parameters (core radius a_e, relative index difference Δ_e, normalized frequency V_e) normalized to the parameters of the corresponding fibers with cladding index variations (core radius a, relative core to inner cladding index difference Δ, normalized frequency V_{13} between core and outer cladding) as functions of the inner cladding relative radius a'/a and of the inner cladding relative height δ.

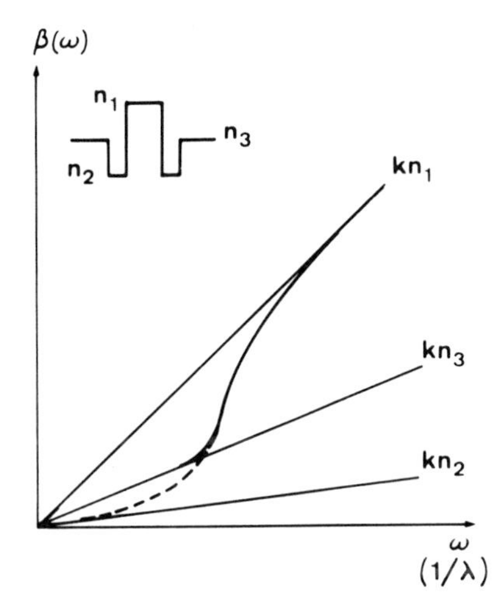

FIGURE 1.23 Behavior of the LP_{01}-mode propagation constant in a depressed inner cladding fiber. The dashed curve represents the expected behavior in the absence of outer cladding, whereas the solid line corresponds to the actual behavior. It is seen that the dispersion ($d^2\beta/d\omega^2$) is strongly increased near cutoff ($\beta \simeq kn_3$).

As usual, the normalized propagation constant b defined this way varies between 0 and 1 for a truly guided mode. A detailed study of b, $d(Vb)/dV$, and $V[d^2(Vb)/dV^2]$ ($V \equiv V_{13}$) has been carried out in Ref. 28, and we give here only some results, which are shown in Figs. 1.24 and 1.25. It can be seen that the fibers with a depressed inner cladding exhibit several differences from step-index fibers. First, as noted in Ref. 25, they can provide a negative value of $V[d^2(Vb)/dV^2]$ in the single-guided-mode region, which is interesting for short-wavelength applications (as M for silica is usually negative below $\lambda = 1.3\ \mu m$, a negative waveguide dispersion will allow compensation of the material dispersion in this spectral region). Second, they can allow a very high positive value of the waveguide dispersion $V[d^2(Vb)/dV^2]$, which is of interest for long-wavelength operation (as M for silica is strongly positive at $\lambda = 1.6\ \mu m$, where the fibers have their minimum loss, a strongly positive waveguide dispersion allows compensation of the material dispersion with moderate index differences). However, the stronger the waveguide dispersion, the tighter the manufacturing and wavelength tolerances, as evidenced by Figs. 1.24 and 1.25.

Finally, a very accurate approximation for b(V) in depressed inner cladding fibers has been described in Ref. 29, which allows

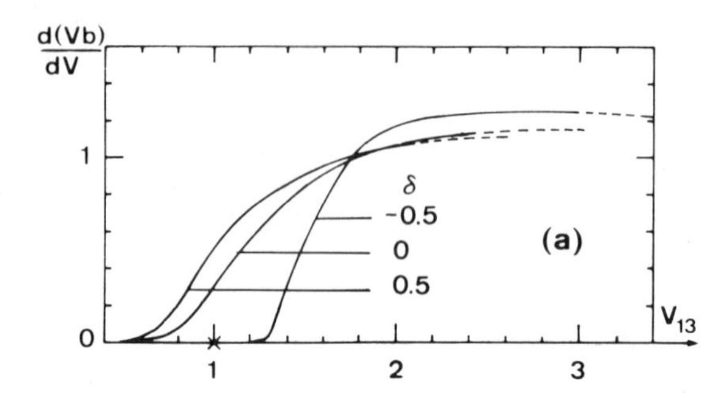

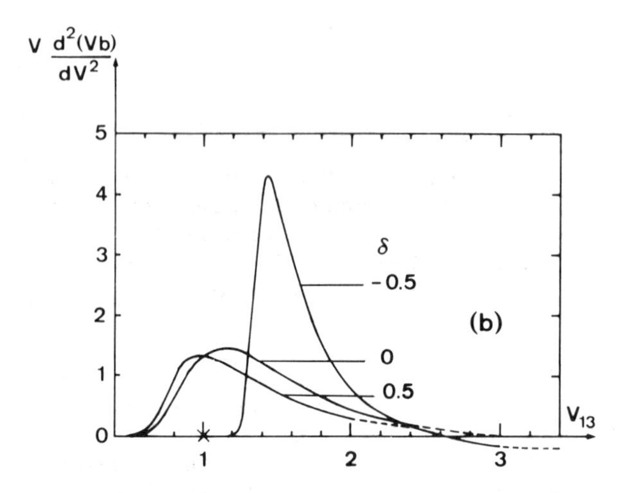

FIGURE 1.24 Dispersion curves for an inner cladding relative radius a'/a = 1.5 and various inner cladding relative heights δ. The region of the single guided mode corresponds to the full lines, and the corss indicates the LP_{01}-mode cutoff. (From M. Monerie, Propagation in doubly-clad single-mode fibers, *IEEE Quantum. Electron.*, Apr. 1982. Reprinted with permission.)

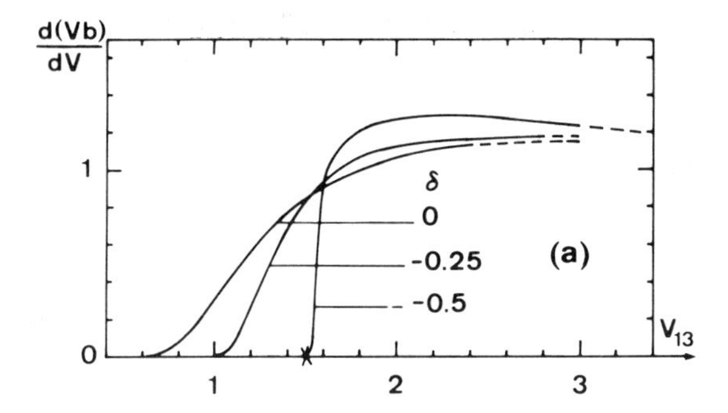

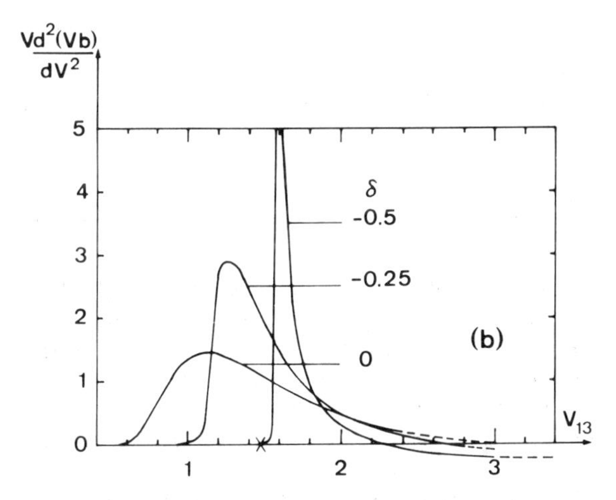

FIGURE 1.25 Dispersion curves for an inner cladding relative radius a'/a = 2 and various inner cladding relative heights δ. The region of the single guided mode corresponds to the solid lines, and the corss indicates the LP_{01}-mode cutoff. (From M. Monerie, Propagation in doubly-clad single-mode fibers, *IEEE Quantum. Electron.*, Apr. 1982. Reprinted with permission.)

an accurate analytical calculation of $V[d^2(Vb)/dV^2]$ with less than 10% error in most practical cases of interest. For this approximation, b(V) is given by

$$b(V) \simeq b_\infty(V) + \frac{Y(V)}{V}[1 + X(V)] \tag{1.71a}$$

$$b_\infty(V) = \frac{b_{st}(V_{12}) + \delta}{1 + \delta} \tag{1.71b}$$

$$Y(V) = 1.4\,\{-\delta + [-\delta(1 + \delta)]^{\frac{1}{2}}\}\,\exp\left[-2V_{12}(-\delta)^{\frac{1}{2}}\left(\frac{a'}{a} - 1\right)\right] \tag{1.71c}$$

$$X(V) = \frac{-2}{[\delta/(1 + \delta)] + b_\infty + 5Y/V}\left\{b_\infty + \frac{2Y}{V} - \left[-\frac{\delta}{1 + \delta}\left(b_\infty + \frac{Y}{V}\right) - \frac{Y}{V}\left(2b_\infty + \frac{Y}{V}\right)\right]^{\frac{1}{2}}\right\} \tag{1.71d}$$

In these expressions, V holds for V_{13}, b is defined in Eq. (1.70), $b_\infty(V)$ is the limiting form of b(V) when $V \to \infty$ (and also when $\delta = 0$, or $a'/a \to \infty$), and $b_{st}(x)$ represents the normalized propagation constant of the LP_{01} mode in a step-index fiber with a normalized frequency x [given, for example, by Eq. (1.39)]. The accuracy of b(V) is better than 1% in most cases of practical interest [29].

From Eqs. (1.71), the waveguide dispersion is obtained as

$$V\frac{d^2(Vb)}{dV^2} \simeq \frac{1}{1 + \delta}V_{12}\frac{d^2[V_{12}b_{st}(V_{12})]}{dV_{12}^2} + V\frac{d^2[Y(1 + X)]}{dV^2} \tag{1.72}$$

where $V_{12} = V/(1 + \delta)^{\frac{1}{2}}$. The first term on the right-hand side is exactly equal to the waveguide dispersion of a step-index fiber with normalized frequency V_{12}, divided by $1 + \delta$. This term is well known and Eq. (1.51) gives a good approximation for it in the range $1.3 \le V_{12} \le 2.8$. When greater values of V_{12} are required, a very good approximation is

$$V_{12}(V_{12}b_{st})'' \simeq -0.14 + 0.29(V_{12} - 1.56)^{-2} \tag{1.73}$$

for $2.8 \leq V_{12} \leq 6$. The second term on the right-hand side of Eq. (1.72) can be evaluated from Eqs. (1.71) with the help of a pocket calculator. The accuracy of Eq. (1.72) is usually better than 10% [29].

1.4.4 Summary

Fibers with raised inner claddings can give rise to an inversion in the order of appearance of the LP_{11} and LP_{02} modes, the latter becoming the second-order mode in place of the former in some cases. Apart from this particularity, fibers with large inner cladding relative radii ($a'/a \geq 6$), as is the case for most practical raised inner cladding and for some depressed inner claddings, exhibit almost the same practical characteristics as the step-index fiber obtained by ignoring the outer cladding. As compared to the step-index fiber obtained with the core and the outer cladding only, the presence of the depressed cladding leads to a better power confinement in the core, but may also introduce a finite cutoff wavelength for the fundamental LP_{01} mode. Above this wavelength, this mode suffers from leakage losses which increase when the depth of the inner cladding increases (for a constant core-inner cladding index difference), and when the inner cladding narrows. Finally, fibers with depressed sufficiently narrow inner claddings exhibit very interesting and useful waveguide dispersion properties.

APPENDIX 1.1 APPROXIMATIONS FOR $J_0(x)$ AND $J_1(x)$

For $0 \leq x \leq 1.8$:

$$J_0(x) \simeq 1 - 0.25x^2 + 0.025x^3, \qquad \text{error} \leq 2\%$$

$$J_0(x) \simeq 1 - 0.21x^2, \qquad \text{error} \leq 4\%$$

For $0 \leq x \leq 2.5$:

$$J_1(x) \simeq 0.17x(3.7 - x), \qquad \text{error} \leq 4\%$$

For very small values of x, the relative error becomes larger but the absolute error remains acceptable.

For $0 \leq x \leq 2.5$:

$$J_1(x) \simeq 0.5815 \sin\left(\frac{\pi x}{3.7}\right), \qquad \begin{array}{l} \text{error} \leq 0.5\%,\ 1.3 < x < 2.5 \\ \text{error} \leq 1\%,\ 0.8 \leq x \leq 1.3 \\ \text{error} \leq 1.5\%,\ 0 \leq x \leq 0.8 \end{array}$$

APPENDIX 1.2 COMPARISON BETWEEN DIFFERENT EQUIVALENT STEP-INDEX-FIBER METHODS

The comparison will concern the method described in Sec. 1.3.2 and the methods described in Refs. 16, 18, 19, and 20. Although some of these methods do not involve the characteristics that we are interested in, or do not explicitly use a perturbation technique, we will see that it is possible to obtain the results given by these methods by starting from Eqs. (1.56). The results of Ref. 16 can be obtained by setting $\Delta_e = \Delta$ and $E_0(r) = E_1(r)$ in Eqs. (1.56). It is then clear that this voluntarily loses one degree of freedom, and that the results cannot be accurate for both the LP_{11}-mode cutoff and the LP_{01}-mode propagation constant, as these two modes have very different field shapes.

The expressions for V_e and a_e given in Ref. 18 can be obtained from Eqs. (1.56) by using $E_0(r) = 1$ and $E_1(r) = (a/r)^{\frac{1}{2}}$ in the core, and vanishing everywhere else. As Eq. (1.55) is stationary in $E(r)$ as well as in β, the results of Ref. 18 are accurate except for central index dips (see below), as the expression above for $E_1(r)$ is unrealistic for $r \to 0$.

Finally, Refs. 19 and 20 are interested only in the LP_{11}-mode cutoff and set $a_e = a$, but use accurate expressions for the field. Table 1.2 compares the cutoff normalized frequencies for various index profiles obtained by exact calculations or ESI methods. As expected from the discussion above, the results of Ref. 16 are less accurate than those of Ref. 18 and the present work, which are within a few percentage points of the exact value.

We turn next to the propagation constant, the evanescent field decay rate, and the mode field radius. The normalized propagation constant $b = (v/V)^2$ is computed exactly (within $\mp$ 0.2%) by the methods described in Refs. 12 and 13, and $b_e\Delta_e/\Delta$ is calculated through the various ESI methods as explained in Sec. 1.3.2. $b\Delta$ being proportional to $(v/a)^2$ gives simultaneously a comparison of the evanescent field decay rates. For the field radius, we compute w_0/a with the field shapes obtained from Refs. 12 and 13 and compare it to $(w_0/a_e) \times (a_e/a)$ obtained from the various ESI methods. Figures 1.26–1.28 shows the results for three index profiles.

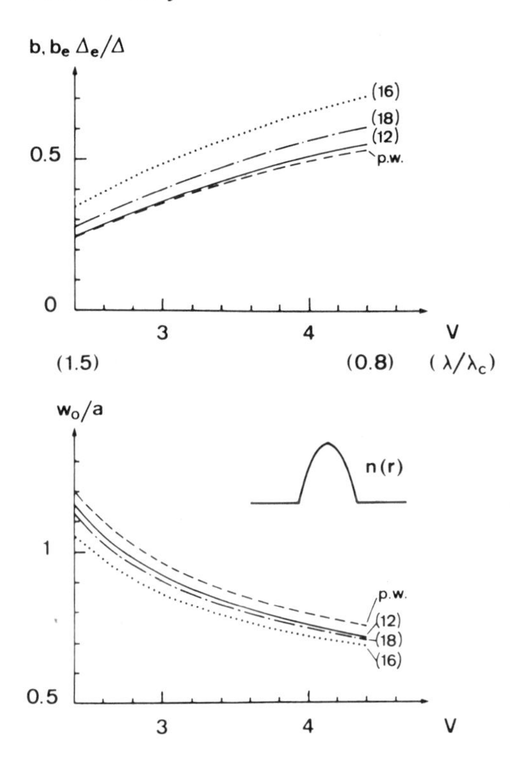

FIGURE 1.26 Exact normalized propagation constant b (Ref. 12, solid line) and equivalent normalized propagation constants $b_e\Delta_e/\Delta$ given by the ESI methods of Refs. 16 and 18 and the present work for a parabolic index fiber $n^2(r) = n_2^2 \{1 + 2\Delta[1 - (r/a)^2]\}$. Exact normalized mode field radius w_0/a [12] and equivalent normalized field radius $[= (w_0/a_e) \times (a_e/a)]$ given by the ESI methods of Refs. 16 and 18 and the present work for the same fiber as above.

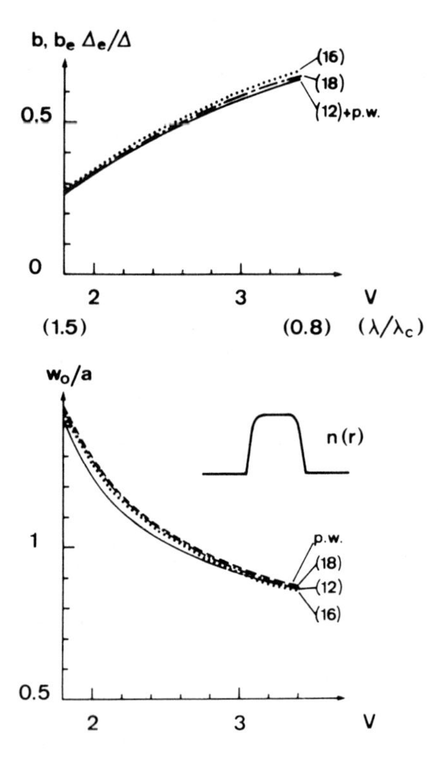

FIGURE 1.27 As Fig. 1.26, but for an index profile $n^2(r) = n_2^2 \{1 + 2\Delta[1 - (r/a)^8]\}$.

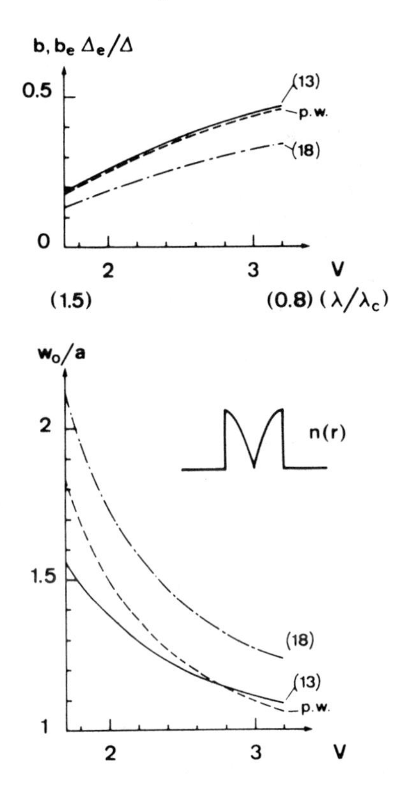

FIGURE 1.28 As Fig. 1.26, but for an index profile $n^2(r) = n_2^2 \{1 + 2\Delta[1 - (1 - r/a)^2]\}$. Exact values from Ref. 13; Ref. 16 is not applicable.

The preceding qualitative discussion of the accuracy of the various ESI methods is confirmed by the quantitative results. It is seen, for example, that the results of Ref. 18 are satisfactory for index gradients due to a rounding of the core–cladding boundary but become inaccurate for a central index dip. Globally, the method presented in Sec. 1.3.2 seems to provide an acceptable accuracy for all cases of practical interest.

ACKNOWLEDGMENT

During the preparation of this chapter, very fruitful discussions with M. Monerie and C. Vassallo were of great help and are sincerely acknowledged here.

REFERENCES

1. D. Gloge, *Appl. Opt.* *10*(10):2252–2258 (1971).
2. D. Marcuse, in *Theory of Dielectric Optical Waveguides, Quantum Electronics: Principles and Applications* (Y.-H. Pao, ed.), Academic, New York, 1974, pp. 60–78.
3. K. Okamoto and T. Okoshi, *IEEE Trans. Microwave Theory Tech.* MTT-26(2):109–114 (1978).
4. H. D. Rudolph and E. G. Neumann, *Nachrichtentech. Z.* *29*(4): 328–329 (1976).
5. R. A. Sammut, *Electron. Lett.* *15*(19):590–591 (1979).
6. D. Marcuse, *Bell Syst. Tech. J.* *56*(5):703–718 (1977).
7. I. H. Malitson, *J. Opt. Soc. Am.* *55*(10):1205–1209 (1965).
8. D. Gloge, *Appl. Opt.* *10*(11):2442–2445 (1971).
9. F. M. E. Sladen, D. N. Payne, and M. J. Adams, *Proc. 4th Eur. Conf. Opt. Commun.*, Genova, Italy, Sept. 12–15, 1978, pp. 48–57.
10. H. Hung-Chia and W. Zi-Hua, *Electron. Lett.* *17*(5):202–204 (1981).
11. D. Marcuse, *Appl. Opt.* *18*(17):2930–2932 (1979).
12. W. A. Gambling and H. Matsumura, *Opt. Quantum Electron.* *10*:31–40 (1978).
13. W. A. Gambling, H. Matsumura, and C. M. Ragdale, *Opt. Quantum Electron.* *10*:301–309 (1978).
14. D. Marcuse, *J. Opt. Soc. Am.* *68*(1):103–109 (1978).
15. A. W. Snyder and R. A. Sammut, *J. Opt. Soc. Am.* *69*(12): 1663–1671 (1979).
16. E. Brinkmeyer, *Appl. Opt.* *18*(6):932–937 (1979).
17. H. Matsumura and T. Suganuma, *Appl. Opt.* *19*(18):3151–3158 (1980).
18. W. J. Stewart, *Electron. Lett.* *16*(10):380–382 (1980).
19. R. A. Sammut and A. K. Ghatak, *Opt. Quantum Electron.* *10*: 475–482 (1978).
20. K. Hotate and T. Okoshi, *Electron. Lett.* *14*(8):246–248 (1978).
21. A. W. Snyder and R. A. Sammut, *Electron. Lett.* *15*(10):269–271 (1979).
22. C. D. Hussey and C. Pask, *Electron. Lett.* *17*(18):644–645 (1981).
23. S. Sentsui, K. Yoshida, Y. Furui, T. Kamiya, T. Kuroha, A. Kawana, and T. Miya, *Proc. 6th Eur. Conf. Opt. Commun.*, York, England, Sept. 1980, *IEE Publ. 190*, pp. 41–44.
24. S. Tanaka, H. Yokota, M. Hoshikawa, T. Miya, and N. Inagaki, *Proc. 6th Eur. Conf. Opt. Commun.*, York, England, Sept. 1980, *IEE Publ. 190*, pp. 37–40.
25. S. Kawakami and S. Nishida, *IEEE J. Quantum Electron.* *QE-10*(12):879–887 (1974).

26. S. Kawakami and S. Nishida, *IEEE J. Quantum Electron.* *QE-11*(14):130–138 (1975).
27. R. A. Sammut, *Opt. Quantum Electron.* *10*:509–514 (1978).
28. M. Monerie, *IEEE J. Quantum Electron.* *QE-18*(4):535–542 (1982).
29. M. Monerie, *Electron. Lett.* *18*(9):386–388 (1982).

2

Birefringence Properties

2.1 INTRODUCTION

Throughout all of Chap. 1 we assumed the fiber to be perfectly circular with a circularly symmetric refractive index distribution, and this led to the perfect degeneracy of the two LP_{01} (or HE_{11}) modes: both the x-polarized ($E_y = 0$) and y-polarized ($E_x = 0$) LP_{01} modes had the same propagation constant. However, actual fibers generally exhibit some ellipticity of the core and/or some anisotropy in the refractive index distribution due to anisotropic stresses. This results in two different propagation constants for the x-polarized (β_x) and y-polarized (β_y) LP_{01} modes, leading to perturbations of the state of polarization (SOP) of the light transmitted by the fiber. In this chapter we study the different sources of birefringence (birefringence will be defined here as $\beta_y - \beta_x$) and the effects on the SOP of transmitted light.

Before entering into the mathematical details of the birefringence properties of single-mode fibers, let us briefly desvribe the origins of this phenomenon. The origin of birefringence when anisotropic stresses are present in the fiber is that, under such conditions, the dielectric permittivity [expressed as n_j^2 in Eqs. (1.2)] is no longer a scalar but becomes a tensor. Equations (1.2) have thus to be written in a vectorial form with different elements in the tensor expressing the dielectric permittivity. This leads to different propagation constants for TE and TM modes (or for LP_{01y} and LP_{01x} modes), even with a circularly symmetric waveguide.

If we come back to the original slab waveguide shown in Fig. 1.1 (with an isotropic scalar dielectric permittivity), we observed in Sec. 1.1.1 that Maxwell's equations admitted two kinds of solutions:

the TE modes ($E_x = E_z = 0$) and the TM modes ($H_x = H_z = 0$). Using Eqs. (1.2a), (1.2c), and (1.3), together with the requirement that the tangential field components be continuous at the dielectric discontinuity, we found that the even TE modes were characterized by

$$u^2 + v^2 = V^2 \tag{2.1a}$$

$$v = u \tan u \tag{2.1b}$$

If we look in detail at the resolution of Eqs. (1.2d), (1.2f), and (1.4) for even TM modes, with the same boundary conditions, it is straightforward to obtain that they are characterized by Eq. (2.1a) and

$$v' = \frac{n_2^2}{n_1^2} u \tan u \tag{2.1c}$$

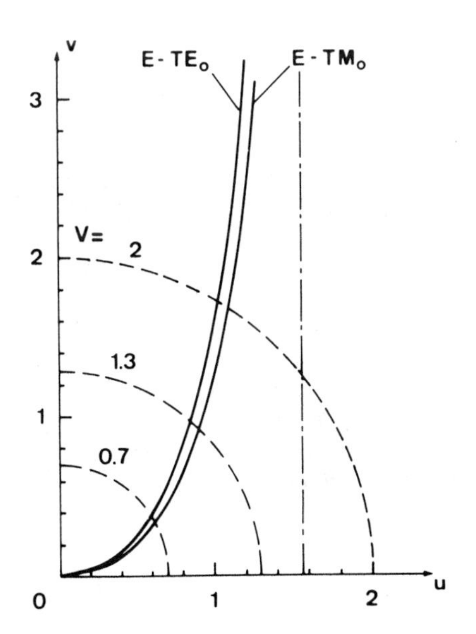

FIGURE 2.1 Graphical resolution of the characteristic equations (2.1a) and (2.1b) for the even TE_0 mode, and (2.1a) and (2.1c) for the even TM_0 mode. The slab waveguide is assumed to have an index difference such that $n_2^2/n_1^2 = 0.8$. The parameter V is given as $V^2 = u^2 + v^2$.

The resolution of Eqs. (2.1a) and (2.1b) or (2.1a) and (2.1c) is illustrated in Fig. 2.1, which shows that there indeed exists a difference between the v parameters of the lowest-order (even TE_0 and even TM_0) modes. With the relationship between the propagation constant β and the parameter v [Eq. (1.6)], we deduce that there exists a difference between the propagation constants of the y-polarized (TE) and x-polarized (TM) modes.

Simple observation of Fig. 2.1 and Eq. (1.6) shows that this birefringence ($\beta_y - \beta_x = \beta_{TE} - \beta_{TM}$) is positive and that, when normalized to the average propagation constant, it vanishes at both zero and infinite frequencies.

2.2 INTRINSIC BIREFRINGENCE

Intrinsic birefringence is that present in the fiber because of built-in anisotropies, either intentional or not.

2.2.1 Shape Birefringence

Birefringence. Let us consider the fiber core shown in Fig. 2.2a and the slab limit shown in Fig. 2.2b, where the major axis is $2a_y$ and the minor axis is $2a_x$ (for the slab, $a_y \to \infty$). In such structures the x-polarized and y-polarized LP_{01} modes will have slightly different propagation constants, leading to a birefringence $\delta\beta = \beta_y - \beta_x$ and a normalized birefringence $B = \delta\beta/\beta << 1$ (β denotes the mean value between β_x and β_y). It can be shown [1] that the normalized birefringence can be written

$$B = \Delta^2 F(e, V_x) \tag{2.2}$$

where Δ is the relative index difference as defined in Chap. 1, we assume a step-index fiber, e is the ellipticity ($e = [1 - (a_x/a_y)^2]^{\frac{1}{2}}$), and V_x is the normalized frequency as defined in Chap. 1 with a_x as "core radius." In the limit of small ellipticities (nearly circular fibers), Eq. (2.2) can be further simplified to [2]

$$B \simeq \Delta^2 e^2 f(V), \qquad e \to 0 \tag{2.3}$$

where the subscript x in V disappears because $a_x \simeq a_y$. For large ellipticities, $F(e, V_x)$ is more complicated, but comparisons between Eq. (2.3) and complete results [3] show that Eq. (2.3) can be used up to $e = 0.3$. Figure 2.3a shows f(V) and $F(1, V_x)$, which is the

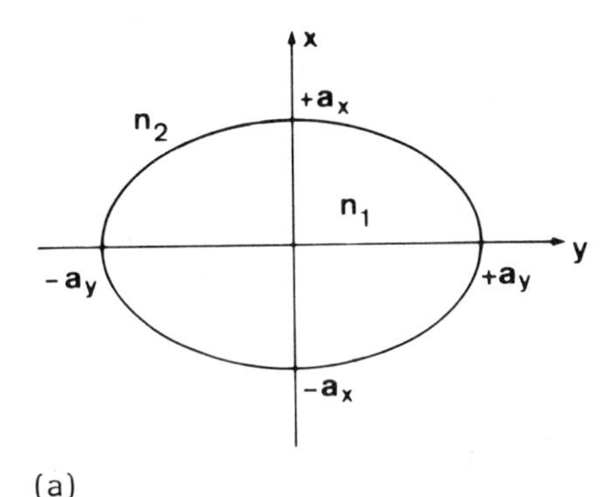

(a)

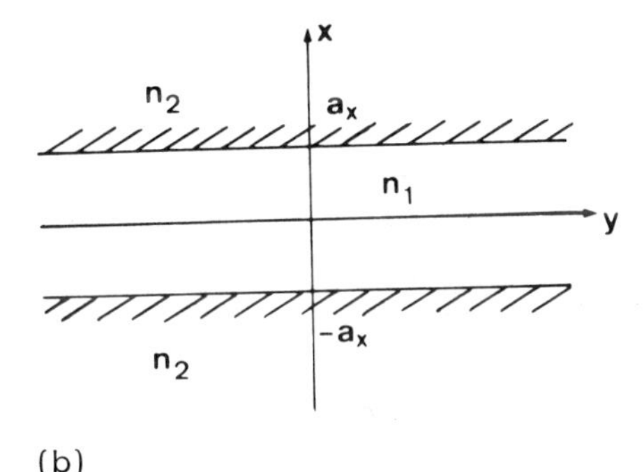

(b)

FIGURE 2.2 Elliptical core fiber (a) and slab limit (b).

limiting value obtained for the slab. It should be noticed that the corresponding birefringence is weak because it arises through the product of the longitudinal field components and the slight nonuniform polarization of the transverse field components [4].

Cutoff. The second-order mode cutoff in elliptical core fibers has received some controversial attention [5–7], and it is a very complicated mathematical problem involving an infinite set of Mathieu functions. However, a recent treatment somewhat analogous to some equivalent step-index-fiber considerations has made possible a fairly simple and accurate approximation [8]. Using the method of Ref. 8, we find that for a step-index elliptical core fiber with semiminor axis a_x, the second-order mode cutoff normalized frequency V_{xc} is approximated to better than a few percentage points by

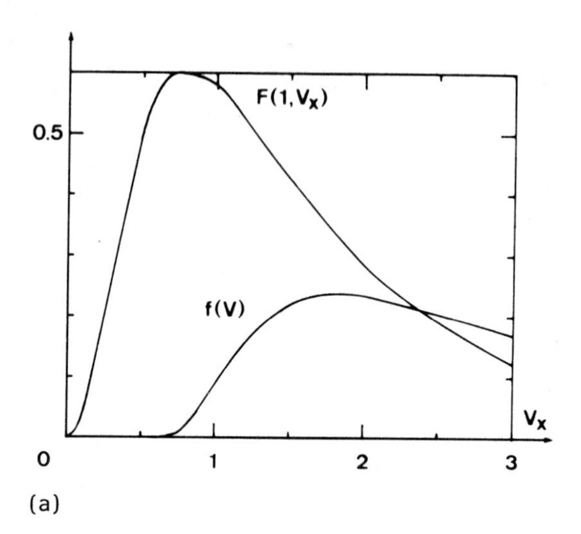

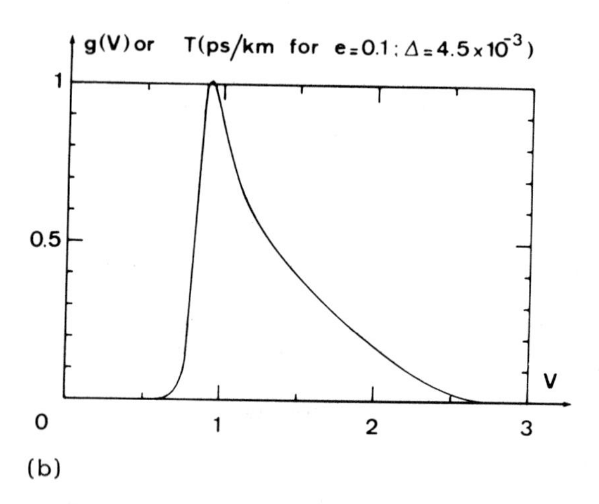

FIGURE 2.3 Effects of fiber ellipticity on modal birefringence. (a) Shape birefringence factors: f(V) for fibers with small ellipticities and $F(1, V_x)$ for the unlimited slab waveguide. [f(V) from Ref. 2; $F(1, V_x)$ from Ref. 3.] (b) Shape birefringence dispersion factor g(V) for small ellipticities, or group delay difference in picoseconds per kilometer for an ellipticity of 10% and a relative index difference of 0.45%.

$$V_{xc} \simeq 2.405\left(\frac{a_x}{a_y}\right)^{\frac{1}{2}}\left(1 + 5\left\{1 - \frac{1 + (a_x/a_y)^2}{[(3/2) + (a_x/a_y)^2 + (3/2)(a_x/a_y)^4]^{\frac{1}{2}}}\right\}\right)^{-\frac{1}{2}} \tag{2.4}$$

Actually, this cutoff is an average of the cutoff frequencies of the two orthogonally polarized second-order modes; however, the birefringence is generally small enough to make this notion valuable.

Birefringence Dispersion. As soon as the two LP_{01} modes have different propagation constants, they also may have different propagation group delay times. This means that the power carried by each mode will suffer different delays, and thus a short pulse of light launched into the fiber will generally be broadened. The difference T between the two group delays of the two polarizations is given by

$$T = \frac{1}{c}\frac{d(\delta\beta)}{dk} \simeq \frac{n_1}{c}\Delta^2 G(e, V_x) \tag{2.5}$$

where n_1 is the core refractive index, c the light velocity in vacuum, any effect of material dispersion has been neglected, and $G(e, V_x)$ is given by

$$G(e, V_x) \simeq F(e, V_x) + V_x \frac{\partial F(e, V_x)}{\partial V_x} \tag{2.6}$$

In the limit of small ellipticities ($e \to 0$), Eq. (2.3) leads to

$$G(e, V_x) = e^2 g(V) \tag{2.7a}$$

$$g(V) \simeq f(V) + V\frac{df(V)}{dV} \tag{2.7b}$$

For large ellipticities $G(e, V_x)$ can be found in Ref. 3, and for small ellipticities, g(V) is plotted in Fig. 2.3b. It should be noticed that in the present case of shape birefringence, use of the simple formula $T \simeq (n_1/c)B$ [9] would lead to large errors, in contrast to the case of stress birefringence, for which it holds (see Sec. 2.2.2).

2.2.2 Stress Birefringence

Because the materials used to manufacture optical fibers have different thermal expansion coefficients, it is possible to build anisotropic stresses into the fiber which lead to birefringence through the photoelastic effect.

Birefringence. Following the work in Ref. 10, we consider the slab model shown in Fig. 2.2b, and again denoting $\delta\beta = \beta_y - \beta_x$, we have (for the stress birefringence alone)

$$B = \frac{\delta\beta}{\beta} = -n_1^2 \frac{p_{11} - p_{12}}{2} (s_2 - s_1)(T_a - T_g) \tag{2.8}$$

where $p_{11} = 0.12$ and $p_{12} = 0.27$ are the usual photoelastic coefficients of silica, s_2 and s_1 are the thermal expansion coefficients of the cladding and of the core, respectively, and $T_a - T_g$ is the difference between the ambient temperature and the manufacturing temperature of the fiber (i.e., the temperature of glass softening). For silica doped with a dopant d, we can write

$$s[(d)] = (d)s_d + [1 - (d)]s_0 \tag{2.9}$$

where (d) is the molar concentration of the dopant d, s_d is its thermal expansion coefficient, and s_0 is the thermal expansion coefficient of silica. Typical values are [10]

$$s_0 = 5 \times 10^{-7}/°C$$

$$d = GeO_2, \qquad s_d = 7 \times 10^{-6}/°C$$

$$d = P_2O_5, \qquad s_d = 14 \times 10^{-6}/°C$$

$$d = B_2O_3, \qquad s_d = 10 \times 10^{-6}/°C$$

Simple considerations show that the anisotropic strain induced along 0y is equal to the tensile strain induced along 0z, and the maximum birefringence allowed by Eq. (2.8) is thus limited by the tensile strength of the fiber. The strain is simply equal to $|(s_2 - s_1)(T_a - T_g)|$ and it should be limited to 2% for laboratory experiments and to 0.4% for long-mean-time-between-failure (MTBF) systems [11]. The corresponding limits of the normalized birefringence are 0.3% and 0.06%, respectively.

Instead of the birefringence itself, it is common to represent it by the *beat length* L_b, which corresponds to the fiber length such that $\delta\beta \times L_b = 2\pi$. One drawback is that L_b usually depends on the light wavelength (except for the twist; see Sec. 2.3.3), but on the other hand, it can be directly observed in some cases (see Chap. 5). For the normalized birefringence limits given above, we have $L_b \simeq 0.2$ mm and $L_b \simeq 1$ mm, respectively, at a wavelength of 1 μm.

Birefringence Dispersion. Again leaving the degeneracy of the two LP_{01} modes leads to a difference in their propagation delay times given by $T = (1/c)(d(\delta\beta)/dk)$. Starting from Eq. (2.8), we simply obtain

$$T = \frac{n_1}{c} B \left\{ 1 - \lambda \frac{d[n_1^3(p_{11} - p_{12})]/d\lambda}{n_1^3(p_{11} - p_{12})} \right\} \tag{2.10}$$

It can be seen in Ref. 12 that the second term in the braces is smaller than 0.1 for $0.6\ \mu m \leq \lambda \leq 1.6\ \mu m$, and we can thus neglect it. This shows that the simple relationship given in Ref. 9 for $T\ (\simeq n_1 B/c)$ holds for this kind of stress-induced birefringence.

2.2.3 Special Manufacturing Conditions

The manufacture of fibers with a controlled amount of intrinsic birefringence has received detailed attention, as the SOP of light transmitted by the fiber may be an important parameter for many applications. It is assumed here that the reader is already familiar with the classical preform manufacturing methods [axial vapor deposition (AVD), outside vapor deposition (OVD), and various types of chemical vapor deposition (CVD)] and fiber drawing processes, which have been described extensively in the literature.

Strong Shape Birefringence. Obtaining a very strong shape birefringence requires both a large ellipticity and a high index difference, as evidenced by Eqs. (2.2) and (2.3) and Fig. 2.3a. This point has been checked experimentally]3,13] and the highest normalized birefringence reported for these structures is about 0.06% with an index difference $\Delta n = 6.5\%$ and an ellipticity $e = 0.9$, in good agreement with theory [3]. Figure 2.4 shows the experimental point on a B or L_b (at 1 μm) versus Δ scale, together with the theoretical maximum values for two ellipticities. The other points and scales will be explained later. However, such large index differences are probably not suitable for very low loss fibers for

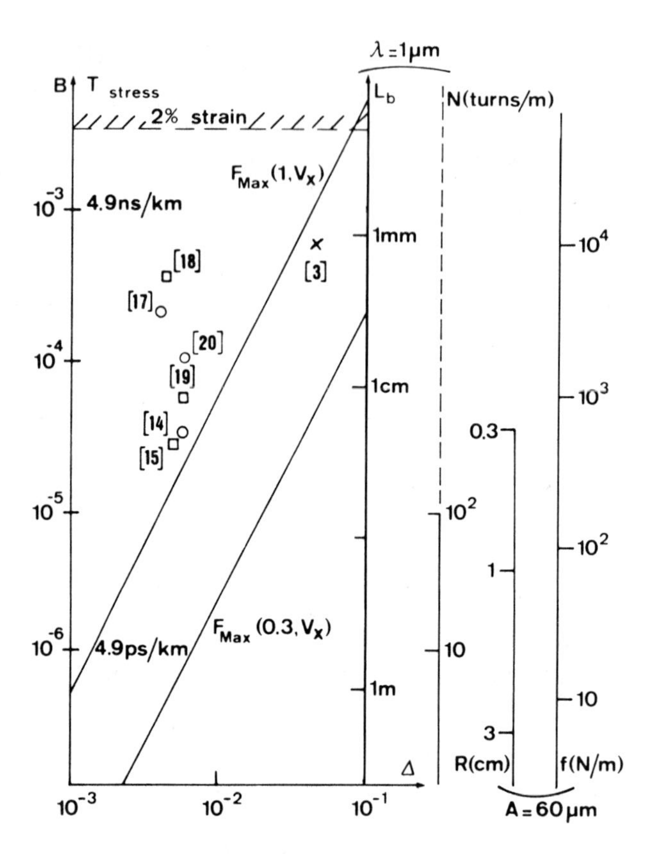

FIGURE 2.4 Plot of normalized birefringence B as a function of relative index difference Δ. Experimental points: cross, pure shape birefringence; open box, pure stress birefringence; open circle, others. Labels within brackets refer to the corresponding reference. T scale: birefringence dispersion or group delay time difference between the two LP_{01} modes, for pure stress birefringence [Eq. (2.10)]. L_b and N scales: beat length [$L_b = 2\pi/(B \times \beta)$] and number of turns per meter of twisted fiber (N), corresponding to a normalized birefringence B (circular when fiber is twisted), calculated at $\lambda = 1$ μm [Eq. (2.18) for N]. R and f scales: fiber radius of curvature (R) or lateral force per unit length (f) required to obtain a normalized birefringence B in a fiber with outside radius A = 60 μm [Eq. (2.12) for R and (.216) for f].

long-distance applications (see Chap. 3). Figure 2.5a shows a cross section of such a fiber, and Fig. 2.5b shows a measurement of its beat length at 0.633 μm wavelength (for a description of the principles of beat length measurement, see Chap. 5).

Structures with asymmetrical side pits of the refractive index (Fig. 2.6) have been proposed [14] but, as seen on Fig. 2.4, the normalized birefringence exceeds the maximum slab limit, and it is thus likely that stress birefringence plays a role in this fiber. (The index difference retained for Fig. 2.4 is the index difference between the core and the pits.)

Although almost no details have been given on the manufacturing methods for all these fibers (except that they are obtained through the MCVD method), one can infer that they have been obtained through controlled depression of the tube during collapse or without continuous rotation of the tube during the deposition stage (only incremental rotations being used for each part of the structure).

Strong Stress Birefringence. As noticed in Ref. 13, it is very difficult to obtain a strong birefringence with only the shape effect and reasonable index differences. It has thus been proposed [15] to induce a strong stress birefringence by grinding flats on the substrate tube before processing it through MCVD. The cladding is made of B_2O_3-doped SiO_2, whereas the core is of pure silica, and because of the difference in the thermal expansion coefficients and surface tension effects during collapse, the resulting fiber has a circular core and a circular outer surface but an elliptical inner cladding, the whole structure exhibiting a strong anisotropic stress distribution. A detailed study of these fibers has been published in Ref. 16, and typical results are a normalized birefringence of 0.003% with a relative index difference $\Delta = 0.5\%$. The corresponding point is plotted in Fig. 2.4 with the label [15]. An alternative approach is to cut the preform after collapse as indicated in Fig. 2.7a. The inner cladding can be doped with GeO_2 [17] or other dopants (in Fig. 2.7b, B_2O_3 was used), and again because of different thermal expansion coefficients and surface tension effects during fiber drawing, the resulting fiber is close to the slab model, as seen in Fig. 2.7b. This kind of fiber combines both shape and stress birefringence, in agreement with Eqs. (2.2) and (2.8), and typical results are $B \simeq 0.022\%$ with $\Delta \simeq 0.4\%$, as shown by point 17 in Fig. 2.4.

However, most of these fibers exhibited some aging properties, strong inhomogeneities, and losses almost twice as high as in conventional fibers. One attempt to reduce these losses consisted of decreasing the inner pressure of the tube during collapse, which resulted in a circular germanosilicate core and elliptical borosilicate inner cladding [18]. Between the elliptical inner cladding and the

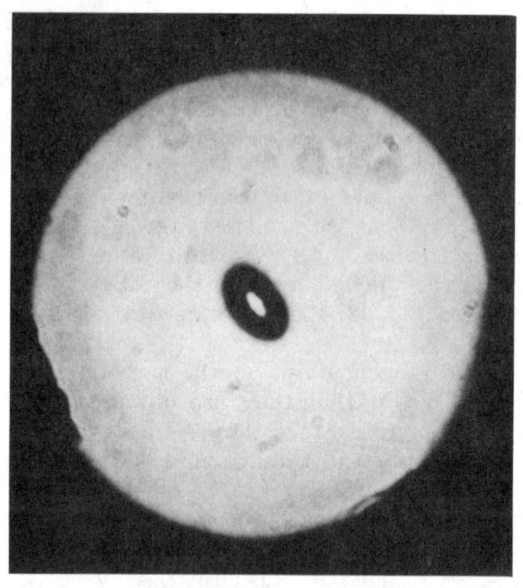

(a)

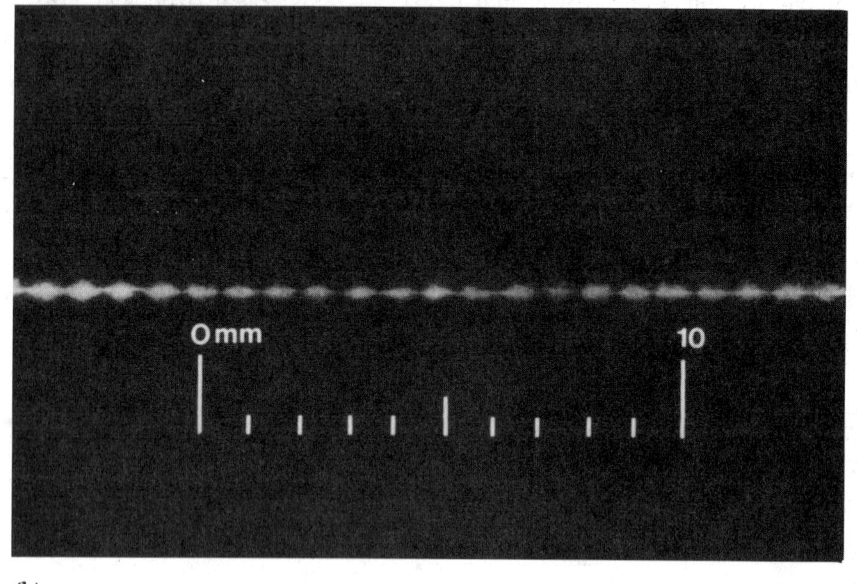

(b)

FIGURE 2.5 Shape birefringent fiber. (a) Cross section of fiber: the bright elliptical spot corresponds to the core, the dark elliptical area to the deposited inner cladding, and the gray area to the starting tube (MCVD). (b) Beat length determination in fiber shown in (a), at λ = 633 nm. (Courtesy of Andrew Corporation, Orland Park, Illinois.)

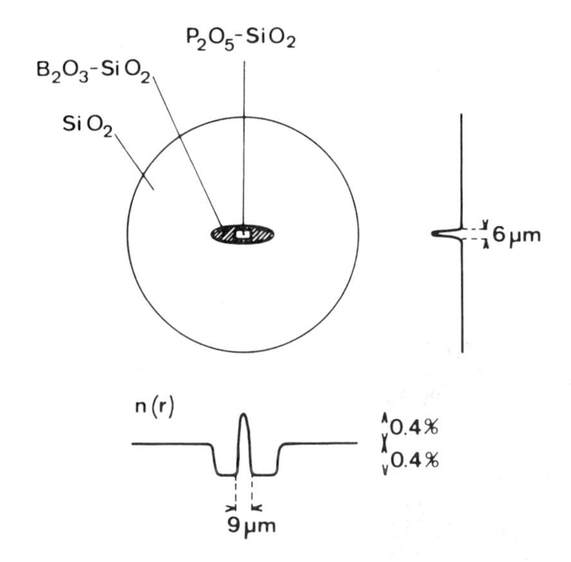

FIGURE 2.6 Shape and stress birefringent fiber with asymmetrical refractive index pits. Parameters are from Ref. 14.

core, an intermediate circular layer of silica was deposited, the idea being to isolate optically the guiding structure from the stress-inducing structure (Fig. 2.8). As seen in Fig. 2.4, strong birefringence was obtained and the minimum loss is about 0.8 dB/km at 1.55 μm [18]. Further improvements have been obtained by manufacturing a GeO_2-doped core through the VAD method and drawing a fiber through the rod in a tube technique with two B_2O_3-doped pits, separated from the core by a large pure-silica region (nine times the core radius) [19]. The birefringence is less than in the previous fiber (see point 19 in Fig. 2.4), but the minimum loss was about 0.6 dB/km at 1.52 μm.

Finally, recent results seem to indicate that it will soon be possible to get, simultaneously, ultralow losses, high birefringence, and (perhaps) zero birefringence dispersion [20]. For this fiber (PANDA fiber), shape and stress birefringence values combine, the stress birefringence being obtained as in Ref. 19; its structure is illustrated in Fig. 2.9. In Ref. 20, the core—cladding relative index difference is 0.6%, the cladding-pits relative index difference is about 0.3% (corresponding to a 14% molar concentration of B_2O_3, from which the thermal expansion coefficient can be obtained), and the separation between the core and the pits is about eight times the core radius. The loss is 0.4 dB/km at 1.54 μm, and the normalized birefringence is about 0.01% (Fig. 2.4). It still seems possible to obtain high birefringences and zero birefringence

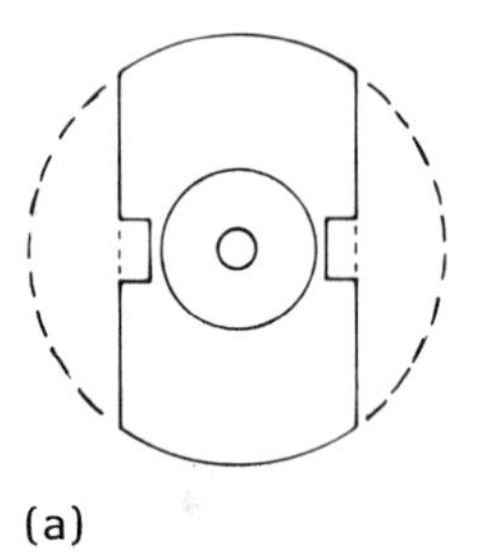

(a)

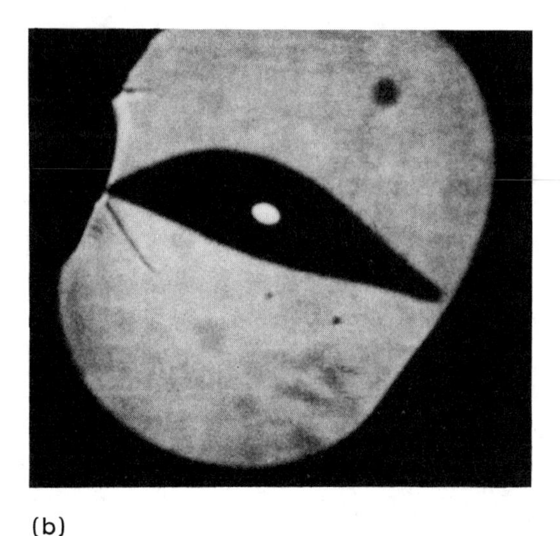

(b)

FIGURE 2.7 Preform (a) and fiber (b) prepared as indicated in Ref. 17, and giving rise to combined shape and stress birefringence. [(a) From Ref. 17.]

dispersion in such fibers, by carefully adjusting the parameters so that the shape and the stress birefringence values combine, whereas their respective dispersions cancel each other out [20].

Small Birefringence. Some applications, such as the Faraday magnetic field sensor (see Chap. 8) or circular polarization-maintaining fibers (see Sec. 2.4.3), require that the fiber exhibit a very small intrinsic linear birefringence. As seen above, this requires an almost perfectly circular core and a good match between the thermal expansion coefficients of the different parts of the fiber structure.

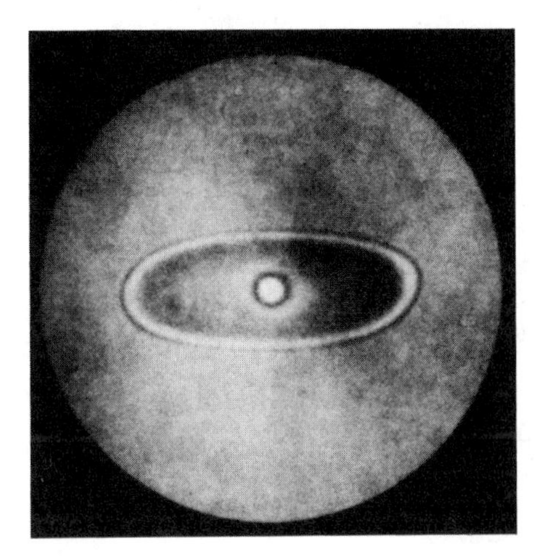

FIGURE 2.8 Stress birefringent fiber of Ref. 18, consisting of a circular GeO_2-SiO_2 core, a circular SiO_2 buffer layer, an elliptical B_2O_3-SiO_2 cladding, and the silica supporting tube. (Courtesy of Hitachi Central Research Laboratory, Tokyo, Japan.)

First progress in this way led to a normalized birefringence of 1.8×10^{-8} with a relative index difference $\Delta \simeq 0.2\%$ [21], and further improvements yielded $B \simeq 4.6 \times 10^{-9}$ with $\Delta \simeq 0.34\%$ [22].

A significant improvement in the manufacturing of low-birefringence fibers has been achieved with the introduction of preform

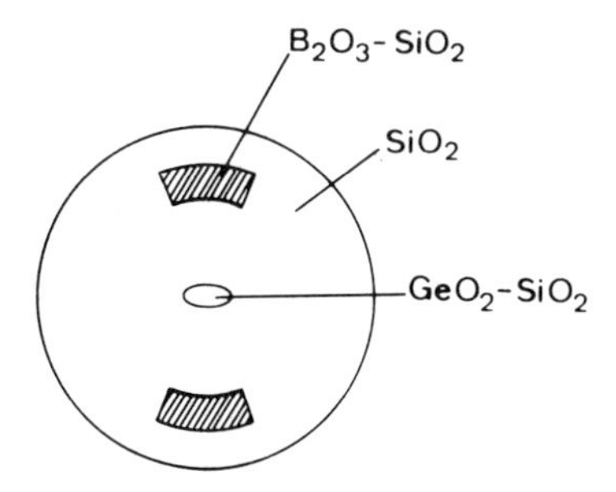

FIGURE 2.9 Structure of the PANDA fiber, combining shape and stress birefringence. (From Y. Sasaki et al., in *Optical Fibers Communication 1982 Digest of Technical Papers*, Optical Society of America, Washington, D.C., April 1982, paper THCC6.)

spinning during fiber drawing [23]. For a typical fiber drawing speed of 0.5 m/s, spin rates of 300 to 1500 rpm have been achieved without a marked increase in loss, thus providing spin pitches of 2 to 10 cm. The theoretical analysis of this effect is analogous to that in Sec. 2.4.2 for fibers with some intrinsic linear birefringence and strong twist, with the simplification that the elastooptic effect is eliminated here [$g = 0$ in Eq. (2.24), because the twist takes place in a viscous part of the glass at the preform tip, and thus no torsional stresses are induced]. From the physical point of view, this can be understood as a succession of birefringent plates with their principal axes of birefringence rotated continuously. The local linear birefringence is unaffected, but there exists an averaging effects, the overall apparent birefringence of a long fiber being reduced to $\delta\beta/2\pi N$, where N is the number of turns per meter. This technique thus allows the reproducible manufacture of very low birefringence fibers (see Sec. 2.5.2).

2.3 INDUCED BIREFRINGENCE

Once the fiber has been manufactured, it is almost impossible to induce any modification of its shape by external means because of the high Young's modulus of silica. But externally applied stresses can lead to induced birefringences through the photoelastic effect. We will consider here that the fiber has no intrinsic linear birefringence and concentrate on the induced birefringence.

2.3.1 Bending

Pure Bending. When a fiber of outside radius A is bent with a radius $R \gg A$ according to Fig. 2.10a, the induced birefringence $\delta\beta = \beta_y - \beta_x$ (for axes orientation see Fig. 2.10a) can be expressed as [24]

$$B_b = \frac{\delta\beta_b}{\beta} = 0.25 n_1^2 (p_{11} - p_{12})(1 + \nu)\left(\frac{A}{R}\right)^2 \qquad (2.11)$$

where ν is the Poisson's ratio of silica ($\nu \simeq 0.16$). With the values $p_{11} = 0.12$ and $p_{12} = 0.27$ for silica, we obtain

$$B_b = -0.093\left(\frac{A}{R}\right)^2 \qquad (2.12)$$

This birefringence is due to a second-order stress compared to the dominant longitudinal stress, which has no effect because of

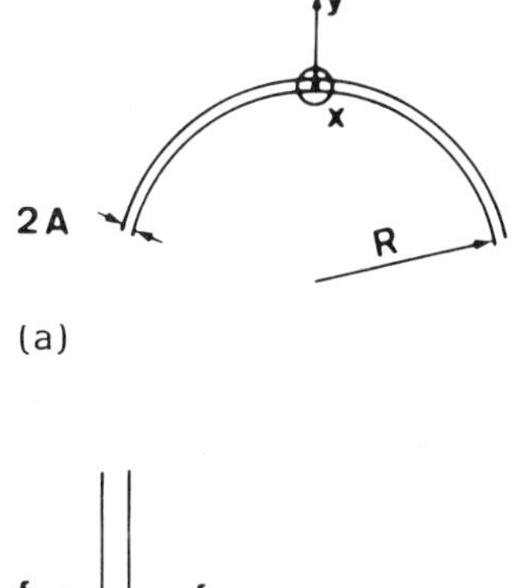

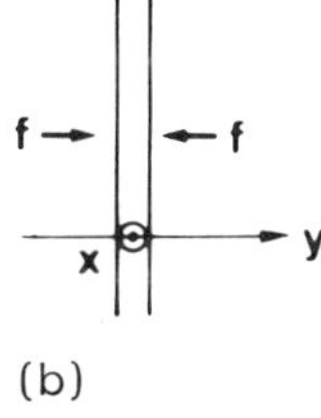

FIGURE 2.10 Externally induced stress birefringence axes. (a) Bending. (b) Lateral force.

symmetry [24]. Again the maximum normalized birefringence that can be induced this way is limited by the tensile strength of the fiber. The strain on the external outer surface of the fiber is given by A/R and should not exceed about 2% for laboratory experiments or 0.4% for long-MTBF systems [11]. The corresponding limits for $|B_b|$ are 3.7×10^{-5} and 1.5×10^{-6}. In Fig. 2.4 we have plotted a scale with the radius of curvature R corresponding to a given normalized birefringence, assuming a fiber with an outside diameter of 120 μm (dividing the outside diameter by K would divide the R scale by K). The scale has been limited at the 2% strain point, which (in terms of normalized birefringence) is independent of A.

Bending Under Tension. If we consider a pure tensile stress applied to the fiber, simple symmetry considerations show that there is no induced birefringence. But if subsequent to the application of the tensile stress, the fiber is wound on a mandrel of radius R that reacts on the fiber surface, the second-order stress effect of bending mixes the two stress components and gives an additional birefringence to the previous one [25]. The built-in tensile stress associated with the manufacturing of stress birefringent fibers does not lead to this effect because there is no reaction of the mandrel against this built-in stress, and this effect is thus restricted to fibers wound under tension. The additional birefringence is given by [25]

$$B_t = \frac{n_1^2}{2}(p_{11} - p_{12})\frac{(1+\nu)(2-3\nu)}{1-\nu}\frac{A}{R}S_{zz} \qquad (2.13)$$

where S_{zz} is the axial tensile strain applied externally, and the axis orientation is the same as in Fig. 2.10a. Using the previous numerical figures, we obtain for the total resultant normalized birefringence in tension-coiled fibers

$$B_{bt} = -0.093\left(\frac{A}{R}\right)^2 - 0.336\,\frac{A}{R}S_{zz} \qquad (2.14)$$

It is thus clear that winding the fiber under tension increases the induced birefringence compared to the case without tension. The maximum tensile strain seen by the fiber is simply $S_{zz} + A/R$, and again considering 2% and 0.4% as maximum values for laboratory experiments or real systems, we obtain maximum values of 4.6×10^{-5} and 1.9×10^{-6}, respectively, for B_{bt}, only 25% higher than in the case of pure bending. Because of the form of Eqs. (2.11) and (2.13) compared to Eq. (2.8), the same formula as that given by Eq. (2.10) for the birefringence dispersion will hold here.

2.3.2 Lateral Force

Let us consider a fiber submitted to a lateral force per unit length f as shown in Fig. 2.10b (the lateral force is assumed to be applied along 0y). The corresponding birefringence $\delta\beta_f = \beta_y - \beta_x$ is given by [26]

$$B_f = \frac{\delta\beta_f}{\beta} = \frac{4n_1^2}{\pi}\frac{1+\nu}{Y}(p_{12} - p_{11})\frac{f}{2A} \qquad (2.15)$$

where $Y = 6.5 \times 10^{10}$ N/m^2 is the Young's modulus of silica. Using numerical values for the various coefficients, we obtain (with f in N/m)

$$B_f = 3.63 \times 10^{-12}\,\frac{f}{A} \qquad (2.16)$$

Considering a 120-μm-outer-diameter fiber, we have plotted a scale of f equivalent to a given normalized birefringence in Fig. 2.4.

Obviously, if the fiber is protected by a soft plastic jacket, the effect is likely to be greatly reduced. Again, the birefringence dispersion is given by Eq. (2.10).

A practical application of these formulas concerns the problem of fiber clamping in polarization-sensitive systems or measurements. One current method consists of fixing the fiber into a V-groove by pressing it inside the V, with the help of a spring or a magnetic clamp. Because of symmetry, one would expect that a V-groove angle of 60° should provide zero birefringence. However, experimental and theoretical considerations [27] show that, because of frictional forces, the minimum birefringence does not vanish but rather goes through a shallow minimum. In any case, the 60° groove angle seems to be the optimum angle for minimizing the residual birefringence.

2.3.3 Twist

Birefringence. We always assume that the fiber has no intrinsic birefringence. Twisting the fiber around its axis with a uniform twist rate $2\pi N$ (rd/m), where N is the number of turns per meter, will induce a shear stress that leads to a circular birefringence, in contrast to all previous cases in which the birefringence was linear. This simply means that the eigenmodes of the twisted fiber are no longer x (resp. y) linearly polarized modes but right (resp. left) circularly polarized modes. The induced birefringence $\delta\beta_c$ is given by [4]

$$\delta\beta_c = \frac{n_1^2}{2}(p_{11} - p_{12})\, 2\pi N = g 2\pi N \tag{2.17}$$

where $\delta\beta_c$ is the difference between the propagation constants of the mode circularly polarized in the same sense as the twist and the mode circularly polarized in the opposite sense. It should be noted that, to first order, $\delta\beta_c$ is independent of λ, contrary to all other cases. Theoretically one would expect $g = -0.16$, but experimentally a value of -0.14 has been found [4,28] and we will use this value hereafter. Also notable is the fact that neglecting the longitudinal components of the fields would make this effect disappear [4]. Starting from Eq. (2.17), we obtain

$$B_c = \frac{\delta\beta_c}{\beta} \underset{-}{\sim} \frac{g}{n_1} N\lambda \underset{-}{\sim} -0.1 N\lambda \tag{2.18}$$

For ensuring acceptable mechanical reliability of the fiber, N should be limited to approximately $100 \times (60/A)$ for laboratory experiments, and to 20 × (60/A) for long-MTBF systems, where A is expressed in micrometers and N in turns per meter [11]. In Fig. 2.4 we have plotted a scale of N equivalent to a given normalized birefringence for $\lambda = 1$ μm. The scale has been limited to the mechanical limit for A = 60 μm and a 2% strain. It can be seen that the maximum circular birefringence approaches the classical linear birefringences built into the fiber.

Birefringence Dispersion. From Eqs. (2.17) and (2.18), we obtain

$$T_c = \frac{\lambda}{c} B_c \left\{ \frac{dn_1}{d\lambda} - n_1 \frac{d[n_1^3(p_{11} - p_{12})]/d\lambda}{n_1^3(p_{11} - p_{12})} \right\} \tag{2.19}$$

The first term in braces is about -1.5×10^4 m^{-1} around $\lambda = 1$ μm (see Ref. 7 of Chap. 1), and the second term is about -10.2×10^4 m^{-1} [12], and we thus obtain at about 1 μm wavelength (but this does not vary significantly for other usual wavelengths):

$$T_c \simeq 8.7 \times 10^4 \frac{\lambda}{c} B_c \simeq -2.9 \times 10^{-5} N\lambda^2 \quad \text{(s/m)} \tag{2.20}$$

with λ in meters. Compared with Eq. (2.10), this shows that for a given normalized birefringence, the birefringence dispersion in twisted fibers is about 10 to 20 times smaller than that in linearly birefringent fibers, depending on the wavelength.

2.3.4 External Fields

Electrical Field. A transverse electrical field disposed like the lateral force in Fig. 2.10b will induce a linear birefringence through the Kerr effect [4]. The corresponding normalized birefringence is given by

$$B_k = K(E_k)^2 \tag{2.21}$$

where E_k is the electrical field amplitude and $K \simeq 2 \times 10^{-22}$ m^2/V^2 is the normalized Kerr effect constant of silica.

Magnetic Field. A magnetic field applied longitudinally along the fiber axis will induce a circular birefringence through the Faraday effect, with

$$\delta\beta_h = 2V_f H_f \tag{2.22}$$

where H_f is the magnetic field amplitude and $V_f \simeq 4.6 \times 10^{-6}$ rad/A is the Verdet constant of silica. It is noteworthy that this effect is nonreciprocal; that is, placing a mirror at the output of the fiber and launching a linear SOP returns a linear SOP rotated by $\delta\beta_h$ times the fiber length.

2.4 PROPAGATION EFFECTS

After examination of the birefringence sources in single-mode optical fibers, it is necessary to look at the influence of these effects on the SOP of the transmitted light. We will first develop the formalisms to be used and then discuss some practical results.

2.4.1 Coupled Modes and Poincaré Sphere

The basic assumption sustaining all the following is that the field at an abscissa z along the fiber is a linear combination of the two fields (E_1 and E_2) composing the local HE_{11} eigenmode. All other modes are neglected, and it is assumed that they do not play any role in the coupling processes, which may be a somewhat risky assumption for propagation over long distances. We also assume that the loss is independent of the SOP. E_1 and E_2 can be linearly, circularly, or elliptically polarized, but they are orthogonal to each other.

Coupled Modes. The general form of the equations governing the evolution of the amplitudes $X_1(z)$ and $X_2(z)$ of the field decomposed on E_1 and E_2 is [29]

$$\frac{dX_1(z)}{dz} = i\,\frac{\delta\beta}{2}\,X_1(z) + \frac{C}{2}\,X_2(z) \tag{2.23a}$$

$$\frac{dX_2(z)}{dz} = -\,\frac{C^*}{2}\,X_1(z) - i\,\frac{\delta\beta}{2}\,X_2(z) \tag{2.23b}$$

where $\delta\beta$ is the difference between the propagation constants of E_2 and E_1, C is a coupling coefficient, and a common phase factor equal to $-i\beta$ has been omitted in diagonal terms. Generally, $\delta\beta$ and C will depend on z, but there is no general solution of Eq. (2.23) in this case. Two important points which are often overlooked should be kept in mind [29]:

1. Whatever the values of $\delta\beta$ and C, *if the light launched into the fiber is strictly monochromatic, there is no depolarization of the light.* This means that at a given instant, the light emerging from the fiber is always perfectly polarized, but its SOP is generally

elliptical. This SOP generally depends on the environmental conditions and on the exact wavelength, and thus depolarization takes place only if the light source has a nonzero line width.

2. *There always exist two orthogonal linearly polarized input states which lead to orthogonal linearly polarized outputs* (for monochromatic light). These particular states depend on the environmental conditions and on the wavelength (except when there is only linear birefringence in the fiber) and should not be confused with eigenstates. An experimental proof of this has been given in Refs. 29 and 30. Before looking at some results, let us briefly discuss the Poincaré sphere representation.

Poincaré Sphere. The Poincaré sphere allows a geometrical representation of birefringence and coupling effects [4]. It consists of a point-to-point mapping of all possible SOPs on the surface of a sphere (see Fig. 2.11). Circular SOPs are on the poles of the sphere, whereas linear SOPs are on the equator. A circular birefringence is represented by a rotation of the point representative of the initial SOP around the vertical axis, with an angle equal to the circular birefringence multiplied by the fiber length. Linear birefringence rotates the initial SOP point around an equatorial axis whose orientation is linked with that of the linear birefringence axes, with an angle equal to the linear birefringence times the fiber length. The combination of several birefringences thus results in the sum of several rotational vectors.

2.4.2 Uniform Birefringences

Equation (2.23) can be solved practically only when all the birefringences under consideration are constant along 0z. This usually depends on the reference axes chosen; for example, a linearly birefringent twisted fiber will have constant coefficients only in its local axes, but not in axes linked to the laboratory (see Fig. 2.12). The respective values of $\delta\beta$ and C will also depend on the nature of the base (E_1, E_2). We give next some values of $\delta\beta$ and C for the usual cases, together with the set of reference axes and the base (E_1, E_2) to be chosen.

Intrinsic Linear Birefringence and Twist. The axes to be chosen are the local (twisted) axes of the fiber (Fig. 2.12). We can choose (E_1, E_2) either as the linearly polarized eigenstates of the fiber in the absence of twist, or as circularly polarized fields with components $(1, \mp i)$ on the local axes. With (E_1, E_2) linear, we have

$$\delta\beta = \delta\beta_\ell, \qquad C = (2 + g)2\pi N \tag{2.24}$$

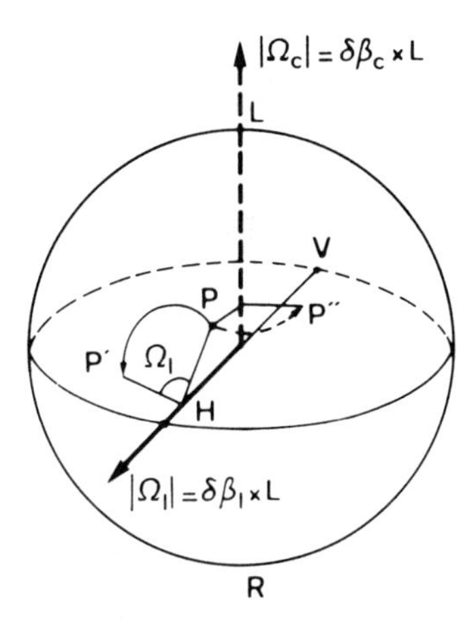

FIGURE 2.11 Birefringence influence on SOP evolution with the Poincaré sphere. Points L and R correspond to left-hand circular and right-hand circular SOPs. Point H corresponds to a linear SOP, along the x axis of the fiber, whereas V corresponds to a linear SOP along the y axis. All points on the V-H circle are linear SOPs. Point P corresponds to an elliptical input SOP. Point P' is obtained from P after a rotation around the H-V axis when the fiber presents only linear birefringence. Point P'' is obtained through rotation around the R-L axis when the fiber presents only circular birefringence.

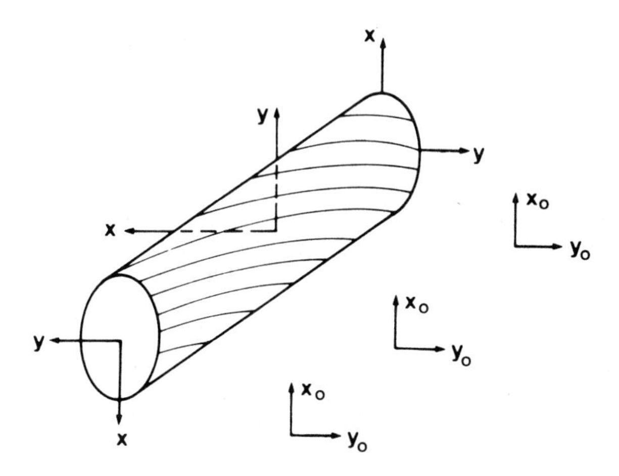

FIGURE 2.12 Twisted fiber with its local birefringence axes (x, y) and laboratory axes (x_0, y_0).

where $\delta\beta_\ell$ is the intrinsic linear birefringence of the fiber, and N and g are as defined in Eq. (2.17). The factor 2 in C arises from the geometrical effect due to the rotation of the local fiber axes.

With (E_1, E_2) circular, we have

$$\delta\beta = (2 + g)2\pi N, \qquad C = i\delta\beta_\ell \tag{2.25}$$

The first choice is best suited when $|\delta\beta_\ell| >> |(2 + g)2\pi N|$, as the resulting eigenstates are nearly linearly polarized, whereas the second choice is better when $|\delta\beta_\ell| << |(2 + g)2\pi N|$, as the resulting eigenstates are nearly circularly polarized [29].

Intrinsic Linear Birefringence and Plane Winding. We assume that the fiber has an intrinsic linear birefringence $\delta\beta_\ell$ and that it is wound on a drum without twist. The bending-induced birefringence $\delta\beta_b$ is given by Eq. (2.11), or (2.14), and we call θ the angle between the intrinsic x axis and the induced x axis (Fig. 2.13). We choose the laboratory axes and (E_1, E_2) the linearly polarized eigenstates of the fiber without bending.

$$\delta\beta = \delta\beta_\ell + \delta\beta_b \cos 2\theta, \qquad C = -i\delta\beta_b \sin 2\theta \tag{2.26}$$

The resulting eigenstates are also linearly polarized [29].

Twisted Fiber and Plane Winding. We assume that the fiber has no intrinsic birefringence and choose (E_1, E_2) circularly polarized, with the laboratory axes.

$$\delta\beta = g2\pi N, \qquad C = -i\delta\beta_b \tag{2.27}$$

If $|\delta\beta| >> |C|$, the eigenstates are nearly circularly polarized.

Helically Wound Fiber. We assume that the fiber is wound around a cylindrical central mandrel in a helical path and that there is no detwisting device, as this is the case in some cabling facilities (Fig. 2.14). The radius of the central mandrel is taken as R_h and the pitch as P_h, and then the resulting radius of curvature R and twist rate $2\pi N$ are

$$R = R_h + \frac{(P_h)^2}{4\pi^2 R_h}, \qquad 2\pi N = \left[\frac{P_h}{2\pi} + 2\pi\frac{(R_h)^2}{P_h}\right]^{-1} \tag{2.28}$$

We assume that the fiber has an intrinsic linear birefringence $\delta\beta_\ell$.

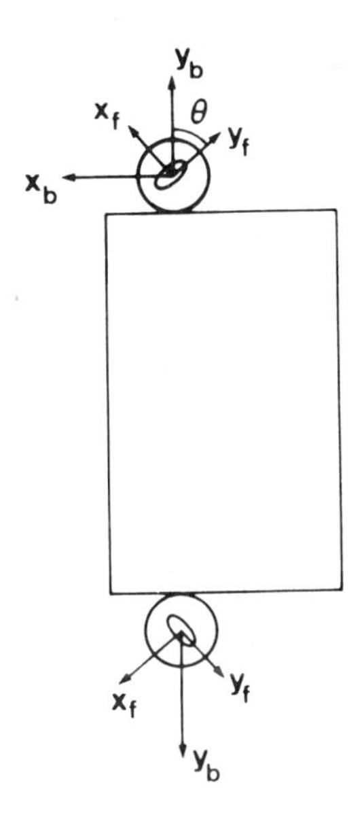

FIGURE 2.13 Fiber with local intrinsic linear birefringence axes (x_f, y_f), wound on a drum without twists. (x_b, y_b) are the bending-induced birefringence axes.

Choosing linear (E_1, E_2) along the axes of the resulting linear birefringence (Fig. 2.14), we have

$$\delta\beta = (\delta\beta_\ell^{\,2} + \delta\beta_b^{\,2} + 2\delta\beta_\ell\,\delta\beta_b \cos 2\theta)^{\frac{1}{2}}$$
$$C = (2 + g)2\pi N \qquad (2.29)$$

where $\delta\beta_b$ is the bending-induced birefringence (with radius R) and θ is the angle between the local intrinsic x axis and the bending induced x axis.

Choosing (E_1, E_2) as circularly polarized eigenstates with the same local axes as above yields

$$\delta\beta = (2 + g)2\pi N$$
$$C = i(\delta\beta_\ell^{\,2} + \delta\beta_b^{\,2} + 2\delta\beta_\ell\,\delta\beta_b \cos 2\theta)^{\frac{1}{2}} \qquad (2.30)$$

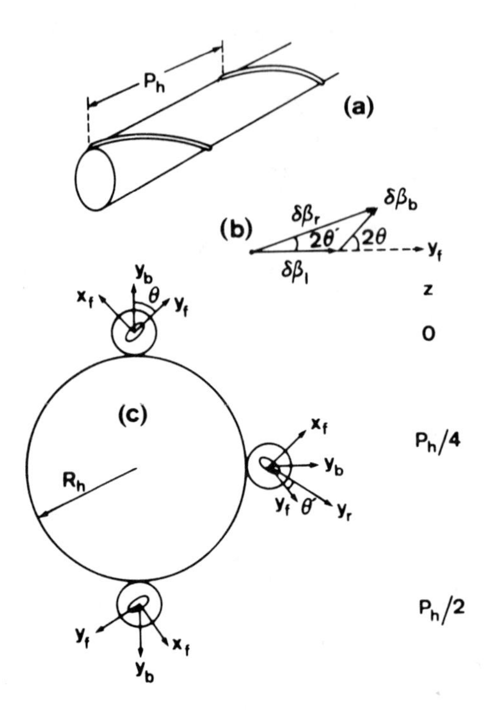

FIGURE 2.14 Fiber with local intrinsic birefringence axes (x_f, y_f) wound helically around a mandrel without detwisting. Inset (b) shows the construction of the linear birefringence vector $\delta\beta_r$, resulting from the combination of the intrinsic linear birefringence $\delta\beta_\ell$ and of the bending-induced birefringence $\delta\beta_b$. In (c), the fiber positions and various axes are represented at various abscissas z along the mandrel. (x_f, y_f) are the intrinsic fiber birefringence axes, (x_b, y_b) are the bending-induced birefringence axes, and (x_r, y_r) are the resulting linear birefringence axes.

It can be seen that this situation is formally analogous to that of intrinsic linear birefringence and straight twist, and the same discussion holds here for the resulting eigenstates.

Influence of a Finite Spectral Width. If the light source has a root-mean-square (rms) spectral width W (in terms of pulsations), there will be a partial depolarization of the light in all cases where two birefringences, with different dispersions, are combined. This is true simply because for a given input SOP, the output SOP will vary with frequency (or wavelength) throughout the source spectrum. If we define ε as the apparent ellipticity of the output SOP,

$$\varepsilon = \frac{I_{max} - I_{min}}{I_{max} + I_{min}} \tag{2.31}$$

where I_{max} and I_{min} are the maximum and minimum intensities passing through a rotating analyzer; when $W \times T \times L >> 1$ (where L is the fiber length), then ε varies as [29]

$$\varepsilon = \frac{|\delta\beta|^2}{|\delta\beta|^2 + |C|^2} |\cos 2\varphi|, \text{ for C real} \tag{2.32a}$$

$$\varepsilon = \left|\sin\left\{2\varphi + \sin^{-1}\left[\frac{|\delta\beta|}{(|\delta\beta|^2 + |C|^2)^{\frac{1}{2}}}\right]\right\}\right|, \text{ for C imaginary} \tag{2.32b}$$

φ is the angle between the linearly polarized input field and the y axis at the input end of the fiber.

Cross-Polarization. A detailed study of the output SOP is given in Ref. 29, and we will concentrate here on the case where a low cross-polarization is desired. We assume that at the fiber input we launch only $X_1(0) = 1$ and $X_2(0) = 0$. After a length L of fiber, $|X_2(L)|^2$ represents the intensity on the unwanted polarization, and when $|X_2(L)|^2 << 1$, it also represents the cross-polarization. From Eqs. (2.23), we get

$$|X_2(L)|^2 = \frac{|C|^2}{|\delta\beta|^2 + |C|^2} \sin^2\left[\left(\frac{|\delta\beta|^2 + |C|^2}{2}\right)^{\frac{1}{2}} L\right] \tag{2.33}$$

We consider as practical cases the last three situations above. Figure 2.15 shows the minimum values of B_ℓ and N (turns per meter) required to obtain a -20-dB cross-polarization for a plane-wound fiber that is either linearly birefringent or twisted [Eqs. (2.26) and (2.27)]. It is assumed that the fiber is wound without tension.

Let us consider now a helically wound fiber, for which we assume that the central mandrel has a radius of 2 mm, as this is a common figure for many types of cables. We consider a pitch varying between 50 and 1000 mm, which according to Eq. (2.28) corresponds to a twist of 20 turns per meter to 1 turn per meter. With Eqs. (2.12), (2.28)–(2.30), and (2.33), we obtain for a -20-dB cross polarization:

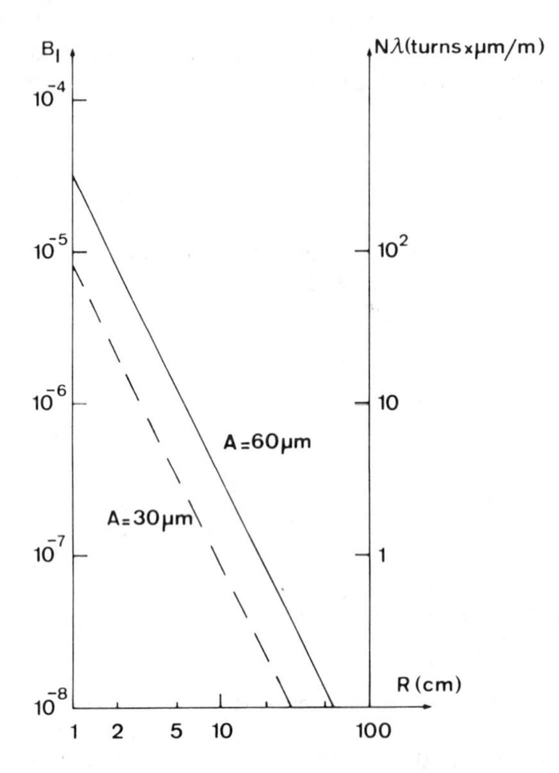

FIGURE 2.15 Minimum fiber birefringence for obtaining a −20-dB cross-polarization when the fiber is wound on a drum with radius R without tension. B scale: minimum intrinsic normalized linear birefringence (linear eigenstates) when the fiber is wound without twist. N scale: minimum number of twist turns per meter of fiber for a fiber without intrinsic linear birefringence (circular eigenstates). A is the outside radius of the fiber.

Linear polarization eigenstates: $B_\ell \geq 12.7\ (\lambda/P_h)$
Circular polarization eigenstates: $B_\ell \leq 0.13\ (\lambda/P_h)$

(The effect of the radius R is found to be negligible.)

Figure 2.16 shows the required minimum or maximum values of B_ℓ for linear polarizations or circular polarizations, respectively. Compared to the results of Sec. 2.2, it can be seen that circular polarizations would be best suited for this type of winding, in contrast to the situation of plane winding with a small radius (see Fig. 2.15). However, the practical use of circular polarization-maintaining twisted fibers requires that the fiber be twisted

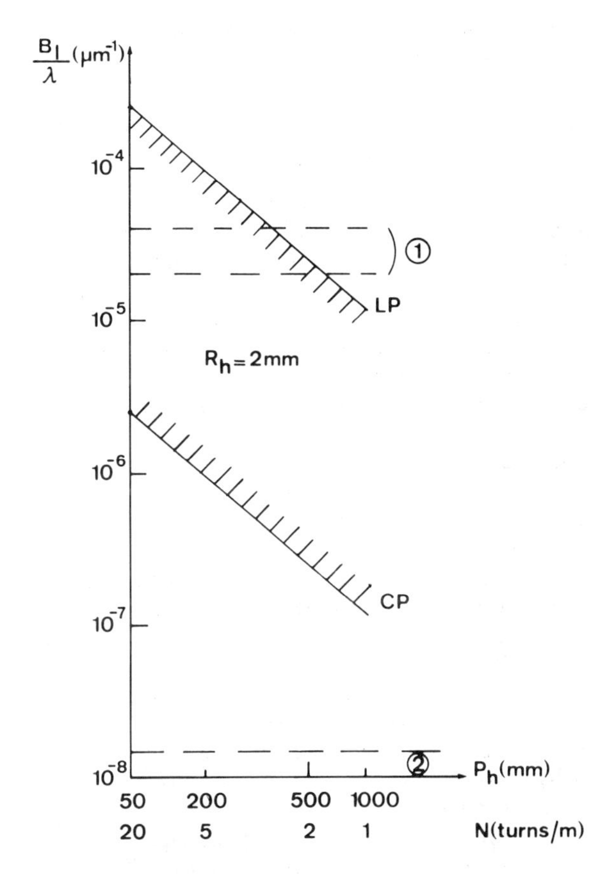

FIGURE 2.16 Minimum and maximum values of intrinsic normalized linear birefringence B_ℓ for −20-dB cross-polarization, as a function of the pitch P_h, for a helically wound fiber designed to maintain either linear (LP) or circular (CP) polarization. Area (1): typical level of highly linear birefringent fibers, as manufactured today (see Sec. 2.2). Area (2): typical level of low-birefringence fibers, as manufactured today (see Sec. 2.2).

everywhere, including where it is clamped (e.g., in connectors). The maximum length of untwisted fiber should be kept well below $2\pi/\delta\beta_\ell$.

2.4.3 Polarization Characteristics of Fiber Links

As seen in the preceding section, when the fiber is wound on a drum (plane winding), linear polarization-maintaining fibers (high intrinsic linear birefringence) are preferable, whereas when the fiber is wound helically with a pitch, as is the case in most cables, circular polarization-maintaining fibers seem to be best suited. Indeed, circular polarization-maintaining cables have been proposed [31], and next we will examine the other problems to be considered when designing a polarization-maintaining single-mode fiber link.

Random Perturbations. Up to now, we have considered only uniform birefringences, as this is the only case where Eqs. (2.23) can be solved practically, and this allowed us to get some physical insight into the effects of birefringence combinations. However, in a real cable the fiber is likely to be disturbed by microbends, which will induce a random coupling term, although the incidence of microbends in modern cables tends to be very low. For a qualitative discussion of these effects, we will refer to a coupled power theory [32] deduced from Eqs. (2.23) by averaging all parameters on a large ensemble of fibers. In the present case, this can also be understood as an average over a large range of environmental conditions, or over a large wavelength range, because for each environmental parameter (e.g., temperature) or for each wavelength we have a slightly different birefringence at each fiber abscissa. For the spectral width, the averaging process can be considered as satisfactory when the fiber length is much greater than 1/(birefringence dispersion × spectral width). Great care should be taken that the results of this coupled power theory not be compared to measurement results carried out on a single fiber, unless the measurements use a large spectral width light source. People familiar with multimode fibers may wonder why the results of the coupled power theory did agree well with experimental results obtained on one multimode fiber but will not agree here with experimental results obtained on one single-mode fiber with a monochromatic source. The reason is simply that here we would measure only one mode at a time and thus have an uncertainty of 100% compared to the theoretical model, whereas in multimode fibers there are always at least 20 modes measured simultaneously (because of limited modal resolution and the use of speckle averaging techniques when monochromatic light is used), leading to about 20% uncertainty [32]. For example, the results of the coupled power theory are in contradiction with the two fundamental points highlighted in Sec. 2.4.1, because they are stated for monochromatic light. Nevertheless, we can get qualitative information from this theory.

In a fiber where the SOP is governed by Eqs. (2.23) with $\delta\beta$ and C randomly varying along 0z ($\delta\beta$ and C are assumed to obey a stationary process with a zero mean value for C, and a mean value of $\delta\beta$ much higher than its fluctuations), the cross-polarization is given by

$$<|X_2(L)|^2> = \frac{L}{4}\int_{-\infty}^{+\infty} <C(x)C^*(x-y)> \exp(i<\delta\beta>y)\, dy \tag{2.34}$$

where < > denotes the average (over an ensemble of fibers, over a large range of environmental conditions, or over a large spectral width), L is the fiber length, and it was assumed that $X_1(0) = 1$; $X_2(0) = 0$; $|X_2(L)| << 1$. The term $<C(x)C^*(x-y)>$ is likely to have an amplitude similar to what is obtained from uniform birefringences, and thus the discussion of the preceding section applies to it, but we observe that we now have its Fourier transform at spatial pulsation $<\delta\beta>$. It is likely that the perturbations have a power spectrum decreasing when $<\delta\beta>$ increases, and thus the larger the $<\delta\beta>$, the smaller the cross-polarization. Typically, the value of $<\delta\beta>$ in the conditions of Figs. 2.15 and 2.16 is 10 times higher with linear polarizations than with circular polarizations, but, on the other hand, the amplitude of $<C(x)C^*(x-y)>$ is about 100 times higher with linear polarizations than with circular polarizations in the case of Fig. 2.16, because region (1) crosses the limiting values of $<\delta\beta_\ell>$ for linear polarizations, whereas we have a margin of about a factor of 10 for circular polarizations between region (2) and the limiting values of $<\delta\beta_\ell>$.

The cross-polarization in the case of plane winding will thus be much smaller with linear birefringent fibers than with twisted fibers, but the case of helical winding is less clear, as we do not have enough information about the power spectrum behavior of C to compare quantitatively the opposite effects of the smaller $<\delta\beta>$ value and smaller $|C|$ value for circular polarizations compared to linear polarizations.

Influence of Joints. Another problem arises from the fact that, at least in terrestrial telecommunication applications, a number of joints (connectors and splices) will disturb the line. Obviously, the case of linear birefringences requires matching the local birefringence axes between successive fibers at each joint, whereas circular polarizations do not need this. It has been shown that the cross-polarization is given by $J<\theta_j^2>$ for linear birefringences [J is the number of joints, and $(<\theta_j^2>)^{\frac{1}{2}}$ is the rms misalignment angle between the birefringence axes], and $(J/2)<[\delta\beta_\ell/(2 + g)2\pi N]^2>$ for circular birefringences [33]. For typical values of 25 joints, a −20-dB cross-polarization requires a rms misalignment of less than 1° for linear polarizations, or a normalized linear birefringence 3.5 times smaller than the values shown in Fig. 2.16 for the circular polarizations. From this point of view, circular polarizations are obviously much more advantageous than linear polarizations.

Typical Performances. If we take into account that circular polarization-maintaining fibers start from standard ultra-low-loss fibers, and that the birefringence dispersion is at least one order of magnitude smaller in these fibers than in linear polarization-maintaining fibers, we can conclude that for telecommunication applications circular polarization-maintaining fibers are better suited than linear polarization-maintaining fibers. The preceding sections give indications of the way to design a cable that is able to exploit this possibility fully. For applications where the fibers are likely to be tightly wound on a drum (such as many sensors), it appears that a high linear birefringence is preferable.

It has been shown in the laboratory that properly twisted fibers maintain circular polarization [28], but no experiment has been carried out on long cabled fibers. For fibers with a high intrinsic linear birefringence, more results have been obtained. The fiber shown in Fig. 2.7b showed a cross-polarization of less than −10 dB after 1.87 km and dispersion measurements showed no effect of mode coupling [34]. The fiber of Ref. 18 showed a cross-polarization of less than −30 dB after 1 km, and the fiber of Ref. 19 less than −20 dB after 500 m and was not affected by plane winding with a radius of a few millimeters.

However, all these results concern bare fibers. No results have yet been reported on cabled and jointed fibers, but it seems likely that the results should obtain at these low cross-polarization levels, provided that cables with ultra-low-microbending contributions are used.

2.5 FIBERS WITH CONTROLLED BIREFRINGENCE

Some very significant applications of single-mode optical fibers require a good control of the state of polarization of the light along the fiber path. The development of high (linear) birefringence fibers has been spurred by their application in the fiber gyro (rotation rate sensor), while low birefringence fibers are of interest to the Faraday current interferometric or polarimetric sensor. We present here some of the specially manufactured fibers available on the market. These descriptions should be understood only as an illustration of the level of industrial activity concerning this type of fiber. They are not intended to provide an exhaustive review of the products offered on the market, nor do they imply any kind of endorsement or recommendation for any particular product. A more detailed review of the various types of fibers with controlled birefringence can be found in Ref. 35.

2.5.1 High Birefringence Fibers

Fibers with a high linear birefringence are manufactured by several companies. Although some manufacturers describe their product as a polarization-maintaining or polarization-retaining fiber, we prefer the term *high birefringence fiber* (HiBi): it can be seen from Sec. 2.4 that a high birefringence is necessary but not sufficient to guarantee polarization-maintaining performances. One has to take some minimal precautions to obtain good performance from this point of view (e.g., jointing two HiBi fibers with a misalignment of their birefringence axes destroys any polarization maintenance possibility; constructing a coil consisting of many layers may induce polarization cross-coupling because of uneven stresses applied on the fiber).

Andrew Corporation (USA) manufactures a highly birefringent fiber with an elliptical core (Fig. 2.5) and a D-shaped outer surface which brings the core very close to the plane surface and combines shape and stress birefringence.

AT&T (USA) has designed a polarization-maintaining fiber of rectangular cross section, with a stress cladding. The rectangular (140 μm × 70 μm) outer cladding allows for mechanical alignment of the polarization axes for launching and jointing purposes and exhibits good winding behavior for gyroscope applications. The fiber is designed to operate at either 1310 or 1500 nm with an attenuation below 1 dB/km. The beat length at 1310 nm is less than 8 mm and the polarization cross-coupling is below −20 dB after 1 km of propagation.

Corning Glass Works (USA) is offering a polarization-retaining single-mode fiber utilizing stress-induced birefringence manufactured through a modified rod-in-tube technique with alumino-silicate glass rods (Fig. 2.17). The fiber drawing process allows the voids to be eliminated in the draw hot zone yielding thus a homogeneous fiber with a circular core and circular cladding. Because of the difference in thermal expansion coefficients between alumino-silicate glass and silica, asymmetric stresses build up when the fiber cools down after drawing, yielding a high linear birefringence. Such fibers are available for operation at either 850 or 1300 nm with losses below 6 dB/km (resp. 2 dB/km), beat lengths below 2.5 mm

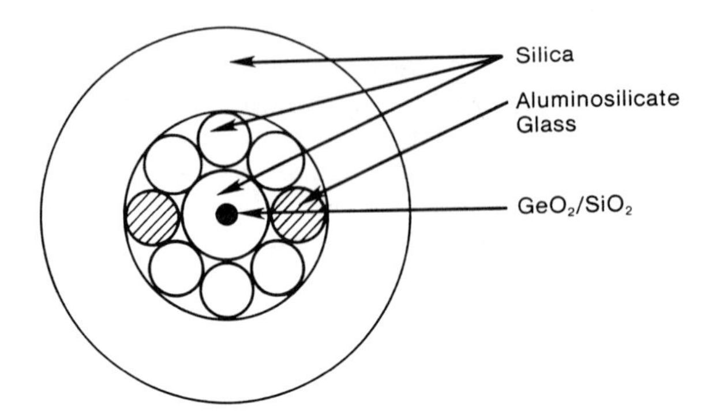

FIGURE 2.17 Modified rod-in-tube design (courtesy of Corning Glass Works, Corning, New York).

(resp. 4 mm), and a typical polarization cross-coupling of −20 dB after 1 km of propagation.

EOTec/3M (USA) produces polarization-maintaining single-mode fibers with a design similar to that illustrated in Figure 2.8. Fibers are available for operation at either 630 or 850 nm with losses of 12 dB/km (resp. 4 dB/km) and a typical cross-polarization of −15 dB after 1 km of propagation. One special version of this fiber is proposed with an elliptical outer surface; this feature is interesting as it enables external identification of the fiber principal axes and promotes better alignment when coiling the fiber (Fig. 2.18).

Fujikura, Furukawa, and Hitachi (Japan) manufacture high birefringence fibers based on the designs illustrated in Figures 2.8 and 2.9

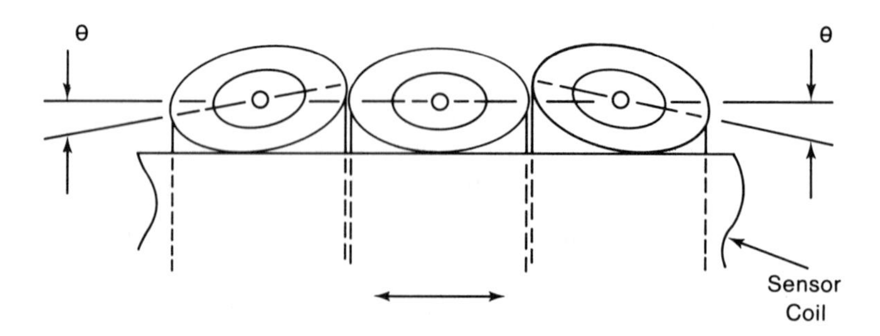

FIGURE 2.18 Elliptically jacketed fibers as they lay on a sensor coil (courtesy of EOTec/3M, West Haven, Connecticut).

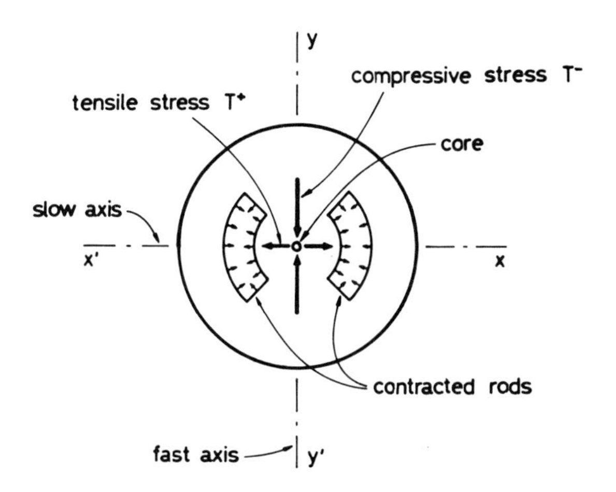

FIGURE 2.19 "Bow-tie" high-birefringence fiber structure as manufactured by York VSOP.

York VSOP (UK) is offering a "HiBi" fiber-utilizing stress-induced birefringence, manufactured through a specially adapted MCVD process. During the preform preparation two longitudinal channels are etched on both sides of the core, yielding a so-called bow-tie structure in a cross section of the fiber (Fig. 2.19). This highly asymmetrical structure produces uneven stresses which induce the high linear birefringence. A full range of fibers are available for operation from 488 nm up to 1500 nm with losses below 5 dB/km at 830 nm and 2 dB/km at 1300 nm and 1500 nm, a beat length below 2 mm at 633 nm, and a typical polarization cross-coupling of −20 dB after 1 km of propagation.

All the manufacturers listed above offer standard products with 125 μm cladding diameter, but they also offer a reduced cladding diameter of 80 μm. This makes it possible to construct more compact fiber coils (sensor applications) without inducing longitudinal strains in excess of the acceptable limit from the fiber lifetime point of view, and without inducing a significant superimposed birefringence which could increase the polarization cross-coupling (see Secs. 2.3.1 and 2.4.2).

2.5.2 Low Birefringence Fibers

York VSOP (UK) is offering an extremely low birefringence fiber manufactured through the preform spinning technique originally developed at Southampton University. Specifications indicate losses below 5 dB/km at 830 nm (2 dB/km at 1300 nm) and beat length figures greater than 360 m (B on the order of 10^{-9} or below).

ACKNOWLEDGMENT

The author is grateful to M. Monerie and C. Vassalo for very helpful discussions during the preparation of this chapter.

REFERENCES

1. T. Okoshi, *IEEE J. Quantum Electron.* *QE-17*(6):879–884 (1981).
2. J. D. Love, R. A. Sammut, and A. W. Snyder, *Electron. Lett.* *15*(20):615–616 (1979).
3. R. B. Dyott, J. R. Cozens, and D. G. Morris, *Electron. Lett.* *15*(13):380–382 (1979).
4. R. Ulrich and A. Simon, *Appl. Opt.* *18*(13):2241–2251 (1979).
5. J. R. Cozens and R. B. Dyott, *Electron. Lett.* *15*(18):558–559 (1979).
6. J. Citerne, *Electron. Lett.* *16*(1):13–14 (1980).
7. S. R. Rengarajan and J. E. Lewis, *Electron. Lett.* *16*(7):263–264 (1980).
8. R. A. Sammut, *Electron. Lett.* *18*(5):221–222 (1982).
9. S. C. Rashleigh and R. Ulrich, *Opt. Lett.* *3*(2):60–62 (1978).
10. I. P. Kaminow and V. Ramaswamy, *Appl. Phys. Lett.* *34*(4):268–270 (1979).
11. A. Zaganiaris, *Proc. 7th Eur. Conf. Opt. Commun.*, Copenhagen, Sept. 8–11, 1981, paper P8.
12. N. K. Sinha, *Phys. Chem. Glass* *19*(4):69–77 (1978).
13. V. Ramaswamy and W. G. French, *Electron. Lett.* *14*(5):143–144 (1978).
14. T. Hosaka, K. Okamoto, Y. Sasaki, and T. Edahiro, *Electron. Lett.* *17*(5):191–193 (1981).
15. R. H. Stolen, V. Ramaswamy, P. Kaiser, and W. Pleibel, *Appl. Phys. Lett.* *33*(8):699–701 (1978).
16. V. Ramaswamy, R. H. Stolen, M. D. Divino, and W. Pleibel, *Appl. Opt.* *18*(24):4080–4084 (1979).
17. I. P. Kaminow, J. R. Simpson, H. M. Presby, and J. B. MacChesney, *Electron. Lett.* *15*(21):677–679 (1979).
18. T. Katsuyama, H. Matsumura, and T. Suganuma, *Electron. Lett.* *17*(13):473–474 (1981).
19. T. Hosaka, K. Okamoto, T. Miya, Y. Sasaki, and T. Edahiro, *Electron. Lett.* *17*(15):530–531 (1981).

20. Y. Sasaki, K. Okamoto, T. Hosaka, and N. Shibata, in *OFC'82 Digest of Technical Papers*, Optical Society of America, Washington, D.C., Apr. 1982, paper THCC6.
21. H. Schneider, H. Harms, A. Papp, and H. Aulich, *Appl. Opt.* *17*(19):3035–3037 (1978).
22. S. R. Norman, D. N. Payne, M. J. Adams, and A. M. Smith, *Electron. Lett.* *15*(11):309–311 (1979).
23. A. J. Barlow, D. N. Payne, M. R. Hadley, and R. J. Mansfield, *Proc. 7th Eur. Conf. Opt. Commun.*, Copenhagen, Sept. 8–11, 1981, paper 2.3.
24. R. Ulrich, S. C. Rashleigh, and W. Eickhoff, *Opt. Lett.* *5*(6): 273–275 (1980).
25. S. C. Rashleigh and R. Ulrich, *Opt. Lett.* *5*(8):354–356 (1980).
26. A. M. Smith, *Electron. Lett.* *16*(20):773–774 (1980).
27. A. Kumar and R. Ulrich, *Opt. Lett.* *6*(12):644–646 (1981).
28. A. M. Smith, *Appl. Opt.* *19*(15):2606–2611 (1980).
29. M. Monerie and L. Jeunhomme, *Opt. Quantum Electron.* *12*: 449–461 (1980).
30. M. Monerie, D. Moutonnet, and L. Jeunhomme, *Proc. 6th Eur. Conf. Opt. Commun.*, York, England, Sept. 1980, *IEE Publ. 190*, pp. 107–109.
31. L. Jeunhomme and M. Monerie, *Electron. Lett.* *16*(24):921–922 (1980).
32. D. Marcuse, in *Theory of Dielectric Optical Waveguides, Quantum Electronics: Principles and Applications* (Y.-H. Pao, ed.), Academic, New York, 1974, pp. 173–193.
33. M. Monerie, *Appl. Opt.* *20*(14):2400–2406 (1981).
34. M. Monerie, P. Lamouler, and L. Jeunhomme, *Electron. Lett.* *16*(24):907–908 (1980).
35. J. Noda, K. Okamoto, and Y. Sasaki, *IEEE J. Lightwave Tech.* *LT-4*(8):1071–1089 (1986).

3
Attenuation

3.1 INTRODUCTION

Among the transmission characteristics of optical fibers, the most important is the attenuation of the light power transmitted by the fiber. The attenuation arises from intrinsic material properties and from waveguide properties. Section 3.2 is devoted to the material attenuation sources (absorption and Rayleigh scattering), and Sec. 3.3 deals with attenuation sources depending on the waveguide properties (microbending and bending loss sensitivity, loss at connectors and splices).

3.2 MATERIAL ATTENUATION

Although new materials such as fluoride glasses may become of interest in the future at wavelengths of either 1700 nm or 3000 to 4000 nm [1,2], we limit the discussion of material attenuation to the case of high-silica glasses, which are the most widely used today for manufacturing low-loss single-mode fibers. We assume the reader to be familiar with the various high-silica fiber manufacturing techniques—inside vapor deposition (IVD), also known as chemical vapor deposition (CVD), and its variants MCVD and PCVD, among others; outside vapor deposition (OVD); and axial vapor deposition (AVD)—and will not describe them here. General reviews of these manufacturing techniques can be found in the literature (e.g., Ref. 3).

3.2.1 Absorption

The absorption loss in high-silica glasses is composed mainly of ultraviolet (UV) and infrared (IR) absorption tails of pure silica, absorptions due to impurities, and IR absorption tails due to the presence of dopants inside the silica.

Pure Silica. The IR absorption tail of silica has been studied in Ref. 4 and it is shown that the vibrations of the basic silica tetrahedron are responsible for strong resonances around 9 to 13 μm with intensities of about 10^{10} dB/km. Overtones and combinations of these fundamental vibrations lead to various absorption peaks at shorter wavelengths, among which the two limiting ones are around 3000 nm (5×10^4 dB/km) and 3800 nm (6×10^5 dB/km). The tail of these various absorption peaks leads to typical values of 0.02 dB/km at 1550 nm, 0.1 dB/km at 1630 nm, and 1 dB/km at 1770 nm. This is responsible for the long-wavelength cutoff of the transmission in high-silica optical fibers around 1800 nm. The UV absorption leads to an absorption tail in the usual wavelengths of interest (Urhach's tail), which has been studied in Ref. 5. The corresponding exponentially decreasing absorption with frequency decrease (wavelength increase) leads to typical values of 1 dB/km at 620 nm and 0.02 dB/km at 1240 nm. Both absorption curves are displayed in Fig. 3.1, labeled IR and UV, respectively.

Effect of Impurities. The usual impurities that lead to spurious absorption effects in the wavelength range of interest are the transition metal ions and water in the form of OH ions. Whereas the concentration of transition metal ions has been reduced to a negligible amount during the past decade, only recently has the presence of water been decreased to a considerable degree. Water thus remains practically the only realistic source of parasitic absorption and still requires careful attention during the manufacturing process.

The presence of OH ions in silica leads to an absorption peak centered at 2730 nm which presents overtones and combinations with Si at 950, 1240, and 1390 nm. For a concentration of 1 ppm, the corresponding attenuations are about 1 dB/km (950 nm), 3 dB/km (1240 nm), and 40 dB/km (1390 nm), and it thus appears necessary to keep the OH concentration at levels below 0.1 ppm if ultra-low losses are desired in the range 1200 to 1600 nm. The spectral loss of a single-mode fiber manufactured by a standard IVD process [6] is shown in Fig. 3.1 and exhibits the OH peaks associated with a typical OH content when no drying action is undertaken. In the IVD manufacturing technique, there are two possible sources of OH contamination: the water directly incorporated into the deposited material, and the OH diffusion from the starting tube toward the core region. For the water diffused from the starting tube during

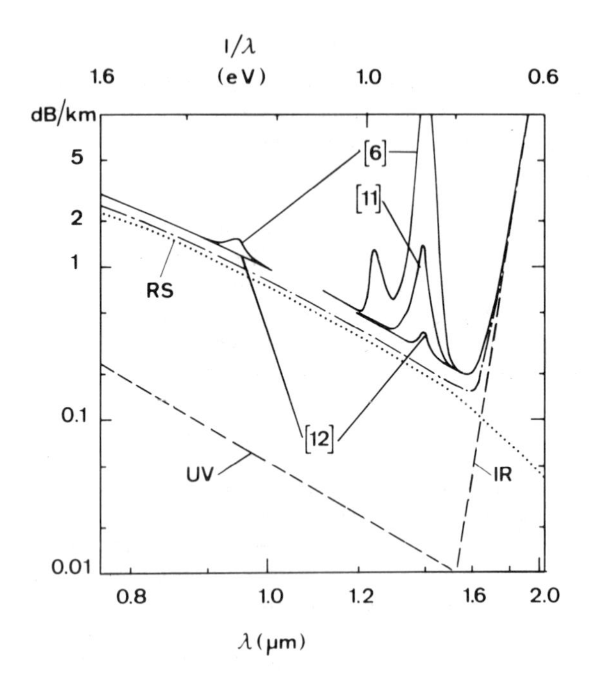

FIGURE 3.1 Fiber loss as a function of wavelength or photon energy. The dashed curves labeled UV and IR refer to pure-silica UV and IR absorption tails. The dotted curve labeled RS corresponds to pure-silica Rayleigh scattering loss. The dashed-dotted curve corresponds to the sum of these three loss factors. The solid lines correspond to experimental loss curves of fibers described in the labeled references (these curves have been interrupted in the region of mode cutoffs).

the deposition stage, the collapsing stage, and the fiber drawing, detailed studies of the OH ion diffusion process [7] combined with the light power distribution (see Fig. 1.12) show that a ratio of 5 or more between the deposited cladding radius and the core radius is necessary in order to get less than 5 dB/km at 1390 nm; practically, most fibers are manufactured with a ratio of 6 or 7. For the water incorporated directly into the deposited material, it has been shown that this almost always occurs during the collapsing stage [8] and that the use of chlorine as a drying agent could efficiently reduce this contamination [9]. Using a $GeCl_4$ gas flow during the collapsing stage makes it possible to reduce the 1390-nm OH absorption loss to about 4 dB/km [10], whereas the use of

10% Cl_2 in O_2 during the collapse decreased the peak to about 1.5 dB/km [11]. Figure 3.1 shows the spectral loss of such a fiber compared to a classical IVD fiber, where the OH peak at 1390 nm reaches about 20 dB/km.

In the AVD and OVD manufacturing techniques, the OH contamination arises almost solely as a result of the flame hydrolysis. As this is unavoidable, it has been necessary to subject the porous blank to a dehydration process before it is sintered into a transparent preform. In the AVD method, the dehydration process uses a chlorine and helium gas mixture at about 1200°C [12] and yields fibers with 1390 nm OH peaks as low as 0.1 dB/km, as shown in Fig. 3.1. The OVD method recently yielded a fiber with ultra-low losses [13], as its spectral loss curve is almost identical to the dashed-dotted curve in Fig. 3.1, except for a 1390-nm OH peak of about 2 dB/km.

Effect of Hydrogen. The first observations of loss due to the presence of hydrogen in optical cables occurred during field trials, when it was found that the loss increased when the cables were flooded with water [14,15]. It was found from comparison with published H_2 absorption spectra that the sources of added loss were the presence of dissolved hydrogen or newly formed OH. An extensive discussion of interactions of hydrogen with silica optical fibers can be found in Ref. 16.

The general conclusion of the various studies conducted on the effect of hydrogen on fiber loss is that at ambient conditions, a properly designed and installed fiber cable should not show any loss increase. Countermeasures to hydrogen formation in the cable and to permeation of hydrogen into the fiber should be taken [16], and a brief summary is given here (see also Sec. 7.3.2). A first step involves avoiding liberation of hydrogen through electrolysis between dissimilar metals present in a cable. For this purpose, cables should not contain aluminum together with either iron or stainless steel. However, copper and stainless steel lead to insignificant hydrogen generation. If silicones are used in the cable or as fiber coating, one should ensure that the silicone material has been completely cured or contains only negligible quantities of SiH.

Effect of Dopants. Dopants such as GeO_2, P_2O_5, F, and B_2O_3 are incorporated into the deposited silica to decrease the manufacturing temperature (especially useful in the IVD process, with P_2O_5) and to modify its refractive index profile (and thus design the waveguide structure). Early studies [4] show that GeO_2 has little effect on the IR absorption tail, but it has been shown more recently that care must be taken with the fiber drawing temperature if a high GeO_2 content is used [11].

The presence of P_2O_5 brings an absorption peak at 3800 nm with an intensity of 10^6 dB/km for a 7 mol % concentration, whereas B_2O_3 brings absorption peaks at 3200 and 3700 nm with intensities of 10^5 and 4×10^6 dB/km, respectively, for a 5 mol % concentration [4]. It is thus apparent that obtaining ultra-low losses at $\lambda \geq 1300$ nm requires that B_2O_3 not be present at radial distances smaller than five times the core radius, and P_2O_5 should be kept out of the core itself.

3.2.2 Rayleigh Scattering

The Rayleigh scattering loss, which is due to microscopic inhomogeneities of the material, shows a λ^{-4} dependence, and is found to be approximately 0.75 dB/km at $\lambda = 1000$ nm in bulk silica samples [5]. The addition of dopants into the silica leads to an increase in the Rayleigh scattering loss, because the microscopic inhomogeneities

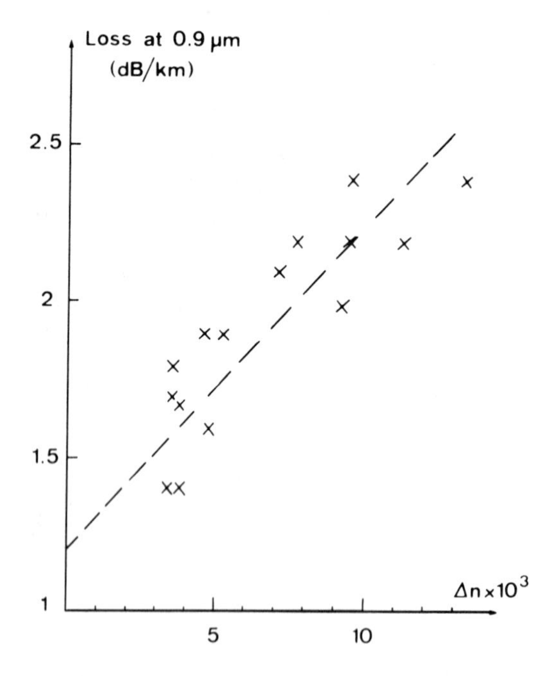

FIGURE 3.2 Fiber loss at 900 nm as a function of index difference due to GeO_2 doping (multimode fiber). Crosses represent measurement results, and the dashed line corresponds to Eq. (3.1) plus the UV absorption tail.

become more important. As the most useful dopant for the core (where most of the guided energy is confined) is GeO_2, measurements of loss in fibers with variable GeO_2 contents have been carried out at 900 nm [17], and the result is shown in Fig. 3.2. It is observed that the loss increases linearly with the index difference, and after taking into account a 0.08-dB/km UV absorption loss at 900 nm (see Sec. 3.2.1), it is found that the measured data can be fitted by a Rayleigh scattering loss term α_s of the form

$$\alpha_s \text{ (dB/km)} = (0.75 + 66\,\Delta n_{Ge})\lambda^{-4} \tag{3.1}$$

with λ in micrometers. Δn_{Ge} indicates that the index difference to be considered is only that due to GeO_2 doping, as the other dopants seem to have much less influence on this loss factor. The corresponding curve is shown in Fig. 3.2.

3.3 WAVEGUIDE ATTENUATION

The losses due to the waveguide structure arise from power leakage (in fibers with depressed inner cladding), bending, microbending of the fiber's axis, and defects at joints between fibers.

3.3.1 Power Leakage

It was shown in Chap. 1 that fibers with depressed inner claddings can present a long-wavelength cutoff for the lowest-order LP_{01} mode. It was shown theoretically that above the LP_{01}-mode cutoff wavelength, the fiber loss increases more or less sharply, depending on the exact fiber design. Very few experimental results on this point have yet been reported, and we present some of them here as an illustration of the power leakage effects in depressed inner cladding fibers. For a clearer understanding of the figures, let us recall that this leakage loss occurs through a tunneling effect between the oscillating field confined in the core and the oscillating field (above the LP_{01} cutoff) spreading in the outer cladding. This leakage loss will thus decrease when the inner cladding thickness increases (everything else being constant) or when the inner cladding depth decreases (for a constant core—inner cladding index difference) (see Sec. 1.4.1 and Fig. 1.20).

The experimental curves in Fig. 3.3 correspond to the loss of various depressed inner cladding fibers with the same core and cladding diameters but varying index differences. There is a qualitative agreement with the theoretical discussion of Sec. 1.4.1 and the curves in Fig. 1.20, and it is confirmed that the long-wavelength

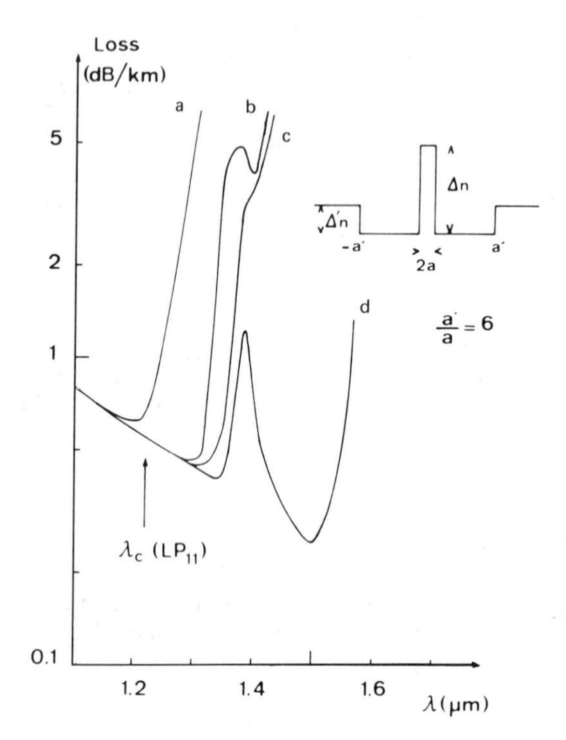

FIGURE 3.3 Experimental loss of various depressed inner cladding fibers with a'/a = 6. Fiber a: $\Delta = 0.51\%$, $\Delta' = -0.23\%$; fiber b: $\Delta = 0.55\%$, $\Delta' = -0.23\%$; fiber c: $\Delta = 0.47\%$, $\Delta' = -0.19\%$; fiber d: $\Delta = 0.45\%$, $\Delta' = -0.17\%$. (From P. D. Lazay and A. D. Pearson, in *Optical Fiber Communication 1982 Digest of Technical Papers*, Optical Society of America, Washington, D.C., Apr. 1982, paper THCC2.)

loss edge of these fibers is very sensitive to the exact fiber design. Figure 3.4 shows transmission curves of fibers with constant index differences, but varying inner cladding diameter-to-core diameter ratios (for all these fibers, the LP_{01}-mode cutoff is about 1.1 μm, as given by the theory). Again it is observed that increasing the cladding thickness decreases the LP_{01} leakage loss. However, the overall sensitivity of the leakage loss to the exact design of depressed inner cladding fibers indicates that practical fiber design needs empirical tests and cannot rely only on theoretical calculations.

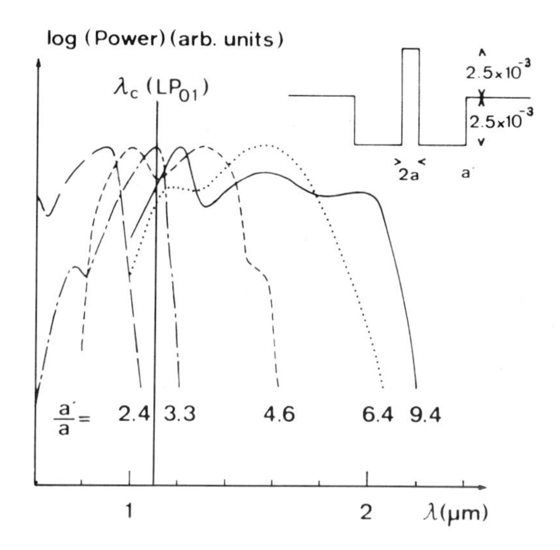

FIGURE 3.4 Experimental transmitted power of various depressed inner cladding fibers with the index profile shown in the insert (absolute index differences). The fiber length is 10 m. $\lambda_c(LP_{01})$ indicates the theoretical cutoff wavelength of the LP_{01} mode (almost insensitive to a'/a when a'/a > 2). (From B. J. Ainslie et al., in *Optical Fiber Communication 1982 Digest of Technical Papers*, Optical Society of America, Washington, D.C., Apr. 1982, paper THCC3).

3.3.2 Bending Losses

The simplest qualitative description of the bending losses in a fiber can be obtained by assuming that in the bent fiber, the field is not significantly changed compared to the field in the straight fiber. The plane wavefronts associated with the guided mode are pivoted at the center of curvature of the bent fiber, and their longitudinal velocity along the local fiber axis increases with the distance from the center of curvature [18]. As the phase velocity (phase front local longitudinal velocity) in the core is slightly smaller than that of a plane wave in the cladding, there must be a critical distance from the center of curvature, above which the phase velocity would exceed that of a plane wave in the cladding. The electromagnetic field resists this phenomenon by radiating power away from the guide, causing radiation losses.

A more quantitative explanation can be obtained by using the fact that a curved fiber can be considered as a straight fiber with a distorted refractive index distribution [19]:

$$n'(r) = n(r)\left(1 + \frac{r}{R}\cos\varphi\right) \qquad (3.2)$$

where n(r) is the index profile of the actual fiber, and the curvature of radius R is assumed to take place in the y0z plane (see Fig. 1.4). This equivalent index profile is shown in Fig. 3.5. A given mode has an effective index given by β/k, and we find a region where this effective index is smaller than the equivalent refractive index. As discussed in Chap. 1, this is equivalent to saying that the mode is at cutoff and experiences radiation losses.

Both of these qualitative explanations allow us to state intuitively that the bending losses will dramatically increase when the radius of curvature decreases (as the field decay in the cladding is exponential-like, decreasing the radius of curvature quickly increases the radiated power proportion); also, a mode close to cutoff will be affected more than a mode far from cutoff (because in the latter case, the line β/k is higher in Fig. 3.5); finally, a high index difference will be of primary importance for decreasing the bending losses, as this increases the value of β/k.

Practical Results. A detailed study of this type of loss has been carried out in Ref. 20 and leads to the distinction between transition losses and pure bending losses. Transition losses occur only in fiber sections following a change in the curvature and are due to coupling between the LP_{01} mode and the leaky modes. As these modes are not lost immediately, they couple back to the LP_{01} mode, and the transition losses appear as an oscillatory function of the fiber length after the curvature change. Because of the resonance-like phenomenon of the coupling, they are also an oscillatory function of the wavelength. The worst case (abrupt change of curvature between a straight fiber and a radius R) leads to an average transition loss

$$A_R\ (\mathrm{dB}) = -10\log\left(1 - k^4 n_1^4 \frac{w_0^6}{8R^2}\right)$$

$$\simeq -10\log\left(1 - 890\frac{w_0^6}{\lambda^4 R^2}\right) \qquad (3.3)$$

where w_0 is the mode field radius as defined in Sec. 1.2.2. In fact, the actual loss oscillates around this average value as a function of fiber length, but tends toward this limit for long fiber lengths. Whereas this loss can be strong (several decibels) in extreme situations (a fiber with a small index difference, well above cutoff), it usually remains below 0.5 dB in most practical cases.

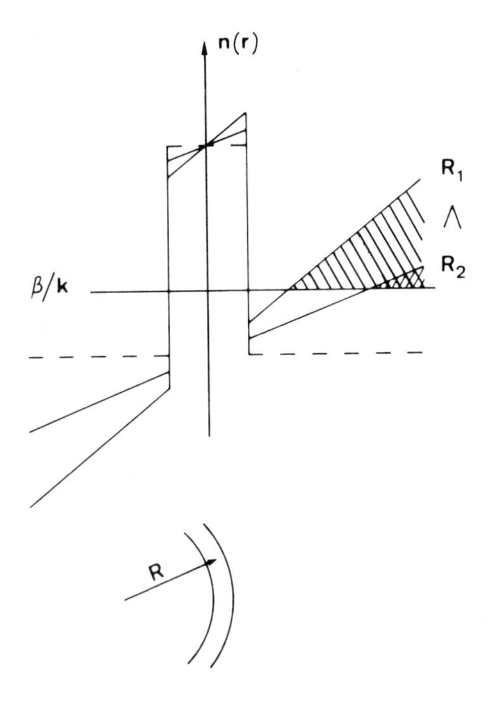

FIGURE 3.5 Index profile (dashed line) and equivalent straight index profiles (solid line) for a curved fiber. The horizontal β/k indicates the mode's effective index, and the shaded area indicates the region where the mode experiences radiation losses.

The pure bending loss is linked with a constant radius of curvature and is due to the leakage of the LP_{01} mode above a given radial distance when the fiber is curved. For a radius R, the loss per unit length is given by [20]

$$\alpha_c = A_c R^{-\frac{1}{2}} \exp(-UR) \tag{3.4a}$$

$$A_c = \frac{1}{2}\left(\frac{\pi}{av^3}\right)^{\frac{1}{2}} \left[\frac{u}{VK_1(v)}\right]^2 \tag{3.4b}$$

$$U = \frac{4\,\Delta n\, v^3}{3aV^2 n_2} \tag{3.4c}$$

where a and Δn are the core radius and core–cladding index difference, and u, v, and V and the LP_{01}-mode parameters and normalized frequency defined in Chap. 1. Extensive calculations of bending losses can be found in Refs. 20 and 21; however, here we concentrate our discussion on obtaining simplified formulas. Analytical approximations can be obtained by using Eqs. (1.14), (1.16), and (1.26):

$$U \simeq 0.705 \frac{(\Delta n)^{3/2}}{\lambda} \left(2.748 - 0.996 \frac{\lambda}{\lambda_c}\right)^3 \quad m^{-1} \tag{3.5}$$

The accuracy is better than 3% for $0.8 \leq \lambda/\lambda_c \leq 2$. On the other hand, a detailed study of the behavior of $v^{-3/2}u^2/V^2K_1{}^2(v)$ shows that it can be approximated by $3.7(\lambda_c/\lambda)^2$, and we thus get

$$A_c \simeq 30(\Delta n)^{1/4}\lambda^{-1/2}\left(\frac{\lambda_c}{\lambda}\right)^{3/2} \quad dB/m^{\frac{1}{2}} \tag{3.6}$$

The accuracy is better than 10% for $1 \leq \lambda/\lambda_c \leq 2$, which is good enough, as A_c is outside the exponential term.

As can be seen from Eqs. (3.4) and (3.5), the pure bending loss depends strongly on the radius of curvature, the index difference, and the ratio λ/λ_c. For example, if two fibers have the same cutoff wavelength but index differences differing by a factor of 2, the fiber with the highest index difference will support a radius of curvature three times smaller than the other one, for the same curvature loss at the same wavelength. Typical cases correspond to fibers with index differences of about 0.5%, operated around 1000 nm, with a cutoff wavelength 10% smaller; this leads to typical values of 3×10^5 dB/km$^{\frac{1}{2}}$ for A_c and 1000 m^{-1} for U. In these conditions, a radius of curvature of 20 mm will bring about a 1-dB/km loss, but 10 mm brings a loss of 10^4 dB/km. This very steep rise obviously comes from the exponential part of α_c, and allows us to define a critical radius of curvature R_c for given index difference, operating wavelength, and cutoff wavelength. When the actual radius of curvature approaches R_c, the bending loss increases sharply from negligible values to intolerably high values. An approximate expression for R_c, valid in the usual wavelength range (around 1000 nm), is obtained from Eqs. (3.4)–(3.6) as

$$R_c \simeq 20 \frac{\lambda}{(\Delta n)^{3/2}} \left(2.748 - 0.996 \frac{\lambda}{\lambda_c}\right)^{-3} \tag{3.7}$$

Once a fiber is characterized by its index difference and its cutoff wavelength λ_c, this simple formula makes it possible to evaluate approximately either the minimum acceptable radius of curvature at a given operating wavelength or the maximum operating wavelength for a given radius of curvature [if the fiber is bent on lengths of more than 1 m, it is safe to double the value of R_c given by Eq. (3.7)]. Reciprocally, if the operating wavelength, cutoff wavelength, and desired critical radius of curvature are given, Eq. (3.7) allows us to calculate the minimum index difference.

Figure 3.6 shows examples of bending losses as a function of the radius of curvature, for fibers with the same cutoff wavelength of 1180 nm, operated at 1300 nm, and with various index differences. The corresponding values of the critical radius of curvature R_c are indicated by arrows. Figure 3.7 shows the variation of the critical radius of curvature R_c as a function of the operating wavelength for fibers with $\Delta n = 0.55\%$ and various cutoff wavelengths.

Finally, Fig. 3.8 shows the experimentally measured added loss as a function of wavelength for a fiber with one turn at a radius $R = 20$ mm, with $\Delta n = 0.27\%$ and $\lambda_c = 900$ nm. The composition between the transition losses (oscillations) and the pure bending loss (general increase) is clearly observable. The dashed curve corresponds to the calculation of the pure bending loss alone, through Eqs. (3.4)–(3.6), after having computed the ESI parameters corresponding to the actual fiber profile (see Chap. 1). Increasing the number of turns would decrease the relative importance of the transition losses, as they are a fixed part independent of the fiber length under curvature.

All of the theory discussed above holds for a step-index matched-cladding fiber, but it has been shown [22] that in other cases, use of the ESI parameters leads to an acceptable accuracy (see Sec. 1.3.2). This is confirmed here in Fig. 3.8, where the parameters used for the calculation are those of the ESI, as the actual index profile involved a central dip and a depressed cladding.

3.3.3 Microbending Losses

Microbendings correspond to a fiber randomly oscillating around its (straight) nominal position with small deviations. Despite the small deviations, the typical periods of the oscillations may be small, and thus the fiber may have sharp local bendings: there are thus two loss cources, one arising from the permanent coupling between the LP_{01} mode and leaky and radiation modes, and another arising from the pure bending loss effect on the LP_{01} mode, which becomes leaky at some radial distance in a curved fiber. Theoretical studies in Refs. 23 and 24 consider only the first phenomenon, whereas the approach taken in Ref. 20 considers both effects, but a

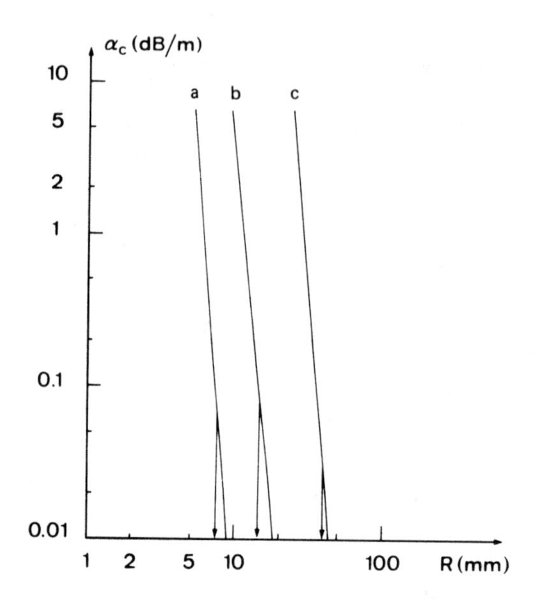

FIGURE 3.6 Pure bending loss as a function of the radius of curvature for fibers operated at 1300 nm, having a cutoff wavelength of 1180 nm and index differences Δn of: (a) 0.825%; (b) 0.55%; (c) 0.275%. The arrows indicate the corresponding critical radius of curvature as given by Eq. (3.7).

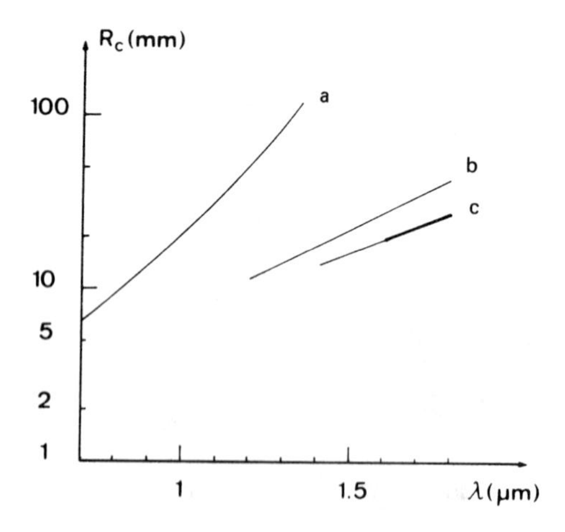

FIGURE 3.7 Critical radius of curvature [Eq. (3.7)] as a function of the operating wavelength for fibers with an index difference Δn = 0.55% and cutoff wavelengths of (a) 700 nm; (b) 1180 nm; (c) 1400 nm. For fibers bent on lengths greater than 1 m, these R_c values should be doubled for safe operation.

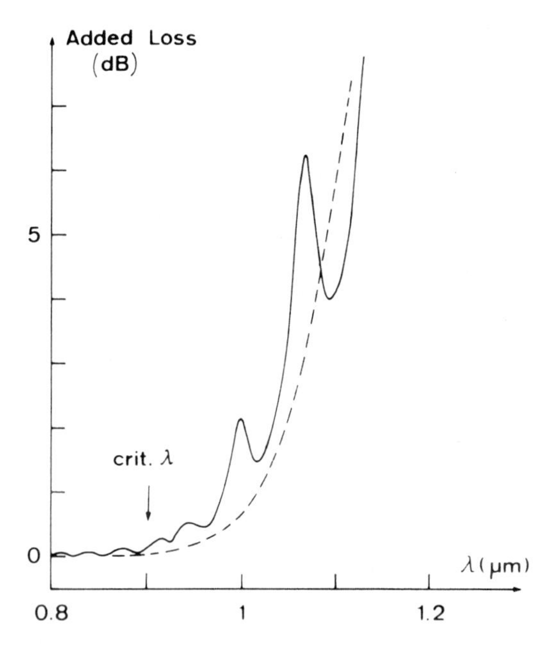

FIGURE 3.8 Added loss in a fiber with one loop under a radius of 20 mm, with $\Delta n = 0.27\%$ and $\lambda_c = 900$ nm. The solid line is the experimental curve, the dashed line corresponds to the theoretical pure bending loss obtained from Eqs. (3.4)–(3.6). The arrow indicates the critical operating wavelength deduced from Eq. (3.7) with $R_c = 20$ mm.

numerical comparison of the predictions of Refs. 24 and 20 shows that the results agree to better than 10% [25] (care should be taken that the attenuation coefficients in Ref. 24 are in terms of field amplitudes and should thus be multiplied by 2). The results of Ref. 24 are the easiest to use and we refer to this approach in the following. It is shown that the microbending loss depends almost entirely on the mode field radius, independently of the refractive index profile. However, the definition of the mode field radius in Ref. 24 shows several differences from the definition used in Chap. 1. The first difference arises from a factor of $2^{\frac{1}{2}}$ due to the use of an e^{-1} or e^{-2} intensity radius in the Gaussian approximation of the LP_{01} field. After correction for this difference, the definition in Ref. 24 can be written

$$w_0'^2 = 2\frac{\int_0^\infty r^3E^2(r)\,dr}{\int_0^\infty rE^2(r)\,dr} \tag{3.8a}$$

where E(r) is the field amplitude and w_0' denotes the mode field radius according to Ref. 24. The definition of w_0 used in Chap. 1 is equivalent to replacing $E^2(r)$ by $E(r)\exp(-r^2/w_0^2)$ in Eq. (3.8a), and it is thus clear that $w_0 \simeq w_0'$ except where the Gaussian approximation of the field is inaccurate. It has been shown [26] that the difference between both definitions is always less than 10% for $0.8 \leq \lambda/\lambda_c \leq 1.8$, and we will thus use the results of Ref. 24 with the definition of w_0 in Chap. 1.

A good approximation of w_0' to better than 0.3% accuracy for $0.85 \leq \lambda/\lambda_c \leq 1.4$ is given by

$$w_0' \simeq 1.064 - 0.726\frac{\lambda}{\lambda_c} + 0.761\left(\frac{\lambda}{\lambda_c}\right)^2 \tag{3.8b}$$

Practical Results. For computing the microbending loss, it is necessary to have a description of the physical fiber axis distortion (at least from a statistical point of view). As this is practically difficult, it seems preferable to refer to the loss encountered by multimode fibers. Microbendings are induced primarily by fiber jacketing and cabling, and it is easy from a practical point of view to determine the microbending loss incurred by a multimode fiber when processed in the same way as the single-mode fibers of interest. If we call α_{mm} the microbending loss of a step-index multimode fiber with numerical aperture NA and core radius a_m, having the same outside diameter and placed in the same mechanical environment as the single-mode fiber under interest, the single-mode microbending loss is given by [24]

$$\alpha_{sm} = 0.05\alpha_{mm}\frac{k^4 w_0^6 (NA)^4}{a_m^2} \tag{3.9}$$

If the reference multimode fiber has the classical parameters NA = 0.2 and $2a_m$ = 50 μm, we get

$$\alpha_{sm} = 2 \times 10^8 \alpha_{mm}(50;\ 0.2)\frac{w_0^6}{\lambda^4} \tag{3.10}$$

with w_0 and λ in meters.

Alternatively, and using the relationships between λ_c, a, and Δn of Chap. 1, we can write Eq. (3.10) as

$$\alpha_{sm} = 2.53 \times 10^4 \alpha_{mm}(50;\ 0.2) \left(\frac{w_0}{a}\right)^6 \left(\frac{\lambda_c}{\lambda}\right)^4 \frac{\lambda_c^2}{(\Delta n)^3} \qquad (3.11)$$

This shows that for given operating and cutoff wavelengths (λ_c/λ and thus w_0/a fixed), the single-mode microbending loss decreases considerably when the index difference increases.

Figures 3.9 and 3.10 show the spectral variation of the microbending loss for single-mode fibers with various index differences and cutoff wavelengths. As expected from the preceding equations, a strong dependence on the index difference is observed, and it is seen that the microbending loss remains low up to about 30% above the cutoff wavelength. Figure 3.11 shows the experimental measurement of the loss added on a depressed-cladding fiber when it is wound with 70 g of tension on its drum (and thus experiences microbends due to multilayer winding), compared to a winding without tension. Also shown are the effect of fiber cabling with step-index single-mode fibers; obviously, these results are indicative only, as quantitative results depend strongly on the exact cable design and structure. Again, the theory holds for step-index matched-cladding fibers, but the use of the ESI parameters or, even better, of the measured mode field radius makes it possible to use these results for other index profiles.

3.3.4 Joint Losses

Losses occurring at joints between fibers have been studied in detail in Ref. 27 using the Gaussian approximation for the LP_{01} mode. This approach is useful, as it causes the results to depend only on the mode field radius and not on the index profile. We assume in the following that the defects occurring at a joint may be:

Fiber tolerances: $w_1 \neq w_2$, where w_1 and w_2 are the mode field radius of both spliced fibers.

Fiber separation: Se, where Se is the distance between the fiber end planes.

Fiber axes offset: d, where d is the lateral distance between the two fiber axes.

Fiber axes misalignment: θ, where θ is the tilt angle between the two axes.

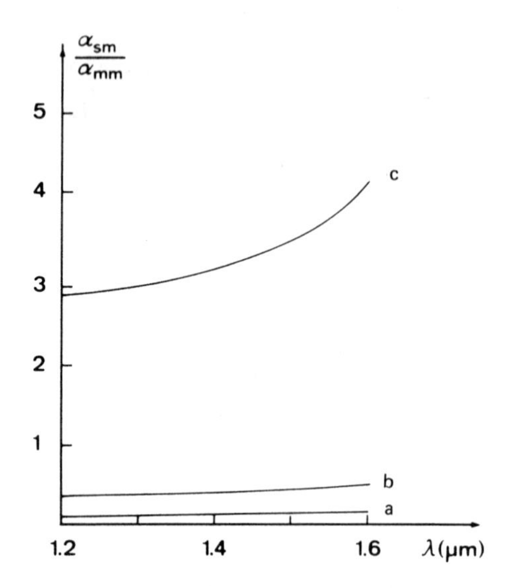

FIGURE 3.9 Single-mode microbending loss α_{sm}, normalized to the microbending loss of a step-index multimode fiber (core diameter 50 μm; NA = 0.2). Both fibers are assumed to have the same outside diameter and to be placed in the same mechanical environment. Single-mode fibers have the same cutoff wavelength of 1180 nm and index differences: (a) 0.825%; (b) 0.55%; (c) 0.275%.

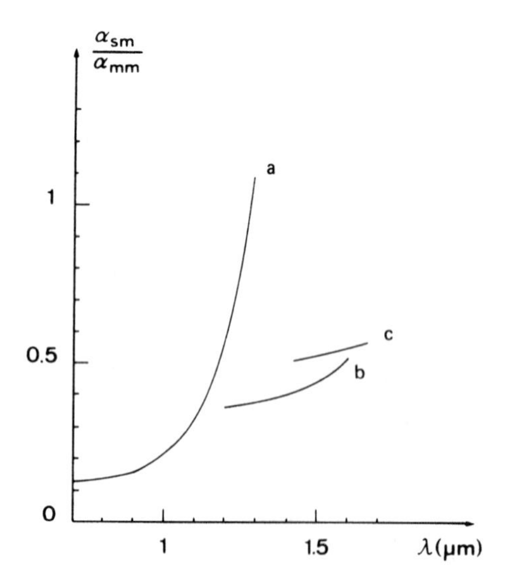

FIGURE 3.10 The same as Fig. 3.9 but for single-mode fibers with the same index difference of 0.55% and cutoff wavelengths of (a) 700 nm; (b) 1180 nm; (c) 1400 nm.

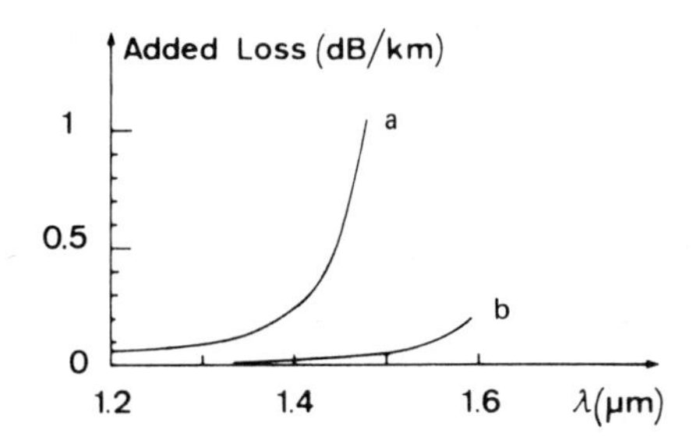

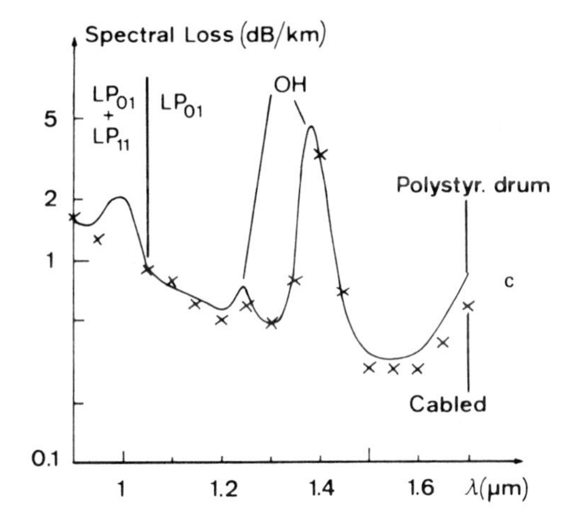

FIGURE 3.11 Microbending added loss as a function of wavelength, as measured experimentally. (a) Depressed cladding fiber with a cutoff wavelength of 1220 nm, a core-inner cladding relative index difference of 0.45%, an outer cladding-inner cladding relative index difference of 0.17%, and an inner cladding diameter-to-core diameter ratio of 6. The fiber is wound with several layers on a drum, with a tension of 70 g (added loss as compared to winding without tension). (After P. D. Lazay and A. D. Pearson, in *Optical Fiber Communication 1982 Digest of Technical Papers*, Optical Society of America, Washington, D.C., Apr. 1982, paper THCC2.) (b) Average of 30 km of fiber, with a cutoff wavelength of 1200 nm and an index difference of 0.32%, assembled in a loose tube cable structure (added loss compared to the bare fiber). [After B. P. Nelson and S. Hornung, *Electron. Lett. 18*(6):270–272 (1982).] (c) Typical effect of fiber cabling in a loose tube cable structure for a step-index fiber with $\Delta n = 0.44\%$, $\lambda_c = 1060$ nm, and an outer diameter of 125 μm. (TSM cable from CLTO; courtesy of R. Jocteur, CLTO-CGE.)

We give next loss formulas for each defect considered as being alone, but it can be shown that the losses corresponding to several simultaneous defects can be approximated by simply adding each loss component, provided that each of these is ≤ 1 dB. It is worth noting that contrary to what happens with multimode fibers, the joint loss here is reciprocal, that is, independent of the light propagation direction.

Fiber Tolerances. For fiber tolerances, the loss is given by

$$\alpha(\text{dB}) = -20 \log\left(\frac{2w_1w_2}{w_1^2 + w_2^2}\right) \tag{3.12}$$

A plot of α as a function of w_1/w_2 is shown in Fig. 3.12a. It is noteworthy that the effect of fiber tolerances is very small. As seen from Fig. 3.12a, a variation of 10% on w produces a loss of 0.05 dB. Using Eq. (1.29), we get

$$\frac{\delta w_0}{w_0} = \left|\ell\left(\frac{\lambda}{\lambda_c}\right)\right| \frac{\delta a}{a} + \left|m\left(\frac{\lambda}{\lambda_c}\right)\right| \frac{\delta(\Delta n)}{\Delta n} \tag{3.13}$$

Figure 3.13 shows $|\ell|$ and $|m|$ as a function of λ/λ_c, and it should be noted that for $\lambda/\lambda_c = 1.285$, the mode field radius is almost insensitive to core diameter fluctuations. If we consider the whole wavelength range $1 \leq \lambda/\lambda_c \leq 1.5$, we have $|\ell| \leq 0.4$ and $|m| \leq 0.7$, and we can deduce the average fiber-to-fiber splice loss with these limiting figures in Eqs. (3.12) and (3.13). The result is shown in Fig. 3.14, and it is clear that fiber tolerances are not critical for obtaining low-loss splices, contrary to what happens in multimode fibers.

Fiber End Separation. For fiber end axial separation, we have

$$\alpha(\text{dB}) = -10 \log\left[\frac{1 + 4Z^2}{(1 + 2Z^2)^2 + Z^2}\right] \tag{3.14}$$

with

$$Z = \frac{Se}{kn_2w_0^2} \tag{3.15}$$

A plot of α as a function of Z is shown in Fig. 3.12b, and the corresponding values of Se in micrometers correspond to $\lambda = 1300$ nm,

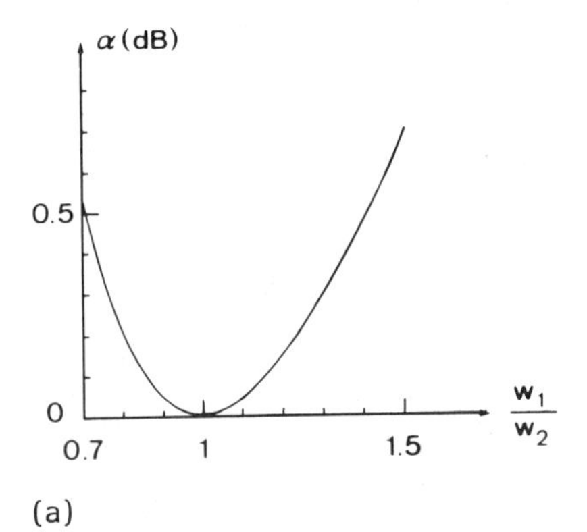

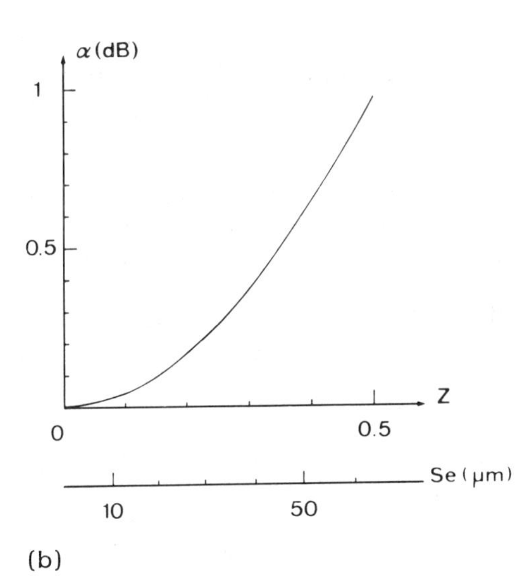

FIGURE 3.12 Joint loss as a function of (a) the ratio of mode field radii in both fibers; and (b) the normalized fiber end faces axial separation Z or the separation Se for $\lambda = 1300$ nm, $\lambda_c = 1180$ nm, and $\Delta n = 0.55\%$.

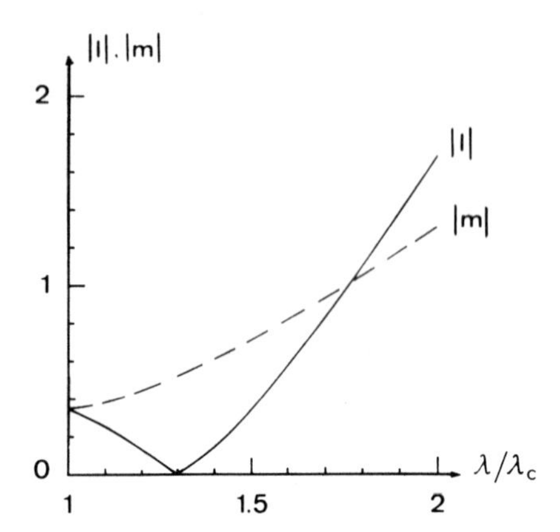

FIGURE 3.13 Sensitivity functions of the mode spot radius w_0 with respect to core radius variations $|\ell|$ or to index difference variations $|m|$.

λ_c = 1180 nm, and Δn = 0.55%. However, this loss source is not present in splices, and as the number of demountable connectors in a system can be minimized, it is not a dominant factor of joint losses.

Axis Offset and Tilts. For a defect involving axis offset and tilts, the loss is given by

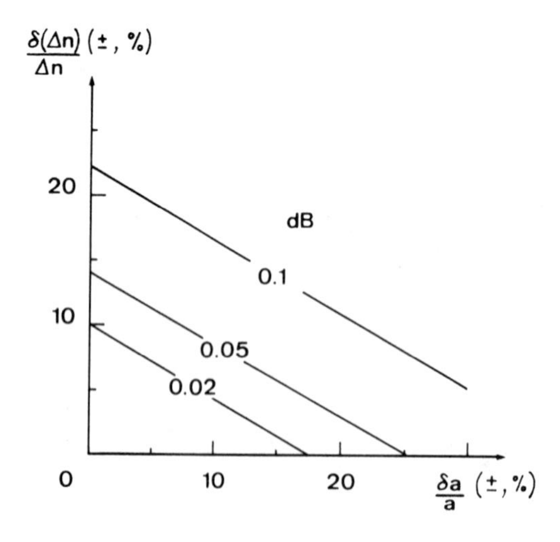

FIGURE 3.14 Fiber-to-fiber core diameter and index difference tolerance for various average joint losses.

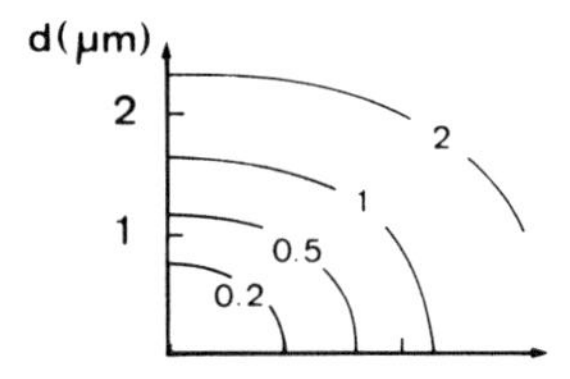

(a)

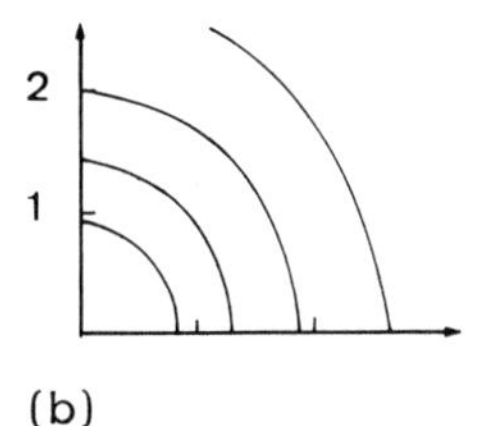

(b)

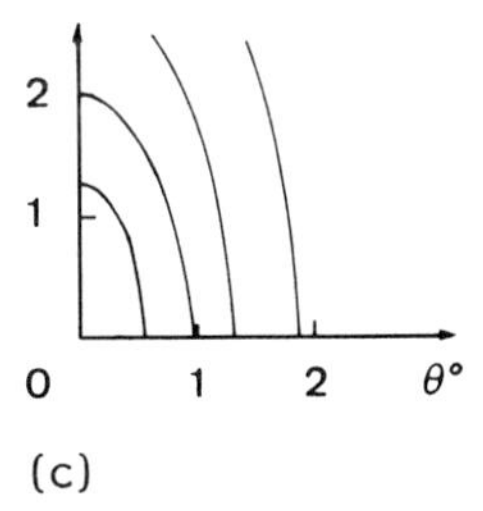

(c)

FIGURE 3.15 Axis offset and tilt angle for joint losses of 0.2, 0.5, 1, and 2 dB and fibers with λ_c = 1180 nm operated at λ = 1300 nm and having index differences of (a) 0.825%, (b) 0.55%, and (c) 0.275%.

$$\alpha(\mathrm{dB}) = 4.34\left[\left(\frac{d}{w_0}\right)^2 + \left(\frac{\pi n_2 w_0 \theta}{\lambda}\right)^2\right] \tag{3.16}$$

This corresponds to a parabolic increase in the loss as a function of either defect parameter. Obviously, an increase in the mode spot radius due to either an increase in the core diameter or a wavelength increase reinforces the contribution of the tilts and decreases that of the offset. Figure 3.15 shows the splice loss as a function of the offset and tilt angle for various fibers.

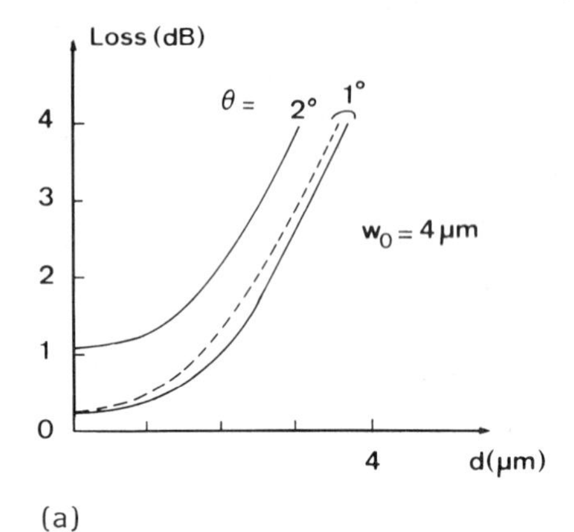

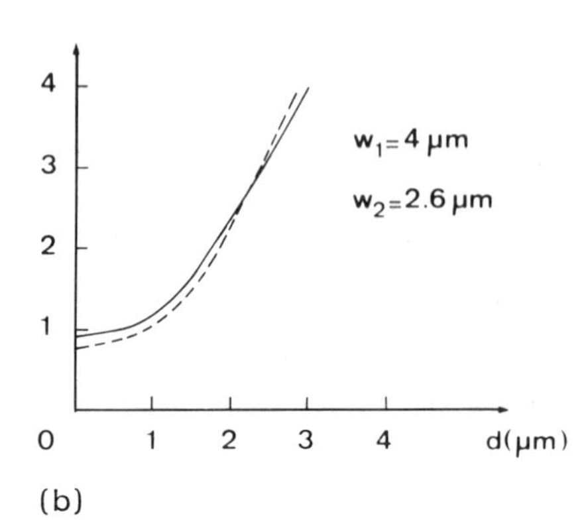

FIGURE 3.16 Experimental (solid lines) and theoretical (dashed lines) splice loss as a function of lateral offset d at 1300 nm. (a) The same fiber, with a spot radius of 4 μm and two values of the tilt angle. (b) Different fibers, with field radii of 4 μm and 2.6 μm. (From R. B. Kummer and S. R. Fleming, in *Optical Fiber Communication 1982 Digest of Technical Papers*, Optical Society of America, Washington, D.C., Apr. 1982, paper THAA1.)

Experimental Results. Figure 3.16 compares theory and experience for splice losses measured on fibers with combined defects; the agreement is found to be good for practical purposes. Finally, it has been mentioned [28] that in connectors, the loss increase as a function of tilt angle is steeper when an index matching fluid is used than when dry connectors are used. On the other hand, as the presence of index matching fluid decreases the intrinsic loss by 0.4 dB, it appears that index matching fluid is of interest only for tilt angles smaller than about 1.5°, an angle above which the dry connectors present a lower loss [this is because in dry connectors, the glass—air interface introduces a refraction effect which divides angle θ in Eq. (3.16) by the refractive index]. However, if very coherent light is used, dry connectors and very small tilt angles may lead to spurious interference effects between the two fiber end faces, giving signal instabilities.

3.4 INFLUENCE OF MANUFACTURING

Besides the influence of manufacturing conditions on the material attenuation sources, the various manufacturing techniques exhibit specific features which influence the overall loss of spliced fibers laid in cables. Again we assume the reader to be familiar with the various high-silica fiber manufacturing techniques (IVD, OVD, AVD, etc.).

3.4.1 Intrinsic Fiber Loss

From this point of view, the VOD and AVD manufacturing processes seem to be almost equivalent and provide a slight advantage over the IVD manufacturing process. In the IVD method, the heat source is outside the silica tube in which the deposition takes place (except for recent plasma-heated IVDs, which do not yet have proven reproducibility), and the temperature is thus maximum on the tube rather than on the deposited material. To avoid tube deformation, it is thus necessary either to maintain a well-controlled slightly positive pressure inside the tube or (easier from the industrial point of view) to incorporate a dopant such as P_2O_5 in the fiber to decrease the working temperature. As we saw in Sec. 3.2.1 that P_2O_5 tends to increase the IR absorption tail, it is anticipated that IVD fibers should always exhibit a slightly higher absorption at 1600 nm than absorptions exhibited by properly processed OVD and AVD fibers. On the other hand, there are probably more data on the IVD process than on any of the others, and Fig. 3.17 shows a histogram concerning the loss of 200 km of IVD fibers obtained with various material deposition rates [29].

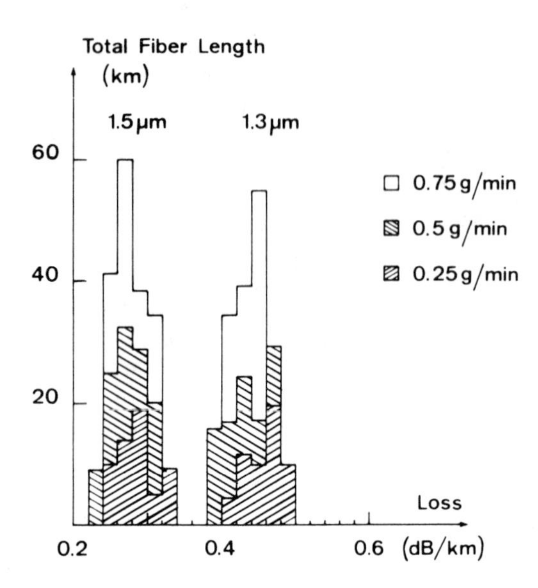

FIGURE 3.17 Histogram of 200-km IVD fiber attenuation at various material deposition rates.

3.4.2 Bending and Microbending Losses

It was shown in Secs. 3.3.2 and 3.3.3 that a high ESF index difference is favorable for decreasing bending and microbending losses. On the other hand, it was shown in Sec. 3.2.2 that a high GeO_2 content increases the Rayleigh scattering loss, and it is thus highly desirable to use the GeO_2 concentration fully to obtain the index difference.

From this point of view, the IVD process is not very well adapted, as it always leaves a central dip in the index profile, which decreases the ESF index difference well below the peak index difference (see Fig. 1.16a); for the same bending and microbending loss sensitivity, IVD fibers exhibit a larger Rayleigh scattering loss. Recent progress has been made in reducing the central index dip by etching the inside layer of deposited material with a CF_4 gas flow during collapse [30]. The sensitivity of the bending loss to the central dip and the efficiency of the dip reduction by this method are illustrated in Fig. 3.18.

The AVD manufacturing process is more efficient, as it is not affected by a central index dip. However, the deposition process leads to a rounding of the index profile at the core–cladding interface which is likely to be more important than in IVD and OVD.

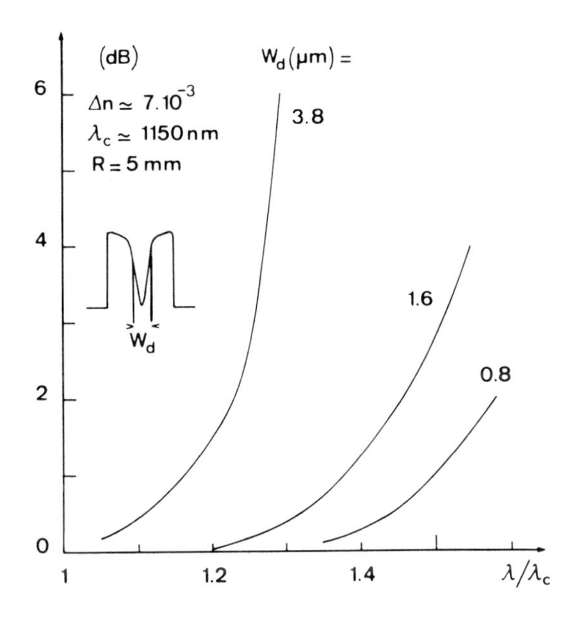

FIGURE 3.18 Experimental bending added loss (one loop with a radius of 5 mm) for IVD fibers with various central index dips (reduced by the method of Ref. 30). All the fibers have the same peak index difference of 0.7% and effective cutoff wavelength of 1150 nm. (From Ref. 17.)

Although the decrease in the ESI index difference compared to the peak index difference is less pronounced with this type of defect, some supplementary loss will probably remain. The OVD process seems to be the best suited, and this is confirmed by the fact that use of this process resulted in the lowest loss ever reported for single-mode fibers [13]. However, very few results have been reported up to now.

3.4.3 Joint Losses

We saw in Sec. 3.3.4 that the lateral axis offset is a dramatic parameter for the splice loss. Part of this lateral offset arises from core–cladding eccentricity, as it is well known that the external surfaces of the fibers tend to align themselves in a fusion splice. From this point of view, the IVD process is the least attractive, probably because the supporting tube defects influence the core–cladding eccentricity. Published values [13,29] report an average eccentricity of less than 0.2 μm for OVD, about 0.3 μm for AVD fibers, and 0.4 μm for IVD fibers, but with some values as high as 1.5 μm.

REFERENCES

1. H. Poignant, J. Le Mellot, and J. F. Bayon, *Electron. Lett.* *17*(8):295–296 (1981).
2. L. Jeunhomme, H. Poignant, and M. Monterie, *Electron. Lett.* *17*(21):808–809 (1981).
3. M. Nakahara, *ITU Telecommun. J.* *48*(11):643–648 (1981).
4. T. Izawa, N. Shibata, and A. Takeda, *Appl. Phys. Lett.* *31*(1):33–35 (1977).
5. D. B, Keck, R. D. Maurer, and P. C. Schultz, *Appl. Phys. Lett.* *22*(7):307–309 (1973).
6. T. Miya, Y. Terunuma, T. Hosaka, and T. Miyashita, *Electron. Lett.* *15*(4):106–108 (1979).
7. M. Horiguchi and M. Kawachi, *Appl. Opt.* *17*(16):2570–2574 (1978).
8. C. Le Sergent, M. Liegois, and Y. Floury, *J. Non-Cryst. Solids* *38–39*:263–268 (1980).
9. K. L. Walker, J. B. MacChesney, and J. R. Simpson, in *IOOC'81 Digest of Technical Papers*, Optical Society of America, Washington, D.C., Apr. 1981, paper WA4.
10. A. D. Pearson, *Proc. 6th Eur. Conf. Opt. Commun.*, York, England, Sept. 1980, *IEE Publ. 190*, pp. 22–25.
11. B. J. Ainslie, K. J. Beales, C. R. Day, and J. D. Rush, *IEEE J. Quantum Electron. QE-17*(6):854–857 (1981).
12. S. Tomaru, M. Uasu, M. Kawachi, and T. Edahiro, *Electron. Lett.* *17*(2):92–93 (1981).
13. G. E. Berkey and A. Sarkar, in *OFC'82 Digest of Technical Papers*, Optical Society of America, Washington, D.C., Apr. 1982, paper THCC5.
14. K. Mochizuki, Y. Namihira, and H. Yamamoto, *Electron. Lett.* *19*:743–745 (1983).
15. N. Uesugi, Y. Murakami, C. Tanaka, Y. Ishida, Y. Negishi, and N. Uchida, *Electron. Lett.* *19*:762–764 (1983).
16. J. Stone, *IEEE J. Lightwave Tech. LT-5*(5):712–733 (1987).
17. C. Brehm and C. Le Sergent, CGE Research Laboratories, unpublished work.
18. E. A. J. Marcatili and S. E. Miller, *Bell Syst. Tech. J.* *48:* 2161–2188 (1969).
19. M. Heiblum and J. H. Harris, *IEEE J. Quantum Electron. QE-11*:75–83 (1975).
20. W. A. Gambling, H. Matsumura, and C. M. Ragdale, *Opt. Quantum Electron.* *11*:43–59 (1979).
21. L. G. Cohen, D. Marcuse, and W. L. Mammel, *IEEE J. Quantum Electron. QE-18*:1467–1472 (1982).

22. H. Matsumura and T. Suganuma, *Appl. Opt.* *19*(18):3151–3158 (1980).
23. D. Marcuse, *Bell Syst. Tech. J.* *55*(7):937–955 (1976).
24. K. Petermann, *Arch. Electron. Uebertrag.* *30*(9):337–342 (1976).
25. P. L. François, CNET, unpublished work.
26. W. A. Gambling, H. Matsumura, and C. M. Ragdale, *Opt. Quantum Electron.* *10* 301–309 (1978).
27. D. Marcuse, *Bell Syst. Tech. J.* *56*(5):703–718 (1977).
28. R. B. Kummer and S. R. Fleming, in *OFC'82 Digest of Technical Papers*, Optical Society of America, Washington, D.C., Apr. 1982, paper THAA1.
29. B. J. Ainslie and C. R. Day, in *OFC'82 Digest of Technical Papers*, Optical Society of America, Washington, D.C., Apr. 1982, paper TUBB1.
30. M. Liegois, G. Lavanant, J. Y. Boniort, and C. Le Sergent, *J. Non-Cryst. Solids* *47*(2):247–250 (1982).

4
Signal Distortion

4.1 INTRODUCTION

Besides the power attenuation by the fiber, the other very important transmission characteristic for telecommunication applications is the signal distortion introduced by the fiber. For very low attenuations, this effect can, in some cases, limit the repeater spacing below what would be possible from the attenuation factor, or conversely, limit the possible bit rate. In this chapter we consider only linear phenomena; that is, we assume the power level launched into the fiber to be less than the threshold for nonlinear effects (see Chap. 9). Under these conditions, the only sources of signal distortion are the difference in propagation delay time between the two LP_{01} modes (birefringence dispersion; see Chap. 2) and the variation of propagation delay time with wavelength (chromatic dispersion; see Chap. 1). We review the effects of both types of dispersion and then discuss the corresponding limitations on the possible bit rates.

4.2 INFLUENCE OF BIREFRINGENCE DISPERSION

As seen in Chap. 2, in a real fiber the two LP_{01} modes (orthogonal polarizations) have slightly different propagation constants and thus different group delay times. As the case where the two polarizations exchange energy (polarization mode coupling) is somewhat complicated, we first discuss the fiber response in the absence of mode coupling.

4.2.1 Without Mode Coupling

In this case, the two LP_{01} modes propagate independently, and thus a short pulse launched into the fiber will emerge as two short pulses after delays $(\tau - T/2)L$ and $(\tau + T/2)L$, respectively, where τ is the mean group delay time of the two LP_{01} modes, T their group delay difference, and L the fiber length (Fig. 4.1a). Expressions for T are given in Chap. 2 for various cases.

The same phenomenon holds for sinusoidally modulated light (in terms of power) and we can define a frequency response for the fiber, to power modulated at pulsation Ω, as the ratio between the output power at pulsation $\Omega(P_s(\Omega, L))$ and the input power at pulsation $\Omega(P_e(\Omega))$. If we assume both modes to be equally excited, we obtain

$$H(\Omega, L) = \frac{P_s(\Omega, L)}{P_e(\Omega)} = \exp(-i\Omega\tau L) \cos\left(\frac{\Omega TL}{2}\right) \tag{4.1}$$

As seen in this expression, it would be difficult to define a -6-dB bandwidth because of the oscillatory nature of $H(\Omega, L)$, and we consider thus the response to a Gaussian pulse launched into the fiber with equal excitation of both modes. As discussed above, the launched Gaussian pulse will emerge as two Gaussian pulses and the rms width of the putput pulse S (in power) is given by

$$S^2 = S_0^2 + \left(\frac{TL}{2}\right)^2 \tag{4.2}$$

where S_0 is the rms width of the input pulse in power. Recall that fibers with $T < 0.05$ ps/km can routinely be manufactured (see Chap. 2), which makes it possible to keep this effect to a practically negligible level.

4.2.2 With Mode Coupling

Uniform Coupling. Consider a fiber with a nominal birefringence $\delta\beta$ and a coupling coefficient C between the two reference polarizations, both parameters being independent of z. It can be seen from Eq. (2.23) that this situation leads to new polarization eigenstates which propagate independently, with a resulting difference in group delay times [1]:

$$T_r = \frac{1}{c}\frac{d}{dk}\left(|\delta\beta|^2 + |C|^2\right)^{\frac{1}{2}} \tag{4.3}$$

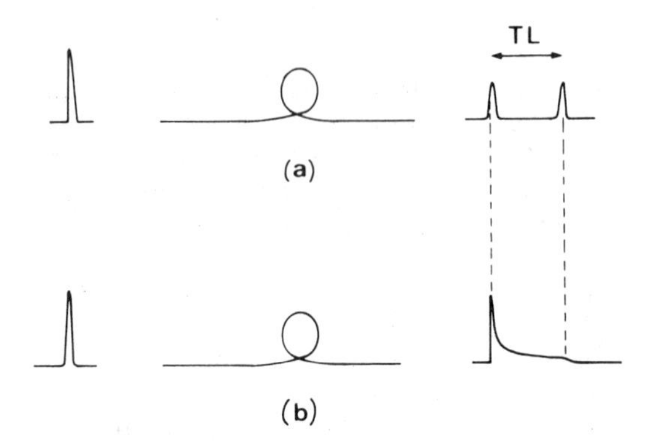

FIGURE 4.1 Pulse distortion effects in birefringent fibers. (a) Without polarization-mode coupling. (b) With random polarization-mode coupling.

where c is the light velocity in vacuum and k is the plane-wave propagation constant in vacuum. Expressions for $\delta\beta$ and C for various cases of interest were given in Chap. 2. If the two eigenstates are equally excited, the discussion of the preceding section simply holds by replacing T by T_r. However, a case of particular interest here occurs when the cross-polarization p is small (polarization-maintaining fibers), which, according to Chap. 2, requires that $|C| << |\delta\beta|$. A Gaussian pulse launched into the fiber will emerge as two Gaussian pulses of relative intensity (1 − p) and p, respectively, separated by T_rL, where L is the fiber length. Because $|C| << |\delta\beta|$, $T_r \simeq T$, and the resulting rms width S of the output pulse (in power) is given by

$$S^2 \simeq S_0^2 + p(1 - p)(TL)^2 \tag{4.4}$$

where S_0 is the rms width of the input pulse in power. We observe here that circular polarization-maintaining fibers are much more favorable than linear ones, as stated intuitively in Chap. 2, because for the same birefringence their dispersion T is about 10 to 20 times smaller. Typical fibers able to maintain linear polarizations were seen in Chap. 2 to have $T \simeq 0.1$ ns/km, and thus for a 50-km-long link with a −20-dB cross-polarization, the second term on the right-hand side of Eq. (4.4) amounts to about $(0.5 \text{ ns})^2$.

Random Coupling. The case of random coupling cannot be solved quantitatively in terms of field equations [Eq. (2.23)], and the power-coupled theory does not yield very useful information, as seen in the following example. Let us consider a fiber with a zero nominal birefringence ($\delta\beta = 0$), perturbed by random twists $N(z)$ (turns per meter, with a sign convention) leading to a random coupling coefficient $C(z) = 2\pi g N(z)$, as shown in Chap. 2. As required by the coupled power theory, we assume that $\langle N(z) \rangle = 0$, where $\langle \ \rangle$ means an average over a large ensemble of such fibers [2]. Because $\delta\beta = 0$, the group delay difference T between the two polarization eigenmodes of the unperturbed fiber also vanishes, and the coupled power theory thus predicts that there is no pulse broadening on average; $\langle \delta t \rangle = 0$ [2], where δt represents the pulse broadening in one fiber of the ensemble. From Chap. 2 we know, however, that in each fiber the local polarization eigenmodes of this type of perturbed fiber are right and left circularly polarized modes, and it thus appears that right and left circularly polarized modes will propagate unaffected (in terms of SOP) through each fiber. Launching a linearly polarized light at the input of the fiber will thus result in the emergence of the two right and left circularly polarized modes, with equal amplitudes and a phase difference given by

$$\delta\phi = \int_0^L C(z)\, dz = 2\pi g \int_0^L N(z)\, dz \tag{4.5}$$

where L is the fiber length. The pulse broadening in each fiber will thus be given by

$$\delta t = \frac{2\pi}{c} \int_0^L N(z)\, dz\, \frac{dg}{dk} \tag{4.6}$$

Obviously, because $\langle N(z) \rangle = 0$, we have $\langle \delta t \rangle = 0$, as predicted by the coupled power theory, but this information does not mean that $\delta t = 0$ in each fiber. This is simply due to the fact that in some fibers the right circular polarization emerges first, whereas in other fibers it emerges last. The coupled power theory does not provide the useful information, as in each fiber there will be a pulse broadening. More useful would be the values of $\langle (\delta t)^2 \rangle^{\frac{1}{2}}$, and it can be shown more generally [3] that provided that $\delta\beta$ is independent of z, we have $\langle (\delta t)^2 \rangle^{\frac{1}{2}} \sim L$, contrary to the expected $L^{\frac{1}{2}}$ dependence from coupled power theory.

Besides this particular example, we can maintain that provided $\langle |C|^2 \rangle \ll \langle |\delta\beta|^2 \rangle$, the relationship given by Eq. (4.4) is likely to hold generally, understanding $(TL)^2$ to be the average value of this parameter. It is thus apparent that high-bit-rate transmission requires either very low birefringence dispersion fibers, or polarization-maintaining fibers with a cross-polarization well below −20 dB. In the latter case, high-birefringence fibers with small birefringence dispersion (such as circular polarization-maintaining fibers or some designs of PANDA fibers; see Sec. 2.2.3) are the most desirable structures.

4.3 INFLUENCE OF CHROMATIC DISPERSION

It was shown in Sec. 1.2.4 that the chromatic dispersion $d\tau/d\lambda$ (where τ is the group delay of the optical wave) is composed of material dispersion terms, waveguide dispersion terms, and mixed terms. Before discussing the effects of the chromatic dispersion on the signal distortion, we discuss the properties of chromatic dispersion, especially the possibilities of dispersion-free fibers.

4.3.1 Chromatic Dispersion Control

We examine here the factors influencing chromatic dispersion and the conditions under which it can be canceled.

Chromatic Dispersion. We assume first that the dopants incorporated into the silica for building the index profile have no influence, as this allows a simpler and more comprehensive discussion, and it will be shown that the corrections introduced by these dopants are generally small. We consider three typical fiber structures, illustrated in Fig. 4.2 with some parameter definitions:

1. The classical step-index matched-cladding fiber (MC).
2. The depressed inner cladding fiber with a thick cladding ($a'/a \geq 5$) (T-DIC). This fiber behaves almost exactly like the matched-cladding fiber for all the waveguide properties (see Chap. 1 for a discussion of its equivalent step-index fiber), but the index difference is obtained with the help of two dopants (generally fluorine for decreasing the index of the inner cladding and germanium for the core). For the same index difference, it thus provides a smaller Rayleigh scattering loss than does the matched-cladding fiber (see Chap. 3).
3. The depressed inner cladding fiber with a narrow cladding ($a'/a \leq 2$) (N-DIC or W). As discussed in Chap. 1, such fibers have very specific waveguide dispersion properties.

It shall be understood that the first two cases actually include graded core fibers, through the use of their ESI parameters. The discussion is qualitatively identical, and only fine tuning of dispersion properties shall be done through specific calculations (see

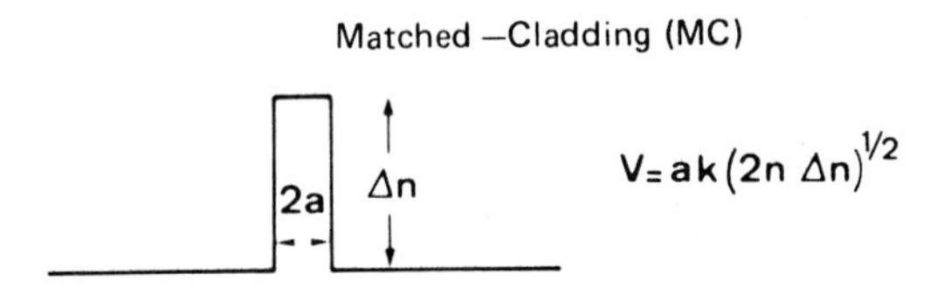

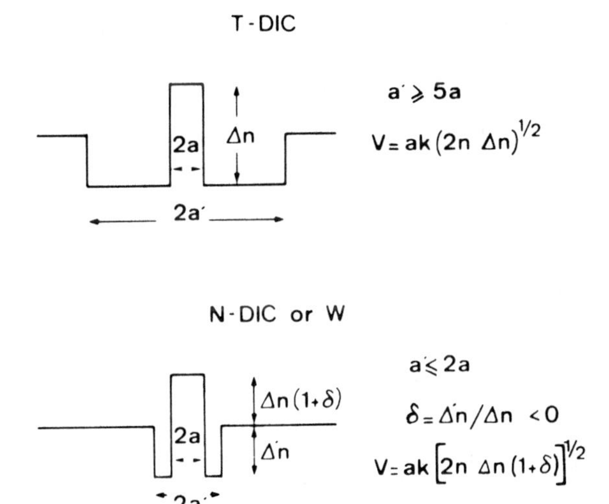

FIGURE 4.2 Fiber structures and parameter definitions for chromatic dispersion studies.

Sec. 4.5.2). Similarly, the discussion of the third case is qualitatively valid for more complicated index profiles having several index rings around the central W structure (see Sec. 4.5.3).

With the parameter definitions shown in Fig. 4.2, and Eq. (1.48), the chromatic dispersion for all these fibers reads essentially as follows (any dopant effects are neglected):

$$C(\lambda) = \frac{d\tau}{d\lambda} \simeq M_2(\lambda) - \frac{\Delta n(1 + \delta)}{c\lambda} V(Vb)'' \tag{4.7}$$

where $M_2(\lambda)$ is the material dispersion of the inner cladding, and Δn, V, and δ are defined in Fig. 4.2 for each fiber structure ($\delta = 0$ for matched-cladding and T-DIC).

Useful approximations of $C(\lambda)$ can be obtained. First, using a simplified model for material dispersion [4] and accurate values of $M_2(\lambda)$ for silica [5], we obtain over a broad wavelength range

$$M_2(\lambda) \simeq 2.66 \times 10^{-2}\lambda - \frac{6.985 \times 10^{10}}{\lambda^3} \qquad (4.8)$$

where $M_2(\lambda)$ is in ps/(nm × km), with λ in nanometers.

Plots of $M_2(\lambda)$ are shown in Fig. 4.3, and the accuracy of Eq. (4.8) is better than 10% from 800 to 1700 nm. In a more restricted wavelength range, we obtain empirically a very useful approximation,

$$M_2(\lambda) \simeq 122\left(1 - \frac{1276}{\lambda}\right) \qquad (4.9)$$

where λ is in nanometers. This approximation gives an accuracy on M_2 better than 0.1 ps/(nm × km) between 1260 and 1660 nm.

From Chap. 1 we approximate the normalized waveguide dispersion for step-index fibers by

$$V(Vb)'' \simeq 0.080 + 0.549(2.834 - V)^2$$

$$\simeq 0.080 + 3.175\left(1.178 - \frac{\lambda_c}{\lambda}\right)^2 \qquad (4.10)$$

where λ_c is the LP_{11}-mode cutoff wavelength. The relative error is less than 5% for $0.9 \leq \lambda/\lambda_c \leq 1.9$ $(1.3 \leq V \leq 2.6)$ (see Fig. 1.13). For W-type fibers, more complicated approximations are given in Sec. 1.4.3. Let us simply recall that these fibers can provide negative or vanishing values of $V(Vb)''$ in the single-mode regime, and very high values of $V(Vb)''$ (see Sec. 1.4.3 and Figs. 1.24 and 1.25).

A general conclusion from the equations above is that at a given wavelength, the chromatic dispersion $C(\lambda)$ decreases when the index difference increases or when the cutoff wavelength decreases [which makes $V(Vb)''$ increase]. This general behavior is illustrated in Fig. 4.4 for step-index fibers. Equivalently, this means that the wavelength of zero chromatic dispersion (λ_D) shifts toward the long wavelengths.

Theory of Dispersion-Free Fibers. The term *dispersion-free fibers* applies to fibers where the second-order dispersion term [$d^2\beta/d\omega^2$ or $C(\lambda)$] vanishes, although the third-order term [$C'(\lambda) = dC(\lambda)/d\lambda$] is generally different from zero. From Eq. (4.7), we find that the

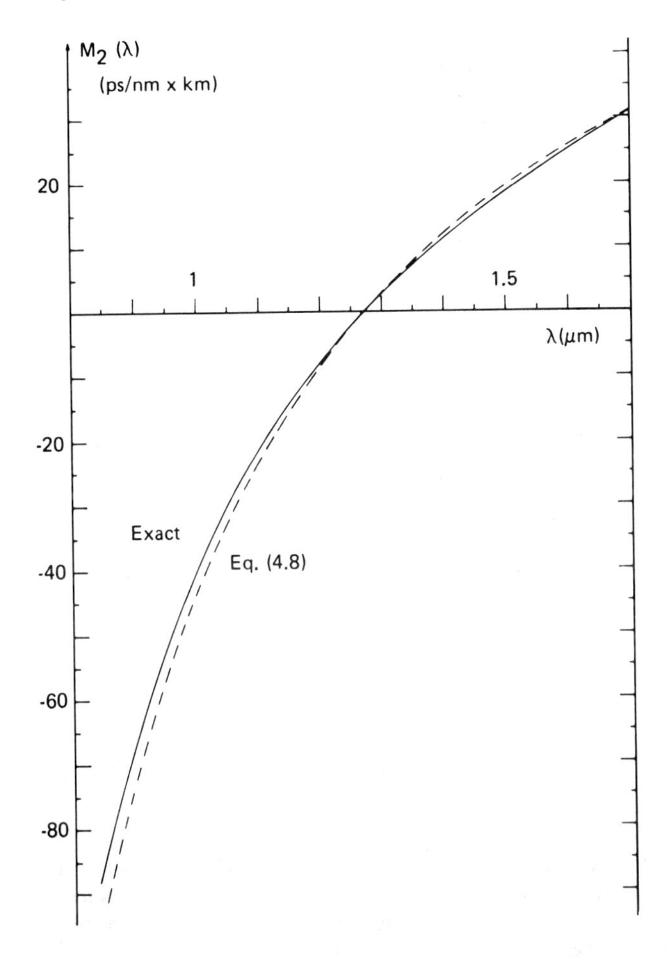

FIGURE 4.3 Material dispersion for fused silica [approximation by Eq. (4.9) is indistinguishable from the exact curve in the range of application].

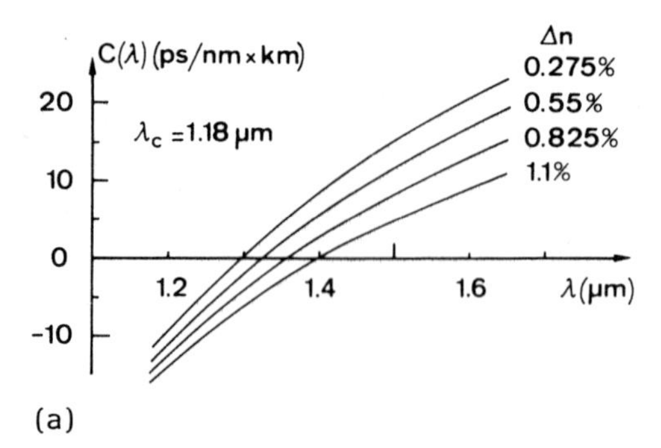

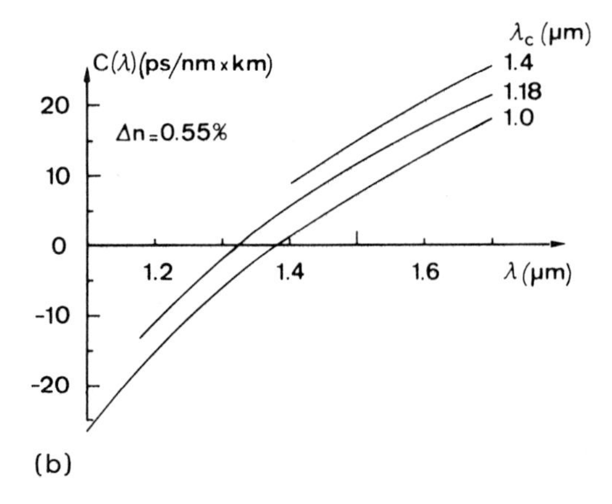

FIGURE 4.4 Chromatic dispersion in step-index (matched-cladding or T-DIC) fibers (dopant effects are neglected). (a) At constant cutoff wavelength and various index differences. (b) At constant index difference for various cutoff wavelengths.

index difference for canceling the chromatic dispersion at wavelength λ_D should satisfy

$$\Delta n_0(1 + \delta) = \frac{M_2 c \lambda_D}{V(Vb)''} \tag{4.11}$$

This relationship is illustrated in Fig. 4.5, which is central to our discussion of dispersion-free fibers.

It can be seen that W-type fibers can provide dispersion cancellation anywhere in the range 1250 to 1700 nm, provided that they are properly designed. Notable is the fact that they are able to provide a dispersion cancellation at 1273 nm $[M_2(\lambda) = 0]$,

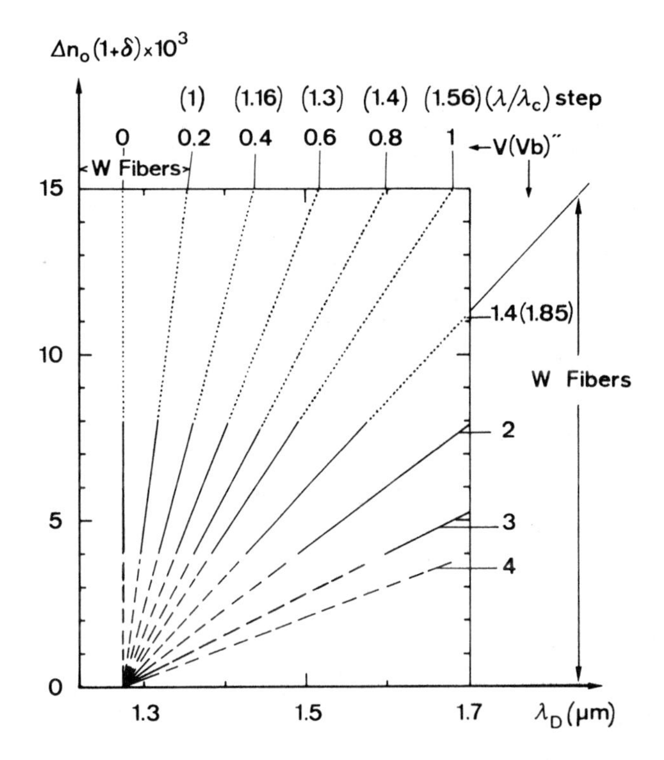

FIGURE 4.5 Index difference $\Delta n_0(1 + \delta)$ (above silica for MC and W fibers, and above inner cladding for T-DIC fibers) for canceling the chromatic dispersion at wavelength λ_D (dopant effects are neglected). Step-index (MC or T-DIC) fibers are limited to the range $0.2 \leq V(Vb)'' \leq 1.4$ (the corresponding values of λ/λ_c are shown in parentheses), whereas W-type fibers cover the range $-0.1 \leq V(Vb)'' \leq 4$ (or even higher, but with unrealistic parameters). The region of solid lines corresponds to a realistic range of index differences (GeO_2 content) for ultra-low-loss MC or W-type fibers. The region of dotted lines should be explored only with T-DIC or linearly graded-core fibers if ultralow losses are desired, and in such a way that Δn_{Ge} (part of Δn due to GeO_2) remains below 0.8%. Finally, the region of dashed lines exhibits high bending and microbending losses.

independently of the index difference, if they are chosen such that $V(Vb)''$ vanishes at this wavelength. T-DIC fibers are slightly more restricted, as they can be dispersion free only between 1300 and 1650 nm (with realistic index differences). Finally, matched-cladding fibers are restricted to 1300 to 1550 nm, which is quite similar to T-DIC. However, for the same value of λ_D, T-DIC fibers can use higher total index differences (for the same GeO_2 content) and smaller values of λ/λ_c, both factors being favorable to lower losses. Practically however, it has been found that canceling the chromatic dispersion around 1550 nm (dispersion-shifting) while still maintaining low losses is best achieved with triangular core index profiles (linear gradient, see Sec. 4.5.2).

Figure 4.6 illustrates the chromatic dispersion of some typical fibers of each kind, all the chosen fibers having approximately the same loss characteristics at their respective λ_D (bending, microbending, joints; see Chap. 3). In addition to the information given in the discussion above, two points are notable: some W-type fiber designs can also cancel $C'(\lambda) = dC(\lambda)/d\lambda$ (W_3 in Fig. 4.6) simultaneously with $C(\lambda)$, and these fibers are very sensitive to manufacturing tolerances (W_2 and W_3).

Tolerances of Dispersion-Free Fibers. We examine here the sensitivity of dispersion-free fibers to the variation of various parameters, such as operating wavelength, core diameter, and index difference. For the sensitivity to the operating wavelength, we get from Eqs. (1.14), (4.7), (4.9), and (4.11),

$$C'(\lambda_D) = \left[\frac{dC(\lambda)}{d\lambda}\right]_{\lambda=\lambda_D} = \frac{122 - K(\lambda_D)}{\lambda_D} \tag{4.12}$$

with

$$K(\lambda) = -M(\lambda)\,\frac{V}{V(Vb)''}\,\frac{d(V(Vb)'')}{d\lambda}$$

Equation (4.9) is approximate mainly because of the use of Eq. (4.7), which neglects any dopant effect. However, within the limit of this approximation, it is valid for any kind of index profile. We may thus use it for the case of step-index core fibers, which are usually designed so that $C(1300\text{ nm}) \approx 0$. We also may use it for triangular core fibers, which are generally used for dispersion shifting and provide $C(1550\text{ nm}) \approx 0$. We obtain typically

$$\begin{aligned} C'_{step}(1300\text{ nm}) &\simeq 0.09\text{ ps}/(\text{nm}^2 \times \text{km}) \\ C'_{step}(1550\text{ nm}) &\simeq 0.06\text{ ps}/(\text{nm}^2 \times \text{km}) \end{aligned} \tag{4.13}$$

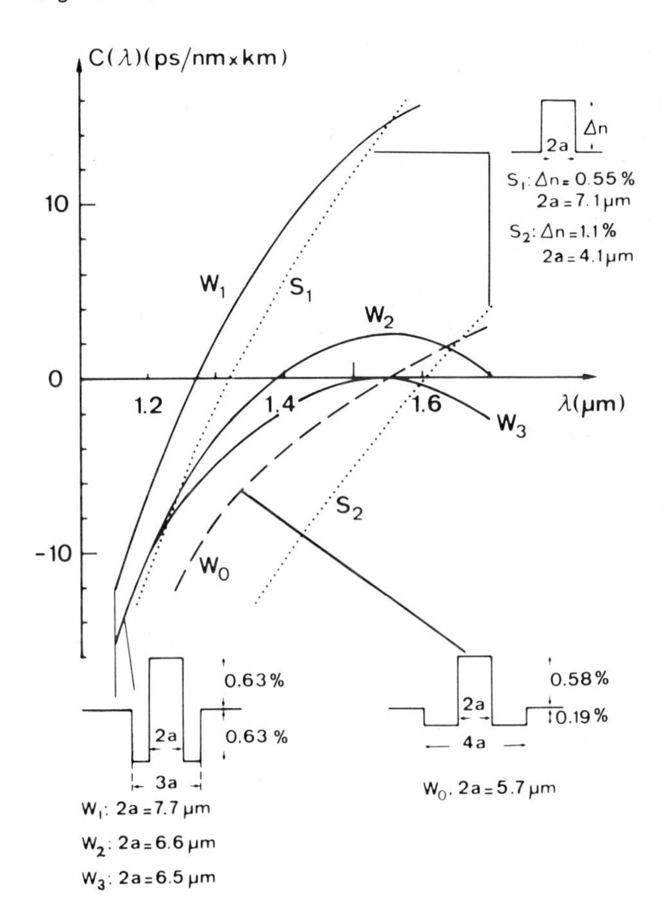

FIGURE 4.6 Examples of dispersion behavior for various fiber designs (dopant effects are neglected; absolute index differences). Structure S_2 is best achieved with triangular core index profile fibers. Note that fiber W_3 provides C = 0 and C' = 0 simultaneously, but that it is very sensitive to the exact design (difference between W_2 and W_3). (Curves W_0, W_2, and W_3 after Monerie [6].)

For W-type fibers, however, Figs. 1.24 and 1.25 show that it is possible to obtain much higher values of $d[V(Vb)'']/d\lambda$. This explains the possibility illustrated in Fig. 4.6 (W_3) of obtaining $C'(\lambda_D) = 0$, and a careful examination of $V(Vb)''$ in Figs. 1.24 and 1.25 shows that W-type fibers may provide the various possibilities illustrated in Fig. 4.7. Obtaining a value of $C'(\lambda_D) = 0$ offers the advantage of a large tolerance on the operating wavelength and opens up the possibility of wavelength multiplexing with low dispersion at all the wavelengths.

For the sensitivity to the core radius a and the index difference Δn, we get from Eqs. (1.14), (4.7), (4.9), and (4.11),

$$dC(\lambda_D) \simeq [122 - \lambda_D C'(\lambda_D)] \frac{da}{a} + \left[6.1 - \frac{\lambda_D C'(\lambda_D)}{2} - M(\lambda_D)\right] \frac{d(\Delta n)}{\Delta n} \tag{4.14}$$

It should be noted that Eqs. (4.9) and (4.14), although approximate, are valid for any index profile.

For step-index and triangular core profile fibers we get typically

$$\begin{aligned} dC_{step}(1310\ \text{nm}) &\simeq -1.1 \frac{d(\Delta n)}{\Delta n} + 4 \frac{da}{a} \quad \text{ps/(nm} \times \text{km)} \\ dC_{tri}(1550\ \text{nm}) &\simeq -7 \frac{d(\Delta n)}{\Delta n} + 29 \frac{da}{a} \quad \text{ps/(nm} \times \text{km)} \end{aligned} \tag{4.15}$$

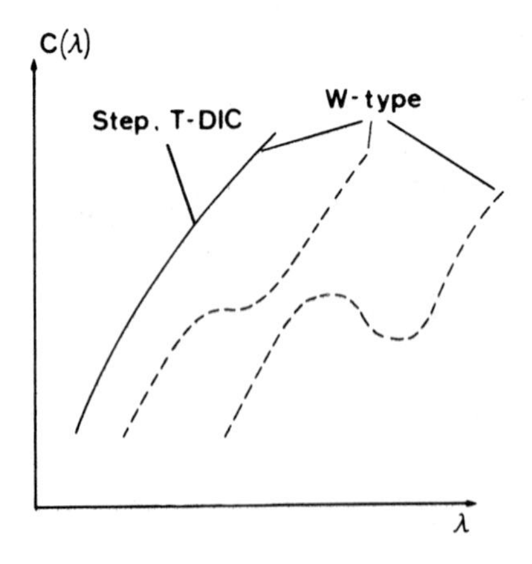

FIGURE 4.7 General behavior of the chromatic dispersion $C(\lambda)$, attainable with various fiber designs.

For tolerances of ±10% on Δn and $\mp 3\%$ on a, this corresponds to variations of ±0.2 ps/(nm × km) at 1310 nm and ±1.6 ps/(nm × km) at 1550 nm. For W-type fibers, the sensitivity to these tolerances will be markedly increased for fibers adjusted to have C(1270 to 1300 nm) = 0 [as they have $V(Vb)'' \simeq 0$] or $C'(\lambda_D) = 0$ (as they usually have very strong values of $d[V(Vb)'']/dV$). This strong sensitivity of W-type fibers is illustrated in Fig. 4.6 (structures W_1, W_2, and W_3). Instead of considering the variation of C with the tolerances, we can consider the variation of λ_D, obtained from Eq. (4.14) through

$$d\lambda_D = -\frac{dC(\lambda_D)}{C'(\lambda_D)} \tag{4.16}$$

For step-index and triangular core profile fibers we typically obtain from Eqs. (4.13) and (4.15),

$$\begin{aligned} &\text{Step;}\ \lambda_D \simeq 1310\ \text{nm} \Rightarrow d\lambda_D \simeq +12\,\frac{d(\Delta n)}{\Delta n} - 44\,\frac{da}{a}\ \text{nm} \\ &\text{Triangle;}\ \lambda_D \simeq 1550\ \text{nm} \Rightarrow d\lambda_D \simeq +117\,\frac{d(\Delta n)}{\Delta n} - 483\,\frac{da}{a}\ \text{nm} \end{aligned} \tag{4.17}$$

For tolerances of ±10% on Δn and ±3% on a, this corresponds to variations of ±2.5 nm around 1300 nm, and ±26 nm around 1550 nm.

Again W-type fibers will be more sensitive to these tolerances, especially those fibers providing $C'(\lambda_D) = 0$. However, the practical significance of these tolerances for long transmission systems composed of many fibers is questionable: $C(\lambda)$ will wander around its nominal value from fiber to fiber, but it has been shown that this could be accommodated for simply by taking the mean value of $C(\lambda)$ into account [7]. This means that for dispersion-free fiber links, operation at V such that $V(Vb)''$ is maximum should be avoided, as it would provide a noncentered (nonzero average) fluctuation of $C(\lambda)$.

Influence of Dopants. If we look back to Eq. (1.48), the complete expression for $C(\lambda)$ is

$$C(\lambda) = M_2(\lambda) - \frac{\Delta n}{c\lambda}\left\{ V(Vb)'' + \frac{\lambda^2}{\Delta n}\,\frac{d^2(\Delta n)}{d\lambda^2}\,\frac{b + (Vb)'}{2} - \frac{\lambda}{n_2\,\Delta n}\,\frac{d(n_2\Delta n)}{d\lambda}\,[V(Vb)'' + (Vb)' - b] \right\} \tag{4.18}$$

An approximation for $[b + (Vb)']/2$ has been given in Eq. (1.49), and we obtain here

$$V(Vb)'' + (Vb)' - b \simeq -1.29 + \frac{5.68}{V} - \frac{2}{V^2} \tag{4.19}$$

Now we consider fibers having a GeO_2-doped core and a fluorine-doped cladding, as these are the most usual cases encountered practically. The total index difference Δn is thus the sum of a contribution from GeO_2 [corresponding to $\Delta n(Ge)$] and from fluorine [corresponding to $\Delta n(F)$]:

$$\Delta n = \Delta n(Ge) + \Delta n(F) \tag{4.20}$$

After combining linearly the behavior of the two dopants for which experimental data have been reported [8], we obtain:

$$\frac{\lambda^2}{\Delta n}\frac{d^2(\Delta n)}{d\lambda^2} \simeq 0.14\,\frac{\Delta n(Ge)}{\Delta n} + 0.1\,\frac{\Delta n(F)}{\Delta n} \tag{4.21}$$

$$\frac{\lambda}{n_2\,\Delta n}\frac{d(n_2\,\Delta n)}{d\lambda} \simeq 1.4 \times 10^{-4}\,\lambda - 0.14\,\frac{\Delta n(Ge)}{\Delta n} - 0.1\,\frac{\Delta n(F)}{\Delta n} \tag{4.22}$$

with λ in nm.

Figure 4.8 shows the dispersion curves of some fibers compared to those obtained when all dopant effects are neglected. It is seen that the corrections introduced in these typical cases are very small, and a more extensive study of the equations above shows that the difference from Eq. (4.7) is always less than 5 ps/(nm × km). The corrections introduced by the dopants are similar to those introduced by classical manufacturing parameter fluctuations. It is thus generally sufficient to ignore the dopant contribution when evaluating the dispersion, at least as a first step.

Figures 4.9 and 4.10 show the value of the index difference Δn_0 which cancels the chromatic dispersion, as a function of the LP_{11} mode cutoff wavelength λ_c, for step-index MC and T-DIC fibers with $\Delta n_F/\Delta n = +0.57$. The strong difference between the two different fiber compositions, for short values of λ_D and λ_c/λ_D close to 1, is due simply to the weakness of the normalized waveguide dispersion coefficient $V(Vb)''$ in these cases. It thus follows that when ignoring the dopant contribution, the absolute error on $C(\lambda)$ or on λ_D remains very small. In the range 1500 nm $\leq \lambda_D \leq$ 1600 nm (Fig. 4.11), the value of Δn_0 is almost insensitive to the fiber

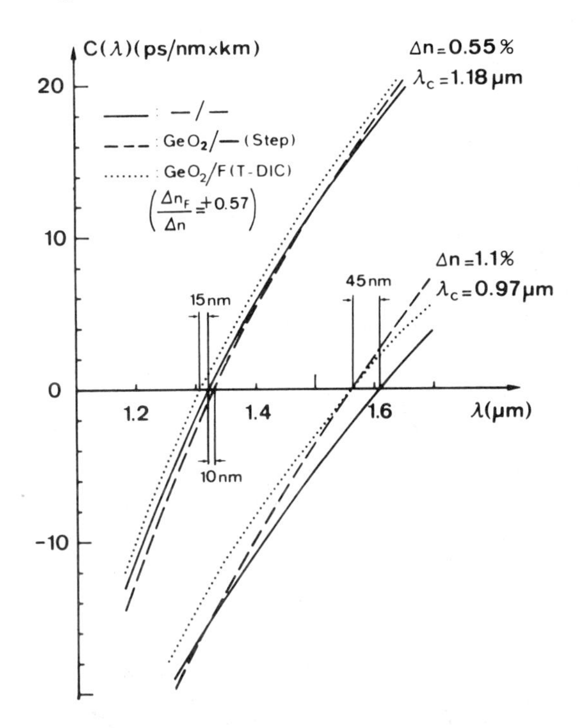

FIGURE 4.8 Chromatic dispersion of various fiber compositions (core dopant/cladding dopant).

composition [because V(Vb)'' is strong], and λ_D depends almost entirely on the core radius.

Practical Realizations. It can be concluded from the previous discussion that the realization of 1300-nm dispersion-free step-index fibers is relatively easy, whereas it appears more difficult to obtain 1550-nm dispersion-free fibers. In Ref. 9, a step-index fiber with $\Delta n \simeq 1.8\%$ and a core diameter of about 4.8 μm was dispersion-free in the 1550-nm region, but with a loss of 2.5 dB/km, far above the ultimate limit. In Ref. 10, a similar fiber was dispersion free around 1500 nm with $\Delta n \simeq 0.74\%$, a cutoff wavelength of 820 nm, and a loss of 1 dB/km. Finally, Fig. 4.11 illustrates some results obtained with W-type fibers, which appear very promising if the manufacturing tolerances can be controlled satisfactorily.

Summary. In summary, several points are of particular importance:

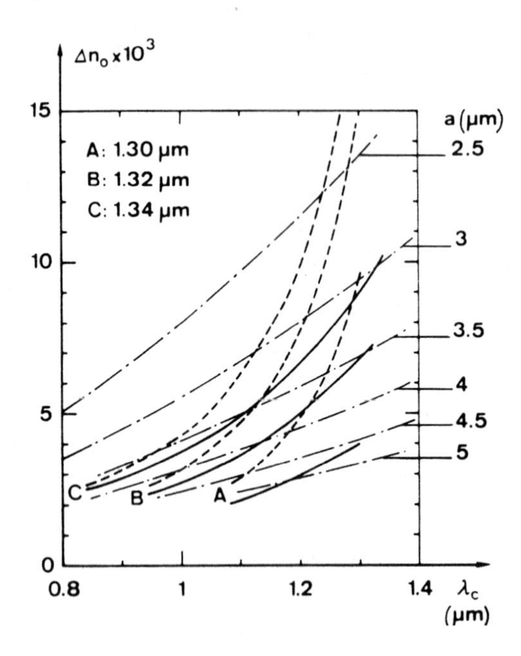

FIGURE 4.9 Index difference Δn_0 for zero chromatic dispersion at various wavelengths (A, λ_D = 1300 nm; B, λ_D = 1320 nm; C, λ_D = 1340 nm) as a function of the cutoff wavelength. Solid lines are for step-index matched-cladding GeO_2-doped-core fibers, and dashed lines are for T-DIC fibers with GeO_2-doped-core and F-doped cladding with $\Delta n_F/\Delta n$ = +0.57. Dashed-dotted lines indicate a constant core radius a.

1. A first theoretical approach to the chromatic dispersion problem, that of ignoring the effects of dopants, is quite simple and gives acceptable accuracy.
2. From a practical point of view, it appears that step-index fibers should be used only for 1300 nm $\leq \lambda_D \leq$ 1400 nm if ultra-low losses are desired. The use of T-DIC fibers makes it possible to cover a wider range, between 1300 and 1600 nm, with step-index core around 1300 nm and triangular core profile around 1550 nm.
3. Although W-type fibers may cause some loss problems, which should be overcome in the near future, they appear attractive, as they can provide one, two, or three values of λ_D anywhere between 1260 and 1700 nm, and also $C'(\lambda_D) = 0$. As an additional advantage, a single preform structure can make it possible to adjust the value of λ_D in the whole range 1300 to 1700 nm

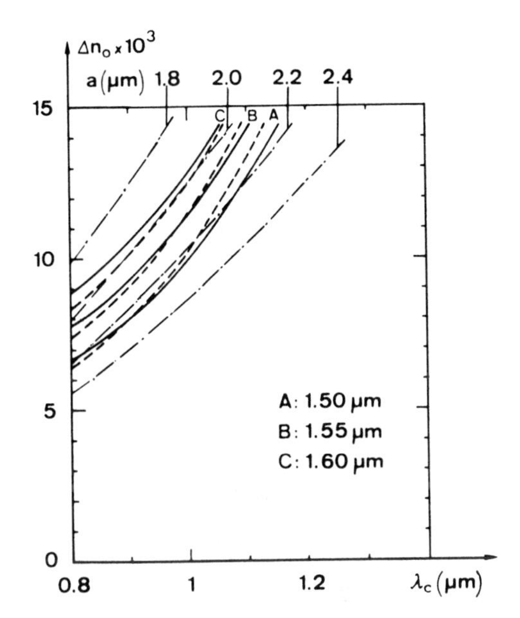

FIGURE 4.10 Same as Fig. 4.9, but for 1500 nm $\leq \lambda_D \leq$ 1600 nm.

by controlling the fiber drawing diameter (see structures W_1, W_2, and W_3 in Fig. 4.6). However, these considerable advantages come at the expense of a very strong sensitivity to the manufacturing tolerances. It is thus likely that W-type fibers will not be used in the first generation of systems operating at 1300 nm, and probably not even in the second generation operating at a single wavelength around 1550 nm.

We now turn our interest to the effect of chromatic dispersion on the transmission characteristics of the fiber. This topic has been studied in detail in Ref. 14, but here we will use a more phenomenological approach. We examine first the limiting case of a polychromatic source (natural line width >> modulation bandwidth), then the limiting case of a monochromatic source (natural line width << modulation bandwidth), and finally, the intermediate case.

4.3.2 Polychromatic Source

We assume here that the natural line width of the source is much larger than the modulation bandwidth and that the light spectrum is unaffected by the modulation. Practically, this corresponds, for

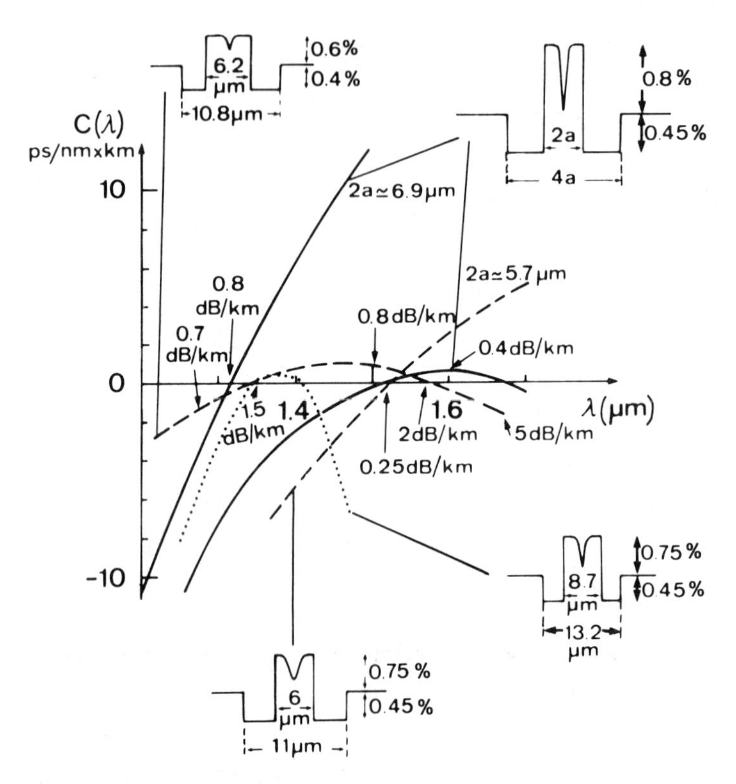

FIGURE 4.11 Chromatic dispersion of some practical W-type fibers, illustrating the various possibilities of these fibers (absolute index differences). Note both that low losses are obtained simultaneously and the accuracy required on the diameter. (Solid line, Ref. 11; dashed line, Ref. 12; dotted line, Ref. 13.)

example, to a multilongitudinal mode semiconductor laser. Following these assumptions, we can write the power launched into the fiber $P_e(t)$ as

$$P_e(t) = \int_{-\infty}^{+\infty} s(\Omega)\ \exp(i\Omega t)\ d\Omega \int_0^{\infty} p(\lambda)\ d\lambda \tag{4.23}$$

where $s(\Omega)$ is the pulsation spectrum of the power modulating signal and $p(\lambda)$ is the spectral power density of the source. As seen in Chap. 1, the propagation delay time along a fiber of length L depends on the wavelength, and thus the wavelength λ will be delayed by $\tau(\lambda) \times L$. The output power $P_S(t, L)$ can thus be written

$$P_s(t, L) = \int_{-\infty}^{+\infty} s(\Omega)\ \exp(i\Omega t) \left\{ \int_0^{\infty} p(\lambda)\ \exp[-i\Omega\tau(\lambda)L]\ d\lambda \right\} d\Omega \tag{4.24}$$

Comparison with Eq. (4.23) shows that we can define the *power frequency response* of the fiber as

$$H(\Omega, L) = \frac{\int_0^{\infty} p(\lambda)\ \exp[-i\Omega\tau(\lambda)L]\ d\lambda}{\int_0^{\infty} p(\lambda)\ d\lambda} \tag{4.25}$$

It follows from Eq. (4.25) that the fiber bandwidth will vary as 1/L. We may use the fact that the spectral line width of the source is much smaller than its center emission wavelength λ_0 and use for $\tau(\lambda)$

$$\tau(\lambda) \simeq \tau_0 + (\lambda - \lambda_0)C_0 + \frac{(\lambda - \lambda_0)^2}{2} C_0' \tag{4.26}$$

with $\tau_0 = \tau(\lambda_0)$, $C_0 = C(\lambda_0)$, and $C_0' = C'(\lambda_0)$.

As a particular case, we consider a source with a Gaussian spectral power density having a rms spectral width $\Delta\lambda$:

$$p(\lambda) = \exp\left[-\frac{1}{2}\left(\frac{\lambda - \lambda_0}{\Delta\lambda}\right)^2\right] \tag{4.27}$$

Using Eqs. (4.25) and (4.26), we find that as long as $|\lambda_0 - \lambda_D| >> \Delta\lambda/2$ [where λ_D is the wavelength of zero chromatic dispersion, $C(\lambda_D) = 0$], the power frequency response is Gaussian:

$$H(\Omega, L) = \exp(-i\Omega\tau_0 L) \exp\left[-\left(\frac{\Omega}{\Omega_0}\right)^2 \ln 2\right]$$

$$\Omega_0 = \frac{(2 \ln 2)^{\frac{1}{2}}}{C_0 L\Delta\lambda} \qquad \text{(}-3\text{-dB optical power bandwidth)} \tag{4.28}$$

More generally, whatever the value of λ_0 (compared to λ_D), a Gaussian light pulse with rms width S_0 (in power) launched into the fiber will emerge with a rms width S (in power) given by

$$S^2 = S_0^2 + (C_0 L\Delta\lambda)^2 + \frac{1}{2}\left[\left(C_0' + \frac{2C_0}{\lambda_0}\right) L(\Delta\lambda)^2\right]^2$$

which can be further simplified by using $\Delta\lambda/\lambda_0 << 1$ to

$$S^2 \simeq S_0^2 + (C_0 L\Delta\lambda)^2 + \frac{1}{2}[C_0' L(\Delta\lambda)^2]^2 \tag{4.29}$$

The output pulse remains Gaussian only when $|\lambda_0 - \lambda_D| >> \Delta\lambda/2$.

4.3.3 Monochromatic Source

Field Frequency Response. We assume here that the natural line width of the source in the absence of modulation is negligible compared to the modulation bandwidth, and that the center emission wavelength is unaffected by the modulation. Practically, this corresponds, for example, to a mode-locked laser or to a single-frequency stabilized laser with external modulation for coherent-type transmission (see Chap. 7). The input field can be written as

$$E(t) = \int_{-\infty}^{+\infty} e(\Omega) \exp(i\Omega t)\, d\Omega \exp(i\omega_0 t) \tag{4.30}$$

where ω_0 is the optical pulsation corresponding to the center wavelength λ_0 and Ω is the modulating pulsation. The corresponding input power is

$$P_e(t) = |E(t)|^2 \tag{4.31}$$

Noteworthy from the practical point of view is that analyzing the light with a scanning Fabry–Perot interferometer followed by a photodetector and an oscilloscope displays $|e(\Omega)|^2$, whereas using a photodetector followed by a spectrum analyzer yields the Fourier transform of $P_e(t)$ (see Fig. 4.12).

The output field is given by

$$S(t) = \exp[i(\omega_0 t - \beta_0 L)] \int_{-\infty}^{+\infty} e(\Omega) \exp(i\{\Omega t - [\beta(\Omega + \omega_0) - \beta_0]L\}) \, d\Omega \tag{4.32}$$

where β_0 is the propagation constant at wavelength λ_0 (pulsation ω_0) and L is the fiber length. The corresponding output power is

$$P_s(t) = |S(t)|^2 \tag{4.33}$$

It follows from Eqs. (4.30)–(4.33) that we can define a *field frequency response* as $\exp[-i\beta(\omega_0 + \Omega)L]$, but that it is not possible to define a power frequency response because of the nonlinearity of Eqs. (4.31) and (4.33). Using a Taylor series expansion allows us to write

$$\beta(\omega_0 + \Omega) \simeq \beta_0 + \Omega\tau_0 - \frac{\Omega^2}{2}\frac{\lambda_0^2}{2\pi c} C_0 + \frac{\Omega^3}{6}\left(\frac{\lambda_0^2}{2\pi c}\right)^2 C_0' \tag{4.34}$$

where C_0 and C_0' are defined in Eq. (4.26) and we used $\Omega \ll \omega_0$. The field frequency response thus arises from a pure phase term which induces a signal distortion through its nonlinear contribution. Note that analyzing the output signal with a scanning Fabry–Perot would show no difference with the input signal, whereas the distortion will be observable on the spectrum analyzer (see Fig. 4.12).

Response to a Gaussian Pulse. We will assume here that $C_0 \gg (\Omega/3\omega_0)\lambda_0 C_0'$, which means that $C_0 \geq 0.01$ ps/(nm × km) for the usual wavelength range 800 nm $\leq \lambda_0 \leq$ 1700 nm and modulation bandwidth (≤10 GHz, or $\Omega \leq 2 \times 10^{10}$ rad/s). In this case, a Gaussian pulse with a rms width S_0 (in power) launched into the fiber will emerge as a Gaussian pulse with rms width S (in power) given by

$$S^2 = S_0^2 + \left(\frac{\lambda_0^2 C_0 L}{4\pi c S_0}\right)^2 \tag{4.35}$$

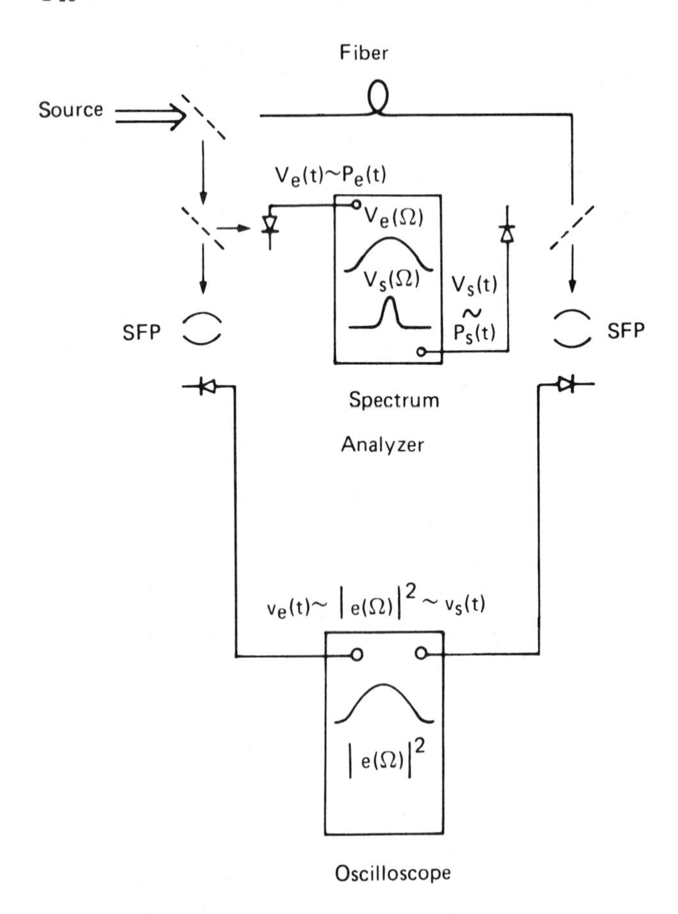

FIGURE 4.12 Observation of spectrum distortion effects in a single-mode fiber with a light source whose line width (in the absence of modulation) is much smaller than the modulation bandwidth. The amplitude optical spectra recorded with scanning Fabry–Perot interferometers (SFP) are unaffected by the fiber (because only the phase of the field is disturbed), whereas the electrical spectra of the detected pulses recorded with a spectrum analyzer show the distortion.

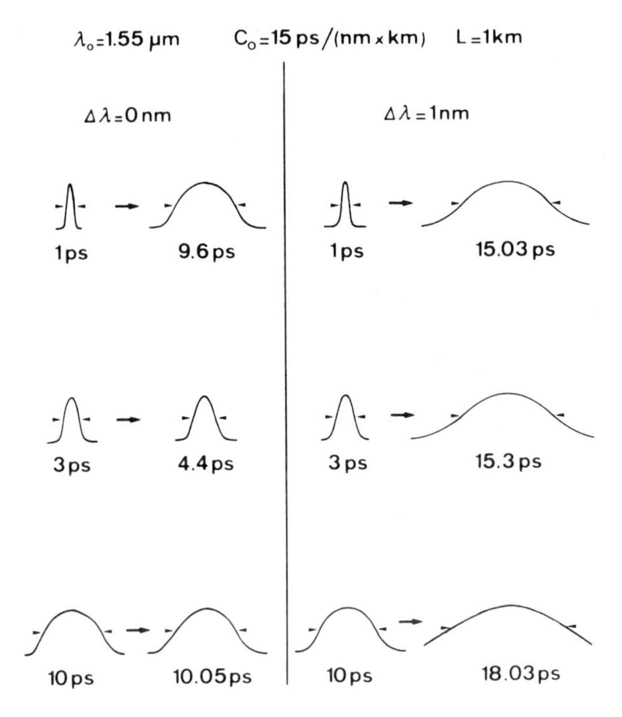

FIGURE 4.13 Examples of pulse distortion effects with a monochromatic source ($\Delta\lambda = 0$ nm) and a polychromatic source ($\Delta\lambda = 1$ nm). Note the presence of an optimum initial pulse width ($\simeq 3$ ps) in the case of the monochromatic source, with respect to the output pulse width.

This equation displays the consequence of the fact that it is not possible to define a power frequency response. *The pulse broadening depends on its initial width*, contrary to the case of a polychromatic source [Eq. (4.29)]. Figure 4.13 illustrates this situation.

From Eq. (4.35) we can minimize the output pulse width by choosing

$$S_0 = (S_0)_{opt} = \left(\frac{\lambda_0^2}{4\pi c} \, |C_0| \, L\right)^{\frac{1}{2}} \tag{4.36}$$

and the optimum output pulse width becomes

$$(S)_{opt} = 2^{\frac{1}{2}}(S_0)_{opt} \tag{4.37}$$

If we consider the field pulse rms width (which has to be considered in coherent transmission systems), we simply have to multiply all rms widths by $2^{\frac{1}{2}}$.

4.3.4 Intermediate Case

In the intermediate case, we consider that the source line width is comparable to the modulation bandwidth, which may correspond to a single longitudinal mode laser with no specific attempt to reduce its width. There is again no way to define a power frequency response, and we consider the case of a Gaussian pulse with a rms width S_0 (in power) launched into the fiber, emitted by a source with a Gaussian power spectral density of rms width W in pulsation. Calling ω_0 the center pulsation of emission (corresponding to λ_0), and assuming that $C_0 >> (W/3\omega_0)(\lambda_0 C_0')$, the output pulse is Gaussian with a rms width S (in power):

$$S^2 = S_0^2 + \left(\frac{\lambda_0^2 C_0 L}{4\pi c S_0}\right)^2 (1 + 4S_0^2 W^2) \tag{4.38}$$

The optimum pulse width $(S_0)_{opt}$ that minimizes S is given by Eq. (4.36) regardless of the value of W [15].

4.4 LIMIT BIT RATES

From all the preceding discussions, we can compute the limit bit rates that can be transmitted into the fiber by adding all the contributions to S and defining the bit rate as BR = 1/4S [15].

4.4.1 Multi Longitudinal Mode Laser

In this case there is no reason to use highly birefringent fibers and we will thus assume that the birefringence is small enough to provide a birefringence dispersion of less than 0.05 ps/km, which has been shown to be realistic (see Chap. 2). We can thus neglect any limitation due to the birefringence dispersion and directly use Eq. (4.29) for S.

For practical telecommunication applications it is desirable that $S \leq S_0 \times 2^{\frac{1}{2}}$ and we thus obtain

$$BR \leq \frac{1}{4} \{2(C_0 L \Delta\lambda)^2 + [C_0' L(\Delta\lambda)^2]^2\}^{-\frac{1}{2}} \tag{4.39}$$

where $\Delta\lambda$ is the rms width of the spectral power density of the source.

C_0 = 15 ps/(nm × km). We choose this value for C_0 as it corresponds approximately to the case of a single-mode fiber designed to have $C_0 = 0$ at 1300 nm, and operated at 1550 nm (see Chap. 7). From Eq. (4.39), we thus get $BR \times L \times \Delta\lambda \leq 12$ Gb × km × nm/s. For sources with $\Delta\lambda = 2.5$ nm, 5 nm, and 10 nm, we obtain $BR \times L \leq 4.8$ Gb × km/s, 2.4 Gb × km/s, and 1.2 Gb × km/s, respectively. (For other values of C_0, these figures are almost inversely proportional to C_0, as long as C_0 remains much greater than $C_0' \times \Delta\lambda$.)

C_0 = 0. When the fiber is free of first-order chromatic dispersion, we are left with the second-order term C_0', which amounts to about 0.1 ps/(nm^2 × km) at 1300 nm, and 0.05 ps/(nm^2 × km) at 1500 nm, as seen in Eq. (4.13). From Eq. (4.39), we obtain

$$BR \times L \times (\Delta\lambda)^2 \leq 2.5 \times 10^3 \text{ Gb} \times \text{km} \times \text{nm}^2/\text{s at 1300 nm}$$
$$BR \times L \times (\Delta\lambda)^2 \leq 5 \times 10^3 \text{ Gb} \times \text{km} \times \text{nm}^2/\text{s at 1550 nm}$$

For sources with $\Delta\lambda = 2.5$ nm, 5 nm, and 10 nm, we get $BR \times L \leq$ 400 Gb × km/s, 100 Gb × km/s, and 25 Gb × km/s, respectively, at 1300 nm, and twice these figures at 1550 nm. These figures can even be greatly enhanced in W-type fibers, where it is possible to obtain simultaneously $C_0 = C_0' = 0$.

4.4.2 Coherent Detection

In this case the light source has a very narrow natural line width and we thus resort to Sec. 4.3.3. As the coherent detection process is linear in terms of the received field amplitude, we have to use the field frequency response and add to the effect of chromatic dispersion the effect of birefringence dispersion with mode coupling. From Eqs. (4.35)–(4.37), we find that for an amplitude-modulated source the optimum field output pulse rms width is

$$(\hat{S})_{opt}^2 = p(1 - p)(TL)^2 + \frac{\lambda_0^2}{\pi c} |C_0| L \tag{4.40}$$

[Although Eq. (4.4) is not directly applicable because we deal here with the field amplitude and a selective detection of the principal polarization state, it turns out that the birefringence dispersion contribution takes the same form.]

We will consider operation at $\lambda_0 = 1550$ nm with an ultimate low-loss fiber [$C_0 \simeq 15$ ps/(nm × km)], as this is the most interesting case for coherent type systems. For the birefringence dispersion, coherent-type systems can be designed with polarization-maintaining fibers, or with low-birefringence fibers and polarization feedback or polarization diversity receivers (see Chap. 7). For polarization-maintaining fibers, according to the discussion in Chap. 2, we will consider a cross-polarization $p = 0.1$ (−10 dB, in terms of power), and the following cases:

Linear polarization, helical winding (LPHW): normalized birefringence $B \simeq 2 \times 10^{-5}$; $T \simeq 0.1$ ns/km
Linear polarization, winding without twist (LP): $B \simeq 2 \times 10^{-6}$; $T \simeq 0.01$ ns/km
Circular polarization, helical winding with 10 turns per meter (CP): $B \simeq 1.6 \times 10^{-6}$; $T \simeq 0.7$ ps/km

For low-birefringence fibers, we consider $p = 0.5$ (power randomly distributed among all possible SOPs) and $T \simeq 0.05$ ps/km. The corresponding limit bit rates are then obtained from Eq. (4.40) and $BR = 1/4(\hat{S})_{opt}$. In the case of linear polarization with helical winding, the bit rate is limited almost entirely by the birefringence dispersion to about 8 Gb × km/s, whereas in the case of low-birefringence fibers, it is limited by the chromatic dispersion to about 40 Gb × km$^{\frac{1}{2}}$/s. All results are summarized in Fig. 4.14. Although lengths of more than 200 km are not directly attainable with coherent detection, because of loss limitations (see Chap. 7), they have to be considered here if direct light amplification without pulse reshaping is used (see Chap. 7).

It is seen in Fig. 4.14 that circular polarization-maintaining fibers and low-birefringence fibers are limited almost entirely by chromatic dispersion effects. In these cases, as the field frequency response is a pure nonlinear phase term, and as the signal photocurrent at the intermediate frequency contains all the phase information, electronic compensation of the nonlinear phase term could be used to increase the allowed bit rate even further. The electronic group delay correction required for the compensation can be deduced from Eqs. (4.32) and (4.34) as

$$\frac{d\tau}{d\nu} \simeq \frac{\lambda_0^2}{c} C_0 L \tag{4.41}$$

where ν is the modulation frequency. With $C_0 = 15$ ps/(nm × km) and $\lambda_0 = 1550$ nm, this amounts to 0.12 ns/GHz for $L = 1000$ km, a correction which is currently achieved in the microwave technology.

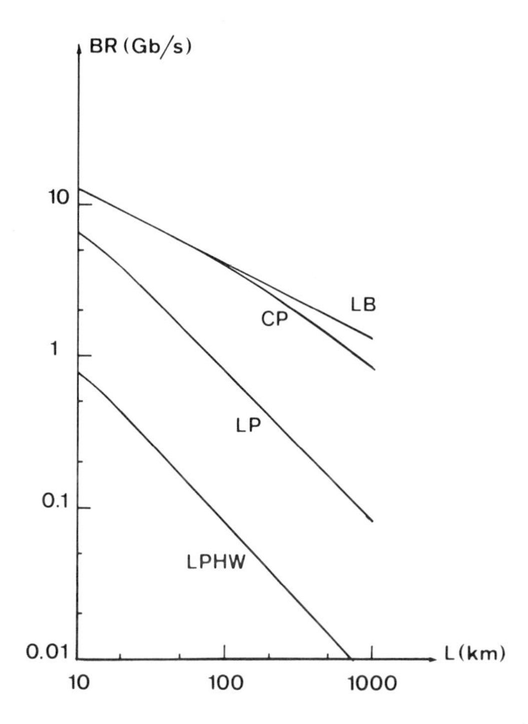

FIGURE 4.14 Maximum possible bit rates in coherent-type systems as a function of link length, for various cases. LPHW, linear polarization-maintaining fibers with helical winding; LP, linear polarization-maintaining fibers with no twist; CP, circular polarization-maintaining fibers; LB, low-birefringence fibers (with polarization feedback or a polarization diversity receiver).

4.4.3 Single Longitudinal Mode Laser

This corresponds to Sec. 4.3.4 for the chromatic dispersion and to low-birefringence fibers. Using Eqs. (4.2), (4.36), and (4.38), we get

$$(S)_{opt}^2 = \left(\frac{TL}{2}\right)^2 + \frac{\lambda_0^2}{2\pi c} |C_0| L + \left(\frac{\lambda_0^2}{2\pi c} C_0 WL\right)^2 \tag{4.42}$$

where W is the rms spectral width (in terms of pulsation) of the laser and T is the birefringence dispersion. Considering a fiber operated at $\lambda_0 = 1550$ nm with $C_0 = 15$ ps/(nm × km) and low birefringence with $T \simeq 0.05$ ps/km, we find that the first term is negligible up to $L = 1000$ km. Considering a laser with a 1000-MHz rms

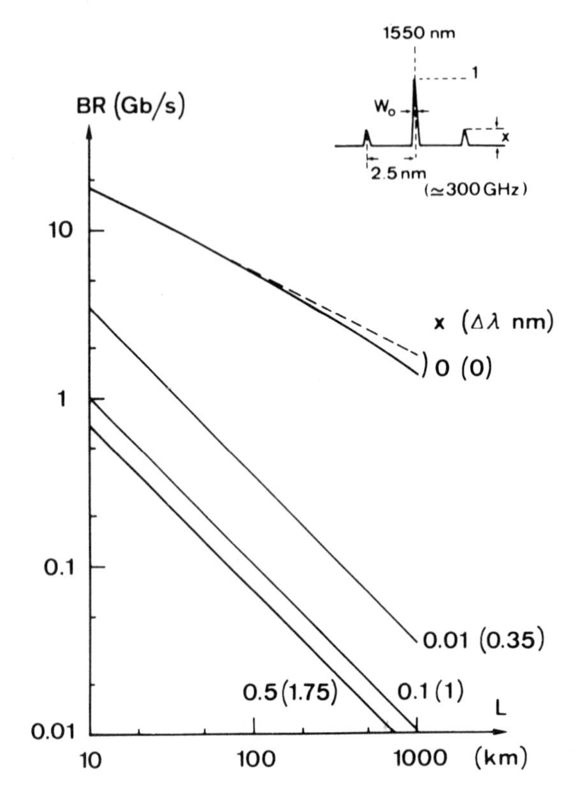

FIGURE 4.15 Maximum possible bit rates in direct detection systems as a function of link length for various laser spectra (shown in the insert), and a dispersion C_0 = 15 ps/(nm × km). x denotes the relative amplitude of the two satellite modes, assumed to have a spacing of 2.5 nm with respect to the central mode at 1550 nm (Δλ is the corresponding rms spectral width). Solid lines are for W (individual modes rms spectral width) of 1 GHz, and dashed lines are for 100 MHz.

spectral width ($W = 2\pi \times 10^8$ rad/s, which is pessimistic), we may also neglect the last term. With the values above and $BR \leq 1/4(S)_{opt}$, we get $BR \times L^{\frac{1}{2}} \leq 57$ Gb $\times$ km$^{\frac{1}{2}}$/s. Note that, contrary to the case of a polychromatic source, the maximum bit rate varies as $L^{-\frac{1}{2}}$, instead of 1/L.

It may happen, however, that a so-called single longitudinal mode laser exhibits some residual power in satellite modes. As the spacing between longitudinal modes is high, especially in lasers intended to be single moded (1 to 2.5 nm at 1550 nm), this leads to a limitation, through a strong increase in the spectral width. This effect is illustrated in Fig. 4.15 and is seen to be very important even for very low levels of the satellite modes. The practical solution to this limitation is to use truly single-mode sources, such as injection-locked lasers [16].

4.4.4 Effect of Mode Jumping

Up to now we have assumed that the source center emission wavelength was unaffected by the modulation. However, it has been observed that when a short current pulse is applied to a semiconductor laser, its center emission wavelength may jump from one mode to its neighbor at a longer wavelength [17]. Whereas this phenomenon will probably have little effect with multi longitudinal mode lasers, it can have spurious effects on pulse broadening with a single longitudinal mode laser. Let us use a qualitative description where we will neglect the pulse broadening due to the finite modulation bandwidth, as the frequency spacing between neighboring longitudinal modes is much greater than the usual modulation bandwidths (as high as 2.5 nm or 200 GHz at 1550 nm). At the beginning of the pulse, the emission wavelength is λ_1; then at some instant during the pulse it jumps to λ_2 ($\lambda_2 - \lambda_1 \leq 2.5$ nm), and it returns back to λ_1 at the end of the pulse. Thus the first and last parts of the pulse will be delayed by $\tau(\lambda_1) \times L$, whereas the central part is delayed by $\tau(\lambda_2) \times L$. Because of the small values of $\lambda_2 - \lambda_1$, we can write

$$\tau(\lambda_2) - \tau(\lambda_1) = (\lambda_2 - \lambda_1)C(\lambda_1) + \frac{(\lambda_2 - \lambda_1)^2}{2} C'(\lambda_1) \qquad (4.43)$$

Thus, depending on the sign of $C(\lambda_1)$, on the value of L, and on the instant at which the jump occurs, the rms width of the pulse may be decreased or increased, whereas the total pulse duration can only increase. The resulting pulse rms width will always be less than the one predicted by assuming that both λ_1 and λ_2 are emitted simultaneously during all the pulse, and we can thus take this figure as an upper limit which will be reached if $[\tau(\lambda_2) - \tau(\lambda_1)] \times L$ is much greater than the pulse duration.

4.5 PRACTICAL DISPERSION-FREE FIBERS

The theoretical aspects of obtaining control of chromatic dispersion have been discussed in Sec. 4.3.1. Practical fiber designs are classified into three dispersion classes (Fig. 4.16):

1. Dispersion-unshifted fibers having a wavelength of zero chromatic dispersion around 1300 nm. These fibers are classified by the U.S. Electronics Industries Association (EIA) as Class IVa, and are subject to Recommendation G.652 issued by CCITT.
2. Dispersion-shifted fibers having a wavelength of zero chromatic dispersion around 1550 nm, in the lowest loss region for silica-based optical fibers. These fibers correspond to EIA Class IVb.
3. Dispersion-flattened fibers exhibiting a zero chromatic dispersion for at least two wavelengths in the range 1270–1600 nm, and a small dispersion (less than, say, 6 psec/nm × km) throughout this wavelength range. They correspond to EIA Class IVc.

4.5.1 Dispersion-Unshifted Fibers

Dispersion-unshifted fibers are the easiest to manufacture as they do not require any specific index profile. They are widely used in long-distance telecommunication systems operating around 1300 nm, but they may also be operated at 1550 nm with lower bit rates.

CCITT Recommendation G.652 indicates that the dispersion properties of these fibers should satisfy:

$1295 \text{ nm} < \lambda_D < 1322 \text{ nm}$
For $1285 \text{ nm} < \lambda < 1330 \text{ nm}$, $|C(\lambda)| \leq 3.5 \text{ ps/(nm} \times \text{km)}$
For $1270 \text{ nm} < \lambda < 1340 \text{ nm}$, $|C(\lambda)| \leq 6 \text{ ps/(nm} \times \text{km)}$
$C(1550 \text{ nm}) \leq 20 \text{ ps/(nm} \times \text{km)}$

Such fibers, produced by several manufacturers, are based on two main designs: all have a step-index core profile, which may be associated with either a matched-index cladding or a thick depressed-index cladding.

Matched-Cladding Designs. These fibers are generally manufactured by OVD or AVD, with a GeO_2-doped core and a pure silica cladding, except for one manufacturer, who uses a pure silica core and a fluorine-doped cladding [18]. The use of pure silica for the

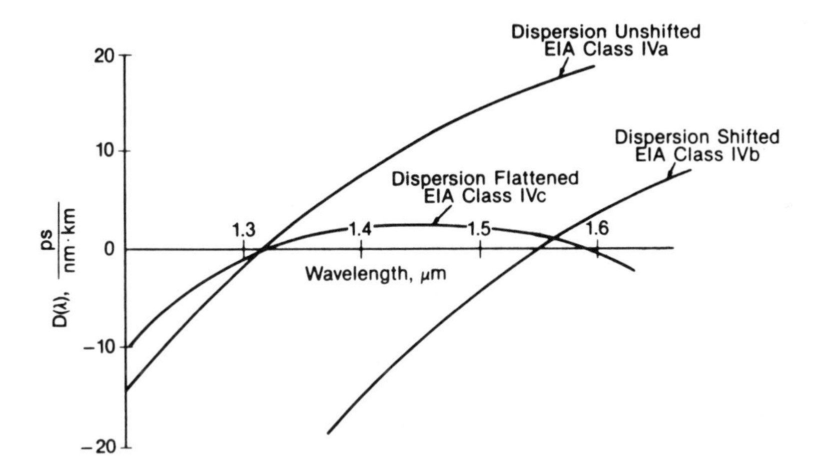

FIGURE 4.16 Dispersion classes for practical single-mode fibers.

core is aimed at providing slightly lower attenuation because of less dopant-induced Rayleigh scattering. Typical design parameters and transmission characteristics for these fibers are given in Table 4.1. These fibers comply exactly with all requirements of CCITT G.652. The dispersion slope around λ_D has been measured for these fibers at a value C' (1310 nm) $\simeq$ 0.093 ps/(nm^2 × km), very close to the theoretical prediction of Eq. (4.13). Referring to Eqs. (4.14) and (4.16), we find for this type of fiber that the zero dispersion wavelength λ_D varies by about 4 nm, for a variation of 12% on Δn, and is almost insensitive to core diameter variations (0.05 nm for 3%).

Depressed-Cladding Designs. These fibers are generally manufactured by IVD, with a GeO_2-doped core and a fluorine slightly doped cladding. This combination of dopants allows a higher total index difference than the matched cladding design, offering better resistance to bending induced losses. The usual design parameters for these fibers is given in Table 4.1, together with their transmission characteristics. These fibers also comply with the requirements of CCITT G.652.

4.5.2 Dispersion-Shifted Fibers

Since the lowest attenuation of high-silica optical fibers is obtained around 1550 nm (called "the third window"), efforts have been made to manufacture single-mode fibers offering simultaneously zero dispersion and minimum attenuation at 1550 nm.

TABLE 4.1 Characteristics of practical dispersion-unshifted fibers

	Step index core profile		
Cladding design	Matched		Depressed
Core dopant	GeO_2	None	GeO_2
$\Delta+$	0.3%	0%	0.27%
Cladding dopant	None	F	F
$\Delta-$	0%	0.3%	0.1%
Total Δ	0.3%	0.3%	0.37%
Core diameter, μm	9	8.5	8.3
$2W_0$ (1.3), μm	10	9.5	8.7
λ_c, nm	1200 ± 70	1200 ± 100	1250 ± 80
λ_D, nm	1310 ± 10	1310 ± 10	1310 ± 10
Average attenuation, dB/km			
at 1300 nm	0.36	0.35	0.38
at 1550 nm	0.20	0.20	0.24

Step-Index Core Fibers. Early attempts for achieving these characteristics with step-index core fibers led to unacceptably high losses, above 1 dB/km at 1550 nm [9,10]. Even after considerable technological improvements, the loss of step-index core, dispersion-shifted fibers remained above 0.35 dB/km at 1550 nm [19,20].

This loss is similar to that obtained at 1300 nm with the usual dispersion-unshifted fibers, making this kind of dispersion-shifted fiber of little interest. However, the minimum loss predicted theoretically for silica fibers around 1550 nm is below 0.2 dB/km. The excess loss above the theoretical value encountered in these step-index dispersion-shifted fibers seems to be associated with the presence of stress at the core–cladding interface because of the high germanium content of the core required by the large index difference. It has been suggested that this mismatch problem may be minimized by grading the refractive index profile of the core, especially for a triangular profile [$g = 1$ in Eq. (1.54a)] [21].

Triangular Core Profile Fibers. Investigations of the influence of the design parameters of triangular core profile fibers have been reported in Refs. 22 and 23. However, it has been shown subsequently that using the ESI parameters approach for such fibers, as described in Sec. 1.3.2, leads to acceptable accuracy [24]. Following this approach, Eqs. (1.58) show that the ESI parameters are given by

$$
\begin{aligned}
a_e &= 0.75\ a \\
\Delta n_e &= 0.59\ \Delta n\ (\text{peak}) \\
V_e &= 0.58\ V
\end{aligned}
\tag{4.43}
$$

Using these parameters together with the theory of Sec. 4.3.1 yields a value of the zero-dispersion wavelength which is smaller than the actually measured value by 10 to 15 nm [24]. This is a systematic error which may thus be taken into account in the calculations. Moreover, it will be seen below that usual manufacturing tolerances yield fluctuations of the zero-dispersion wavelength of about 10 to 20 nm. One may thus conclude that the ESI parameters approach provides a simple tool with an acceptable practical accuracy for designing such fibers. The first successful realization of a fiber approaching the ultimate loss limit simultaneously with minimum dispersion around 1550 nm is reported in Ref. 25. The fiber had a core diameter of 6.4 μm (a_e = 2.4 μm), and a peak relative index difference of 0.8% (Δ_e = 0.47%). The cutoff wavelength observed for the second-order mode was 850 nm (compared to 880 nm obtained from the ESI parameters), the mode field diameter at 1550 nm was measured as 9.4 μm (10.2 μm obtained from the ESI parameters), and the zero-dispersion wavelength was found to be 1560 nm (1550 nm predicted from ESI parameters). The loss at 1560 nm was 0.24 dB/km, about 30% below the usual loss at 1300 nm, making this fiber really attractive.

Dispersion-shifted fibers based on a triangular core profile are now offered by several manufacturers, and their main characteristics are reported in Table 4.2. The simple matched-cladding design has been found to exhibit some sensitivity to bending-induced losses, which can cause trouble in applications where the fiber may be subjected to small bend diameters. This problem is avoided by using a raised index ring in the cladding with a mean radius of about 6.3 μm (twice the core radius), a width of about 1.4 μm, and a refractive index 0.004 above that of the cladding [26]. The effect of adding this index ring is to increase the guiding strength, thus yielding an increase in the cutoff wavelength of the second-order mode and better resistance to bending losses. Obviously, the index profile becomes more difficult to control, and moreover the dispersion slope

TABLE 4.2 Characteristics of practical dispersion-shifted fibers

	Triangular core profile		
Cladding design	Matched	Raised-ring	Depressed
Total Δ	1%	0.9%	1.1%
Core diameter, μm	6.2	6.3	?
$2\,W_0$ (1.55), μm	9.0	9.0	7.0
λ_c, nm	880	1200 ± 100	1100
λ_D, nm	1550	1550	1550
C' (λ_D), ps/(nm^2 × km)	0.066	0.08	0.056
Average attenuation, dB/km, at 1550 nm	0.24	0.21	0.24

and the sensitivity to manufacturing tolerances are increased: Referring to Eqs. (4.14) and (4.16) we obtain for this type of fiber that the zero-dispersion wavelength varies by about 35 nm for a 12% variation in Δn, and by about 1 nm for a 4% variation in a. With the matched-cladding design, the same parameter fluctuations induce a variation in λ_D of 14 nm for Δn and 19 nm for a. As it is much easier to control the core diameter than the index difference, the increased sensitivity to Δn is a drawback of these triangular core fibers. Finally, another approach is used to overcome the bending loss sensitivity, by increasing the peak index difference and using a depressed-index cladding, together with a small core diameter. The drawback here is that the loss seems to be slightly higher, by about 10%.

4.5.3 Dispersion-Flattened Fibers

Fibers exhibiting a low dispersion over a broad wavelength range are of interest for the purpose of increasing the fiber transmission capacity through wavelength multiplexing. Flattening the dispersion curve was first achieved with W-type fibers, also called doubly clad fibers [11]. The typical result obtained is shown in Fig. 4.11. However, it appeared soon that these fibers were very sensitive to bending-induced losses. It has been found that adding a raised refractive index ring surrounded by an index well, around the basic W-type structure, leads to a simultaneous low loss and low dispersion over the entire 1300–1600 nm range [27]. In these fibers, called quadruple clad or segmented core [28], the light leaking outside of the core when the fiber is bent is retrapped by the index ring. This makes it possible to use these fibers in standard cables designed for usual single-mode fibers [29].

REFERENCES

1. M. Monerie and L. Jeunhomme, *Opt. Quantum Electron. 12*: 449–461 (1980).
2. D. Marcuse, in *Theory of Dielectric Optical Waveguides, Quantum Electronics: Principles and Applications* (Y.-H. Pao, ed.), Academic, New York, 1974, pp. 173–193.
3. C. Vassallo, CNET, unpublished work.
4. S. H. Wemple, *Appl. Opt. 18*(1):31–35 (1979).
5. I. H. Malitson, *J. Opt. Soc. Am.* 55(10):1205–1210 (1965).
6. M. Monerie, *IEEE J. Quantum Electron. QE-18*(4):535–542 (1982).
7. D. Marcuse, *Appl. Opt. 20*(4):696–700 (1981).
8. F. M. E. Sladen, D. N. Payne, and M. J. Adams, *Proc. 4th Eur. Conf. Opt. Commun.*, Genova, Sept. 12–15, 1978, pp. 48–57.
9. L. G. Cohen, C. Lin, and W. G. French, *Electron Lett. 15*(12): 334–335 (1979).
10. A. Kawana, T. Miya, N. Imoto, and H. Tsuchiya, *Electron. Lett. 16*(5):188–189 (1980).
11. S. J. Jang, L. G. Cohen, W. L. Mammel, and M. A. Saifi, *Bell Syst. Tech. J. 61*(3):385–390 (1982).
12. C. Brehm, C. Le Sergent, J. J. Bernard, and J. M. Gabriagues, CGE Research Laboratories, unpublished work.
13. T. Miya, L. Okamoto, Y. Ohmori, and Y. Sasaki, *IEEE J. Quantum Electron. QE-17*(6).858–861 (1981).
14. D. Marcuse, *Appl. Opt. 19*(10):1653–1660 (1980).
15. C. Lin and D. Marcuse, *Electron. Lett. 17*(1):54–55 (1981).
16. D. J. Malyon and A. P. McDonna, *Electron. Lett. 18*(11):445–447 (1982).
17. D. Botez, *J. Opt. Commun. 1*(2):42–50 (1980).
18. H. Kanamori, H. Yokota, G. Tanaka, M. Watanabe, Y. Ishiguro, I. Yoshida, T. Kakii, S. Itoh, Y. Asano, and S. Tanaka, *IEEE J. Lightwave Tech. LT-4*(8):1144–1150 (1986).
19. B. J. Ainslie, K. J. Beales, C. R. Day, and J. D. Rush, *IEEE J. Quantum Electron. QE-17*(6):854–857 (1981).
20. T. Tomaru, M. Kawachi, M. Yasu, T. Miya, and T. Edahiro, *Electron. Lett. 17*:731–732 (1981).
21. M. A. Saifi, S. J. Jang, L. G. Cohen, and J. Stone, *Opt. Lett. 7*:43–45 (1982).
22. U. C. Paek, G. E. Peterson, and A. Carnevale, *Bell Syst. Tech. J. 60*:583–598 (1981).

23. K. I. White, *Electron. Lett. 18*:725–727 (1982).
24. J. C. Augé, P. Dupont, and L. B. Jeunhomme, *SPIE Fiber Optics 559*:13–17 (1985).
25. B. J. Ainslie, K. J. Beales, D. M. Cooper, C. R. Day, and J. D. Rush, *Electron. Lett. 18*:842–844 (1982).
26. T. D. Croft, J. E. Ritter, and V. A. Bhagavatula, *IEEE J. Lightwave Tech. LT-3*(5):931–934 (1985).
27. L. G. Cohen, W. L. Mammel, and S. J. Jang, *Electron. Lett. 18*:1023–1024 (1982).
28. V. A. Bhagavatula, M. S. Spotz, W. F. Love, and D. B. Keck, *Electron. Lett. 19*:317–318 (1983).
29. J. Augé, C. Brehm, P. Dupont, L. Fersing, C. Le Sergent, J. Ramos, F. Alard, J. F. Bayon, Y. Durteste, P. L. François, D. Grot, J. Y. Guilloux, P. Lamouler, M. Monerie, and L. Rivoallan, *Proceedings of IOOC-ECOC'85*, Venice, Italy October 1–4, pp. 205–208 (1985).

5 Characterization

5.1 INTRODUCTION

Generally speaking, the characterization of single-mode fibers (SMFs) is of interest to the manufacturer, the theoretician, and the system designer. However, the parameters of primary importance to the former are different from those more specific to the latter, and as the field of single-mode fibers has been growing rapidly, many different measurement techniques are reported in the literature. Instead of giving a detailed description of all these techniques, we will try to give an overview of the various parameters and associated measurement methods, discussing in detail only those techniques applicable to routine characterization and recommended by standardization organizations. However, the usefulness of perhaps more complicated alternative parameters or methods should not be overlooked, as a full understanding of propagation in single-mode fibers often requires the comparison of different results.

The main groups active in standardizing optical fiber measurement procedures are the International Telegraph and Telephone Consultative Committee (CCITT), the International Electrotechnical Commission (IEC), and the U.S. Electronic Industries Association (EIA). The CCITT and IEC are working at an international level, whereas the EIA is active only at a national level in the United States together with the American National Standards Institute (ANSI) and the Institute of Electrical and Electronics Engineers (IEEE).

The study group XV/working party XV-5 is in charge of preparing standards for optical cables and systems in the CCITT. The document describing standard parameters and the corresponding

Chapter 5

TABLE 5.1 Standard measurement procedures recommended by CCITT and EIA[a]

Parameter	CCITT-G.652	EIA-455-xx	Present book
Geometrical parameters (excluding mode diameter)	Transmitted near field (RTM)	—	5.2.2
Refractive index profile	Refracted near field (ATM)	FOTP-44	
Effective cutoff wavelength	Transmitted power (RTM)	FOTP-80 (uncabled fiber)	5.3.1
		FOTP-170 (cabled fiber)	
Mode field diameter	Near field scan (RTM)	FOTP-165	5.3.2
	Offset joint (RTM)	FOTP-166	
	Far field scan (RTM)	FOTP-164	
	Variable aperture (RTM)	FOTP-167	
	Knife edge scan (RTM)	FOTP-174	
Attenuation	Cutback (RTM)	FOTP-78	5.3.3
	Insertion loss (ATM)	FOTP-171	
	Backscattering (ATM)	—	
Chromatic dispersion	Spectral time delay (RTM)	FOTP-168	5.4.2
	Spectral phase shift (RTM)	FOTP-169	
	Differential phase shift	FOTP-175	
	Interferometric (ATM)	—	

[a]RTM = reference test method; ATM = alternative test method.

measurement procedures is CCITT Recommendation G.652, which can be obtained from CCITT's office in Geneva, Switzerland. Table 5.1 lists the various parameters and the corresponding measurement procedures recommended by CCITT G.652. These procedures are described in more detail in Secs. 5.2, 5.3, and 5.4. Some of the procedures are considered by CCITT as alternative test methods (ATM) that can be used as well as reference test methods (RTM) for normal product acceptance purposes. However, when using an ATM, should any discrepancy arise, it is recommended that the RTM be employed as the technique for providing the definitive measurement results. Commission 455 of the EIA is in charge of preparing fiber optics test procedures (FOTP). The corresponding documents are referenced as EIA-455-xx or FOTP-xx and are also listed in Table 5.1.

5.2 STRUCTURAL PARAMETERS

The structural parameters are directly related to the manufacturing conditions and concern the dopant distribution, geometry, and index profile. Although the final product is the fiber itself, characterization at the intermediate preform stage is of primary importance for at least three reasons:

1. The measured parameters on the preform are directly related to the deposition and collapsing process.
2. Determining the structural parameters of the preform allows us to adapt the fiber drawing parameters precisely, thus obtaining specified fiber propagation characteristics (cutoff wavelength, mode spot radius, etc.).
3. The radial dimensions of the preform are about 50 to 100 times larger than those of the fiber, and thus the measurement resolution is much better.

One may doubt whether the fiber drawing process really yields a perfect homothetical transformation of the preform index profile. Whereas this seems to be true for the outside diameter (with almost circular outside surfaces) and for the core–cladding concentricity, it is more questionable for dopant distribution because of evaporation and diffusion phenomena, especially when phosphorus and/or fluorine are used. However, it is generally observed that the agreement between preform and fiber measurement results is satisfactory. It is thus likely that for early process optimization, measurements on preforms are not sufficient, but that for quality control in a well-defined process, they could be used alone.

5.2.1 Preform Characterization

Dopant Concentration. One category of measurement techniques is concerned with the determination of dopant concentration as a function of radial distance. This obviously is directly meaningful to the manufacturer, but it can also be useful to users if the refractive index is assumed to exhibit a linear dependence on dopant concentration, as is generally the case.

One method consists of cutting a preform slice and measuring the germanium distribution by means of a scanning electron microprobe. Other commonly used dopants are not accessible because of their quick evaporation under electronic beam heating, and this method has the serious drawback of being destructive.

A more recent method uses transverse UV fluorescence [1]. It consists of illuminating the preform transversely with a He–Cd laser (325 nm) whose beam is focused so that it is much narrower than the preform core diameter and passes through the core center. At this excitation wavelength, only the germanium oxide shows absorption, and it fluoresces at about 420 nm with an intensity proportional to the Ge concentration. Observation of this fluorescence (with a video camera) at 90° with respect to the beam axis and the preform axis makes it possible to deduce the radial germanium distribution. Again, only germanium can be observed this way, but this method is nondestructive and makes it possible to check the uniformity along the axis and under different radial directions by rotating the preform. Care should be taken, however, as the index profile cannot be deduced reliably from this measurement, because GeO_2 and GeO behave differently under UV excitation [2].

Index Profile. Index profile determination of SMF preforms as well as graded-index multimode preforms has received considerable attention. Besides the straightforward destructive technique, consisting of placing a thin preform slice in an interferometer, various nondestructive techniques using transverse observation have been implemented. They all have several common features, such as the necessity of inserting the preform inside an index matching liquid cell and the possibility of checking the preform uniformity along its axis and of observing the preform in several directions.

In the focusing method [3], the preform is transversely illuminated with a white-light parallel beam, and the beam then impinges on a video camera whose focus is adjusted on the core–cladding interface. This allows monitoring of the focusing properties of the core and, assuming circular symmetry, a somewhat complex mathematical treatment allows index profile reconstruction.

In the ray deflection method [4], a laser beam much narrower than the core diameter traverses the preform perpendicularly to its axis at variable radial distances, and the deflection angle of the beam is measured as a function of the radial distance. Again a mathematical treatment yields the index profile. An experimentally simpler procedure (spatial filtering technique [5]), based on the same principle, uses a collimated beam larger than the preform, and a spatial filtering mask. This method is used in a commercially available preform measurement apparatus. In addition to that, the method can be extended to axially nonsymmetrical preforms [6,7] but increases the computation time considerably.

The only drawback of these various methods is that they require a mathematical treatment before any characteristic parameter of the index profile can be obtained. On the other hand, the transverse interferometric method still retains the advantages of the previous methods but permits direct visualization of the geometry, and in some cases of the maximum index difference. The preform is inserted in one arm of a Mach–Zehnder interferometer (in a temperature-stabilized index matching liquid cell), and is traversed perpendicularly to its axis by a collimated laser beam larger than the preform. The meridian plane of the preform perpendicular to the beam is imaged onto a video camera. All diameters of the various regions can be measured immediately once the optic's magnification is known (the outside diameter can be visualized by slightly changing the liquid temperature and hence refractive index), as well as the concentricity. In addition to that, for a perfectly step-indexed fiber, the core–cladding index difference is deduced from the maximum fringe displacement. In the general case, however, a mathematical treatment is necessary to account for disturbing index gradients [8] (note that the correction used in Ref. 8 due to the ray deflection is negligible in the case of most SMF preforms). Figure 5.1a shows a typical interferogram of a SMF preform, where the displayed part comprises the deposited depressed cladding (in which individual layers processed by the IVD manufacturing method are observable), and the core with its central dip. Figure 5.1b compares the actual fringe pattern and the fringe pattern expected for perfect step transitions at the various interfaces and no central dip but the same index differences.

More recently, a measurement method combining experimental simplicity and accurate two-dimensional index profiling has been described [9]. Figure 5.2a illustrates the experimental arrangement. The preform is immersed in an index matching liquid (index n = cladding index) and is illuminated by a parallel light beam. When a ray crosses the preform end–liquid interface at a point on the preform end plane (whose only requirement is that it be carefully sawed), it is deviated by an angle ϕ related to the relative index difference Δ between that point and the liquid by the linear relationship

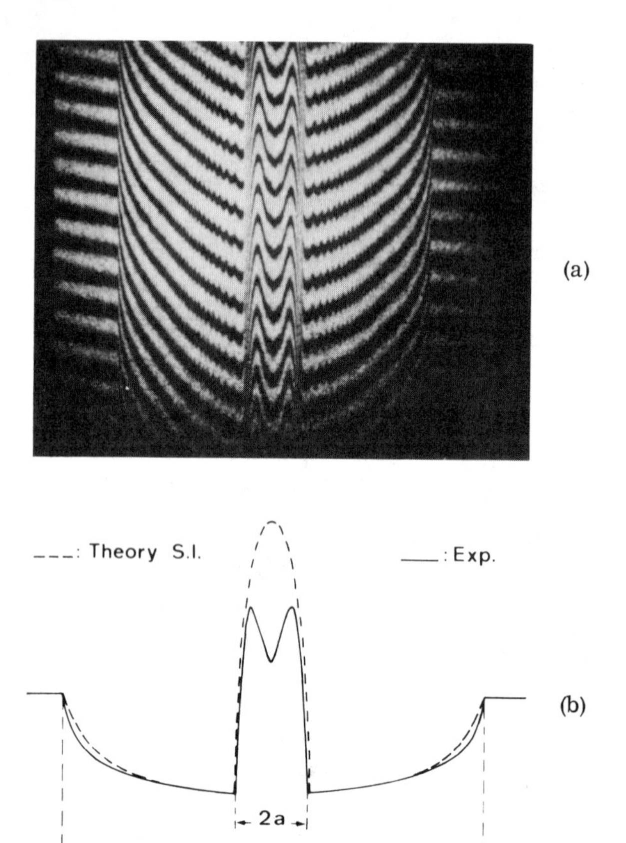

FIGURE 5.1 (a) Transverse interferogram of a single-mode preform manufactured using the IVD process. (b) Comparison of the actual experimental fringe pattern (solid line) with the theoretical fringe pattern (dashed line) expected for perfect step transitions and no central dip but the same index differences.

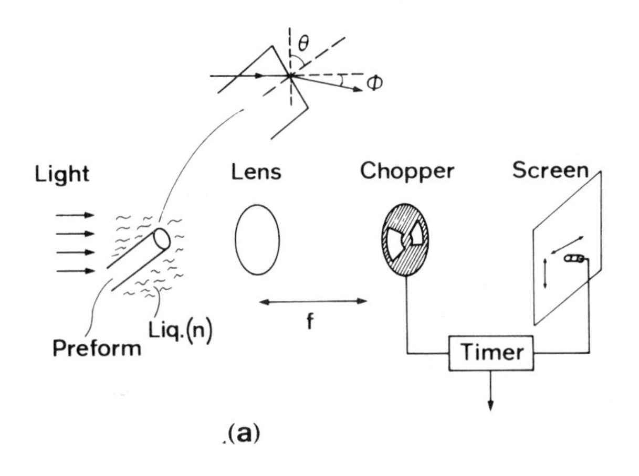

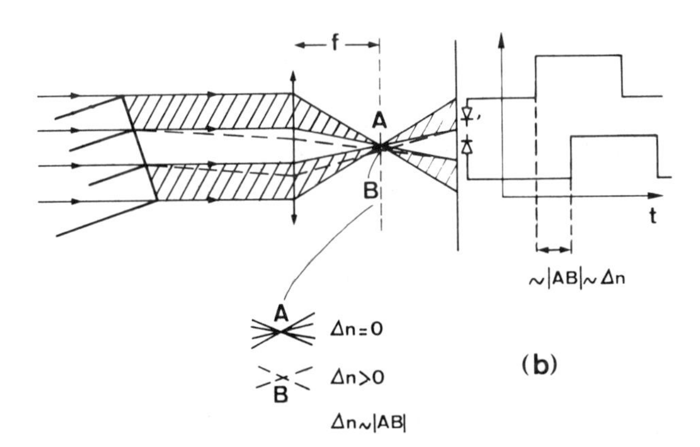

FIGURE 5.2 (a) Experimental arrangement for simple two-dimensional preform index profiling using the spatial filtering technique. Replacement of the chopper by a cylindrical lens allows visual display of the two-dimensional index profile. (After I. Sasaki et al., in *Optical Fiber Communication 1982 Digest of Technical Papers*, Optical Society of America, Washington, D.C., Apr. 1982, paper WAA5.) (b) Data acquisition principle of the spatial filtering technique.

$$\Delta = -\frac{1}{2} n\phi \sin 2\theta \tag{5.1}$$

The lens images the preform end onto a screen, which can be used for visual inspection, or which is replaced by a two-dimensional scanning photodetector for quantitative records of the index profile. For visual inspection, a cylindrical lens is placed in the focal plane of the lens (instead of the chopper of Fig. 5.2a). This has the effect of shifting the image along the preform axis, proportionally to ϕ, thus yielding direct viewing of the two-dimensional index profile, much like a stereoscan picture. For quantitative measurements, the measurement principle of ϕ with a chopper in the focal plane, and a timer counter, is shown in Fig. 5.2b.

If we take as a reference in the focal plane the point (A) corresponding to a zero index difference, all the rays deviated by the same amount ϕ (and thus passing through points of the same index difference) intersect at point B, the distance AB being proportional to the index difference. The chopper simply transforms this geometrical distance into a pulse delay, and the phase difference between the photodetected signals in two different image points is proportional to the index difference between the two corresponding points on the preform end. Thus measuring the phase of the detected signal as a function of photodetector position in the two-dimensional image yields directly a two-dimensional scan of the index profile, even with asymmetrical preforms.

5.2.2 Fiber Characterization

Because of the small size, index profiling on the fiber itself is much more difficult than on the preform. As noted earlier, accurate index profile determinations on the fiber are likely to be useful for detailed investigations at an early stage, but should be unnecessary in the production stage, where approximate index profile determinations should be sufficient (if not totally unnecessary).

Accurate Index Profiling. Various methods are applicable to accurate index profile determinations on the fiber, but some of them have serious drawbacks. The interferometric method on a fiber slice requires a time-consuming sample preparation, and the reflection method seems to be disturbed by chemical modifications on the fiber end face.

The most widely used technique is the refracted near-field technique (RNF) first described in Ref. 10, for which details are given in Ref. 11. This method is recommended as an alternative test method by CCITT-G.652 and EIA/FOTP-455-44. One advantage of this technique is that the same bench can be used for both single-mode fibers and multimode fibers (Fig. 5.3). A laser beam is strongly focused in a small spot on the input end face of the fiber, which is immersed in an index-matching liquid cell. A collecting lens permits measurement of the light escaping laterally from the fiber and records the received power as a function of the radial distance of the launching spot, which scans the fiber diameter. An opaque disk in the collecting beam rejects the leaky modes' contribution and the power curve is simply the opposite of the index profile. The refractive index scale is found by knowing the index of the supporting silica tube and of the liquid, whose temperature should be carefully controlled. For the determination of the various diameters, the decision level on the various refractive index difference interfaces (core–inner cladding, inner cladding–outer cladding, outer cladding–index matching fluid), is defined as 50% of the corresponding index step encountered. By using a reflective elliptical index-matching liquid cell as the collecting optic, the resolution has been improved to 0.35 μm [12]. However, the method requires high-quality fiber end faces and critical beam focusing, and its practical use can be very time-consuming is some situations.

Diameters and Approximate Index Profiling. For single-mode fibers, the classical near-field scanning technique can be modified to yield quick and easy diameter and index profile determination [13]. This method is recommended as the reference test method by CCITT-G.652. The principle is to use the fiber without any cladding mode stripping, which may pose a problem, as some primary coatings have a higher refractive index than that of the silica supporting tube. In this case, the primary coating should be removed on the short sample to be used. Excitation of the fiber should be done using a highly diverging source emitting blue light for better resolution.

The classical near-field theory states that in a multimode fiber in which all modes are equally excited, the power distribution on the output end face is proportional to the index profile except near the core–cladding interface, where leaky modes disturb the power distribution. Under the conditions described above, the whole fiber without a cladding mode stripper can be regarded as a multimode fiber whose core corresponds to the silica part and whose cladding is the surrounding medium (primary coating, or air). It is almost a step-index fiber with only a small perturbation at the center (actually the core of the SMF), and this perturbation, which is the interesting part in our application, will not be disturbed by the leaky modes [14], thus allowing a simple core index profile determination.

Equivalent Step-Index Fiber Parameters. Let us recall that the theoretical approach of ESI parameters in Chap. 1 is aimed at

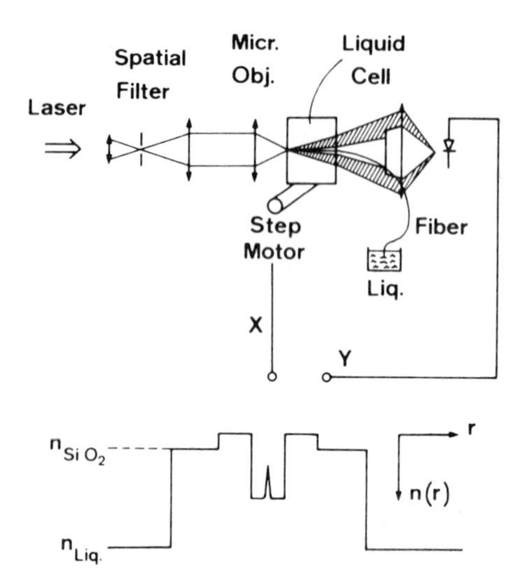

FIGURE 5.3 Experimental arrangement for optical fiber profiling by the refracted near-field technique. (After Ref. 10.)

replacing the complete index profile by step-index fiber parameters (core diameter, index difference) leading to a good approximation of the most important fiber transmission characteristics: mode field diameter and chromatic dispersion. The combination of a measurement of the actual fiber (or preform) index profile together with Eq. (1.57) allows us to determine ESI parameters. Whereas this procedure of predicting the fiber transmission characteristics is not always very accurate, it is very reproducible. It has thus been found to provide an efficient tool for routine characterization of fibers manufactured in a stabilized process.

5.3 STATIC TRANSMISSION PARAMETERS

The static transmission parameters comprise the effective cutoff wavelength, which indicates where the fiber can be used in single-mode applications; the parameters necessary to describe the field shape (mode field radius, ESI parameters), which are useful to predict splicing, microbending, bending losses, and dispersion characteristics; the spectral loss; and the birefringence. However, most of them are not yet uniquely defined and care should be taken to assess the significance of the measured parameters.

5.3.1 Effective Cutoff Wavelength

Theoretical and Effective Cutoff Wavelength. In Chap. 1 the LP_{11}-mode cutoff wavelength was defined as the wavelength above which the mode begins radiating because of a nonzero imaginary part in its propagation constant. Because of this imaginary part, the LP_{11} mode is affected by a very high loss per unit length, and hence only the LP_{01} (HE_{11}) mode propagates on significant fiber lengths. Equations (1.14)–(1.16) relate this theoretical cutoff wavelength to the structural parameters of the fiber.

In practice, however, the fiber is always affected by microbends, core diameter fluctuations, and bends, which together strongly increase the LP_{11} loss for wavelengths slightly below its theoretical cutoff wavelength. This results in introducing an *effective cutoff wavelength* as the wavelength above which, on a *given fiber sample in given conditions*, the LP_{11} mode is no longer transmitted. Obviously, the effective cutoff wavelength is always smaller than the theoretical cutoff wavelength, the difference between both decreasing when the index difference increases, because this decreases the bending and microbending losses (see Chap. 3). Before describing the associated measurement techniques, it seems valuable to discuss briefly the significance, advantages, and drawbacks of using the effective cutoff wavelength.

Because of its practical definition, the effective cutoff wavelength is directly accessible experimentally, whereas the theoretical cutoff wavelength is only an asymptotic limit value of the effective cutoff wavelength in an ideal single-mode fiber. On the other hand, the effective cutoff wavelength is not uniquely defined; it depends on both the fiber characteristics and on the physical environment (length, microbends, bends, etc.). For example, a decrease of about 60 nm per decade of fiber length increase has been observed for the effective cutoff wavelength of usual Telecom grade fibers [15]. For this reason, the effective cutoff wavelength should generally not be used to deduce the structural parameters of the fiber or the V value.

Significance and Usefulness. First of all there is a general tendency to operate practical systems at a wavelength as close as possible to cutoff in the single-mode region. This enhances fundamental mode confinement and reduces bending and microbending losses. On the other hand, if by chance or mistake it happens that the system is operated below cutoff in the bimodal region, this may have very important consequences for system performance: System bandwidth can be reduced by the appearance of multiple pulses corresponding to different propagation delay times among the various modes; modal noise can occur because of interference between the two modes at fiber joints.

These conflicting requirements call for an accurate evaluation of the effective cutoff wavelength for the actual fiber section involved in the system. As this is generally not possible, the measurement method must provide a "safe" value which guarantees effective single-mode operation above the measured value in usual systems configurations. This can be ensured by conditioning the fiber sample used in the test procedure with fewer bending and microbending losses than in the system configuration.

Global Measurement Technique. These measurements are performed on long fiber pieces and yield an effective cutoff. The simplest measurement technique is combined with the spectral loss measurement (Sec. 5.3.3) and consists of simply observing (either on the short reference sample or on the total fiber) the more or less abrupt transmitted power decrease when the wavelength increases and reaches the effective cutoff wavelength. Figure 5.4 shows several curves of the transmitted power on a 2-m-long fiber sample as a function of wavelength, which correspond to different curvatures of the sample. Let us first concentrate on curve 1, which corresponds to a radius of curvature R = 90 mm. In addition to variations in the transmitted power associated with the spectral response of the measurement apparatus, we notice around 1000 mm an abrupt power decrease associated with the LP_{11}-mode leakage. Ideally, this decrease would be a vertical line corresponding to the theoretical cutoff wavelength, and the power drop would be a factor of 3 (3.4 dB) because there are twice as many LP_{11} modes as LP_{01} modes (because of the sin φ and cos φ possibilities in LP_{11}, and provided that all modes are excited equally). Because of microbends, bends, and core diameter fluctuations, the LP_{11} mode is gradually attenuated well before the theoretical cutoff, and this results in a less abrupt decrease and an apparent drop of less than 4.8 dB. The same phenomenon qualitatively holds around 700 nm for the (LP_{02} + LP_{21}) mode cutoff, and the dashed-dotted lines indicate the approximate separations between the different regimes. A simple but approximate method of defining the effective cutoff wavelength is to use the wavelength corresponding to point A in Fig. 5.4. Looking at curves 2 to 5 shows that the effective cutoff wavelength so obtained will depend on the curvature of the sample fiber (the same is true for microbends). This definition will thus be usable only if the curvatures and microbends affecting the sample are realistic from a user's point of view.

Using the foregoing measurement technique in a more sophisticated (and more time-consuming) way consists of plotting the effective cutoff wavelengh for various curvatures (with definitions similar to that of curve 1 through point A in Fig. 5.4), as a function of curvature, as shown in Fig. 5.5. By extrapolating a mean

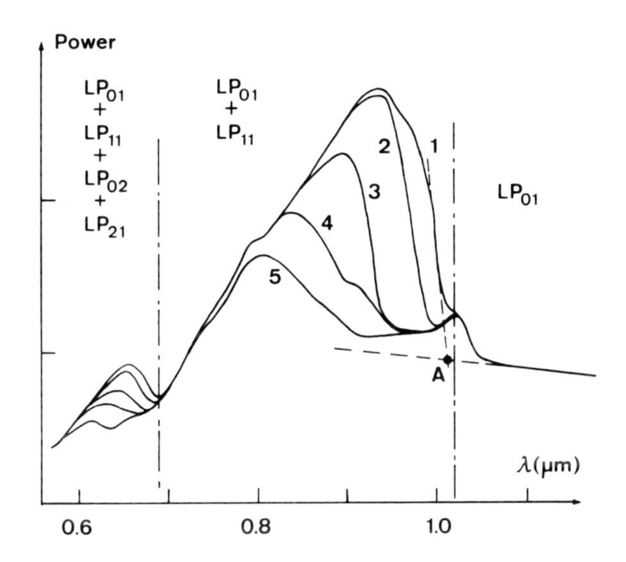

FIGURE 5.4 Transmitted power in a 2-m-long fiber sample for various radii of curvature R, illustrating the determination of the effective cutoff wavelength. 1, R = 90 mm; 2, R = 75 mm; 3, R = 50 mm; 4, R = 40 mm; 5, R = 30 mm.

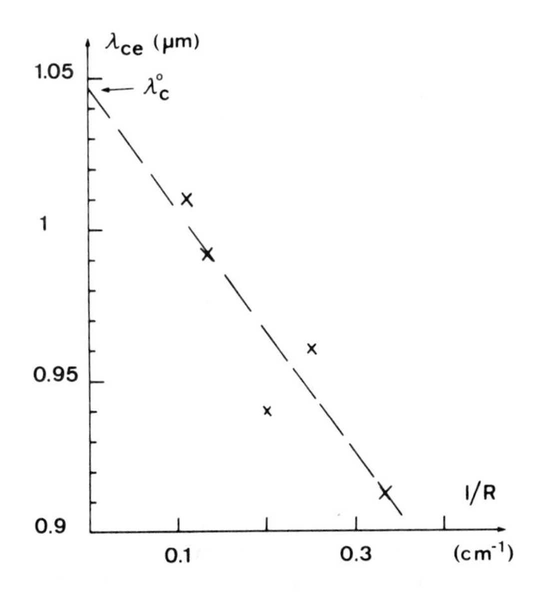

FIGURE 5.5 Effective cutoff wavelength as a function of curvature (from Fig. 5.4), and extrapolation for $\lambda_c{}^0$.

straight line to an infinite radius of curvature, we obtain an asymptotic value λ_c^0. Obviously, λ_c^0 is closer to the theoretical cutoff wavelength than any measured effective cutoff wavelength λ_{ce}, but it is not clear whether λ_c^0 is equal to it, because of the unknown influence of residual microbends. It should also be noted that λ_c^0 is difficult to attain directly using a straight fiber sample, because of the need for an efficient cladding mode stripper, which requires bending the fiber (Sec. 5.3.3).

Local Measurement Techniques. These techniques are aimed at approaching as closely as possible the theoretical cutoff wavelength. The technique described above has the advantage of using the same measurement bench as that used for spectral loss measurements, but its accuracy is questionable. To get closer approximations of the theoretical cutoff wavelength, alternative techniques have been developed which overcome the bends and microbends problems by using very short fiber samples (less than 1 cm).

In one method [16], a 1-cm-long single-mode fiber sample is spliced to a multimode fiber that is excited by a monochromator, and the SMF is surrounded by an index-matching liquid which acts as a cladding mode stripper. The field shape can be determined by observing the near field on the output end of the SMF and the shape change associated with the LP_{11} cutoff when the wavelength reaches the cutoff wavelength (see Fig. 1.9). Although accurate determination of the shape change may be difficult, this method works well.

A more recent technique consists of exciting a single-mode fiber, placed in an index-matching liquid cell, with a monochromator and collecting the power radiated laterally by the first 2mm of fiber [17]. The experimental arrangement can be similar to that of Fig. 5.3, except for the light source (replaced here by a monochromator) and the stepping motor (which is not required here). When the wavelength increases and reaches the cutoff wavelength, the LP_{11} mode begins to radiate and thus there is a corresponding collected power increase. The power increase is not abrupt, possibly because of residual core imperfections, but perhaps also more fundamentally because the imaginary part of the propagation constant of the LP_{11} mode (which governs its radiation loss per unit length) does not abruptly increase to infinity. This is the only method that is directly consistent with the definition of the theoretical cutoff wavelength.

Standard Measurement Procedures. Detailed investigations and comparisons of various measurement methods, such as reported in Ref. 18, have led to the establishment of standard measurement procedures recommended by CCITT-G.652 and EIA-455-80 (for uncabled fibers) and EIA-455-170 (for cabled fibers). These procedures are based on the measurement of transmitted power as a function of wavelength, using a classical spectral loss test system and pertain to the category of global measurement techniques as described above. They lead to the determination of the effective cutoff wavelength on well-defined fiber or fiber cable sections, under specified conditions.

A first measurement of the transmitted power $P_1(\lambda)$ is made through the test sample. For uncabled fibers, the test sample shall be 2 m in length and be bent to form a loosely constrained loop (one full circle) with a diameter of 28 cm. The remaining part of the fiber shall be substantially free of bends. Although some incidental bends of larger radii are permissible, they must not significantly influence the measurement result (see Fig. 5.6). For cabled fibers, the test sample shall have a total length of 22 m and one loop of 76 mm diameter shall be applied to each uncabled fiber length exposed at both ends (see Fig. 5.7). A second measurement is then made of the transmitted power $P_2(\lambda)$ through a reference sample. This reference sample may be either the same fiber piece as above with at least one loop with a diameter of less than 6 cm (as depicted in Fig. 5.6) or a 1- to 2-m-long piece of multimode fiber (as shown in Fig. 5.7). In the first case, the method is called transmitted power/single-bend attenuation; in the second case, it is designated as transmitted power step. Both methods of referencing may be applied to cabled or uncabled fiber measurements.

After the measurements described above have been performed, the logarithmic ratio between transmitted power $P_1(\lambda)$ and $P_2(\lambda)$ is calculated as:

$$R(\lambda) = 10 \log \left[\frac{P_1(\lambda)}{P_2(\lambda)} \right] \tag{5.2}$$

A typical result obtained with the transmitted power/single-bend attenuation method illustrated in Fig. 5.6 is shown in Fig. 5.8. In that case, the effective cutoff wavelength is determined as the largest wavelength at which $R(\lambda)$ is equal to 0.1 dB. This is equivalent to saying that the second-order mode has been attenuated by 19.3 dB. A typical result obtained with the transmitted power step method illustrated in Fig. 5.7 is shown in Fig. 5.9. The effective cutoff wavelength is obtained here as the intersection of a plot of $R(\lambda)$ and a straight line displaced 0.1 dB and parallel to the straight line fitted to the long wavelength portion of $R(\lambda)$. An interlaboratory comparison conducted in 1984 shows that each method has a standard deviation of about 8–10 nm, and that the two

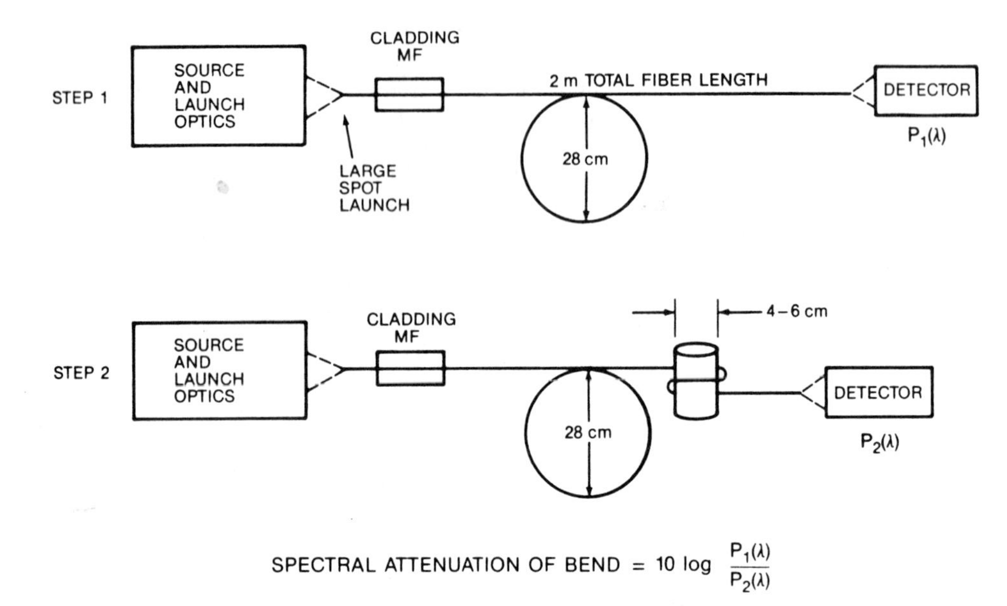

FIGURE 5.6 Standard test setup for the measurement of the effective cutoff wavelength on an uncabled fiber sample: transmitted-power/single-bend attenuation method. (Courtesy of D. N. Franzen, National Bureau of Standards, Boulder, Colorado.)

methods are in very close agreement (less than 2 nm discrepancy) [18]. Moreover, the choice of the 28-cm bend diameter does not seem to be critical.

5.3.2 Mode Field Diameter

Many methods have been published for measuring the mode field diameter in single-mode fibers. Besides a great variety of experimental test setups, the situation is complicated by the fact that the mode field diameter is not uniquely defined. As soon as the mode field shape is not perfectly Gaussian, looking for a single parameter to characterize its width can be done in various ways. With a single set of raw data obtained from a given measurement setup, different mathematical procedures can be applied to this raw data to extract a *mode field diameter*. One may, for example, maximize the overlap integral with a Gaussian function, or directly measure the width between the points at $1/e^2$ relative intensity, or calculate the root mean square (rms) width of the distribution, and so on.

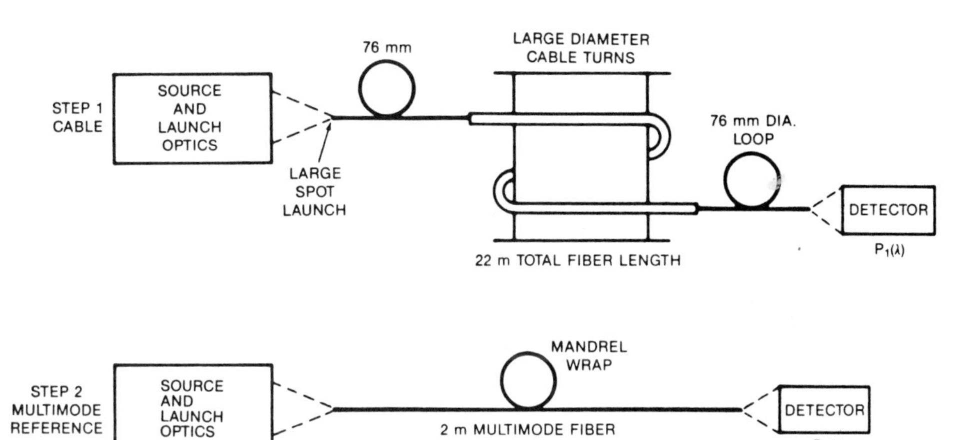

FIGURE 5.7 Standard test setup for measurement of the effective cutoff wavelength on a cabled fiber sample: transmitted-power step method. (Courtesy of D. N. Franzen, National Bureau of Standards, Boulder, Colorado.)

Fortunately, various studies have shown that the differences between the most usual and practical measurement methods lie within less than ±5% even for triangular core fibers at 1550 nm [19,20]. Moreover, there exists a set of properly selected combinations of experimental test setup plus mathematical data reduction procedures, which are directly related one to the other through exact mathematical relationships. Choosing a measurement method among these selected combinations ensures that there is no other source of discrepancy than usual measurement uncertainties. Thus, five different measurement methods have been standardized as reference test methods for the mode field diameter.

These methods are described below in some detail, and one may select the one which fits best with the equipment available or with the acceptable measurement time. A general diagram showing the five standard methods together with the mathematical relationships between them is shown in Fig. 5.10. The values of mode field diameter obtained by the five methods for a given piece of fiber shall thus be strictly identical (from a theoretical point of view). For all measurement methods, a light source operating at the proper wavelength (usually 1300 nm or 1550 nm) shall be used. It is of course

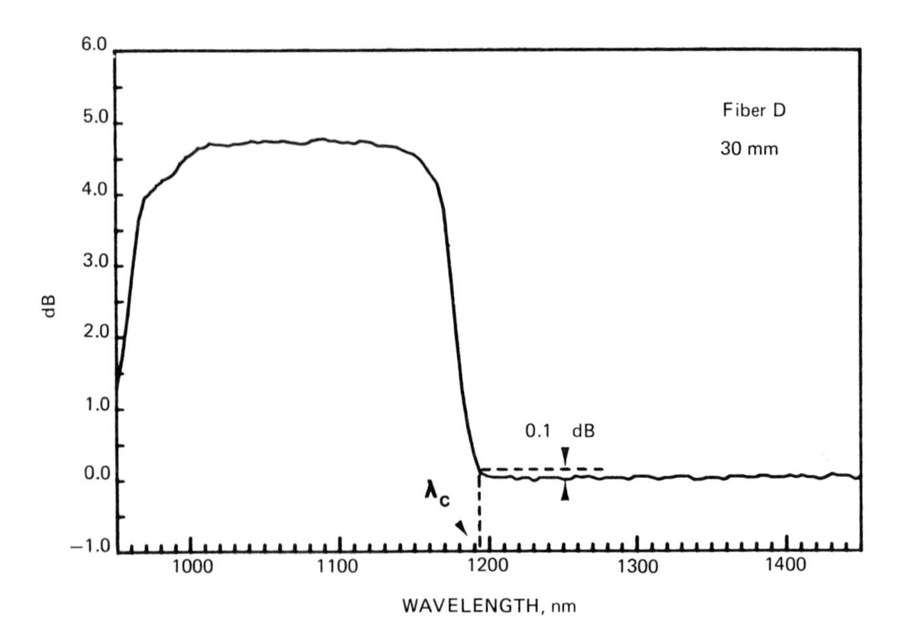

FIGURE 5.8 Determination of the effective cutoff wavelength in the transmitted-power/single-bend attenuation method. (Courtesy of D. N. Franzen, National Bureau of Standards, Boulder, Colorado.)

very important to make sure that cladding light is eliminated with appropriate cladding mode strippers and to eliminate any higher-order modes (other than the fundamental LP_{01} mode). This may be achieved with a small radius loop such as the one used for cutoff wavelength measurements.

The near-field scanning technique is the most straightforward method. The fiber output end is imaged through magnifying optics onto the plane of a scanning photodetector with a pinhole aperture. The numerical aperture and magnification of the optics are selected to be compatible with the desired spatial resolution. Of course, the magnification of the optics is calibrated with a specimen whose dimensions are accurately known. Because only a very small fraction of the total intensity reaches the detector, a laser is generally required as a light source.

Scanning the (imaged) intensity distribution with the detector yields the radial intensity distribution $f^2(r)$. Neither the maximization of the overlap integral with a Gaussian field distribution as in Sec. 1.2.2 nor the rms width of $f^2(r)$ as in Eq. (3.8a) are used for

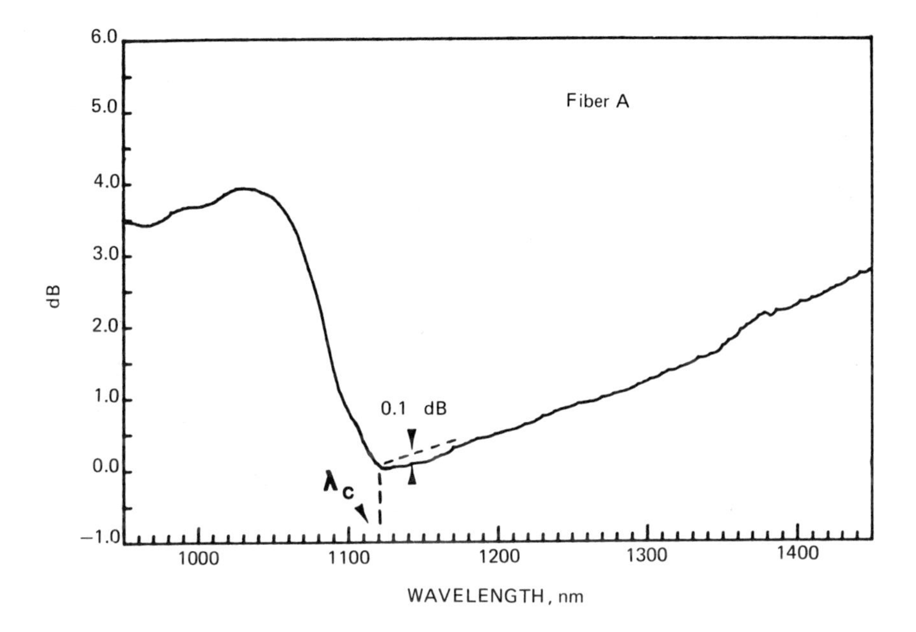

FIGURE 5.9 Determination of the effective cutoff wavelength in the transmitted-power step method. (Courtesy of D. N. Franzen, National Bureau of Standards, Boulder, Colorado.)

obtaining the mode field diameter. A more recent mathematical definition is used instead [21] as it is rigorously related through mathematical relationships to the other standard definitions of mode field diameter [22]. A comparison with the former definitions can be found in Ref. 20. The mode field diameter $2w_0$ is obtained from

$$w_0^2 = 2 \frac{\int_0^{\infty} r f^2(r)\, dr}{\int_0^{\infty} r \left[\frac{df(r)}{dr}\right]^2 dr} \tag{5.3}$$

From a practical point of view, this method has several drawbacks: The experimental setup requires careful adjustment of the focus, which is difficult in the infrared, and the mathematical data reduction procedure involves the derivative of the measured function, which generally decreases the accuracy.

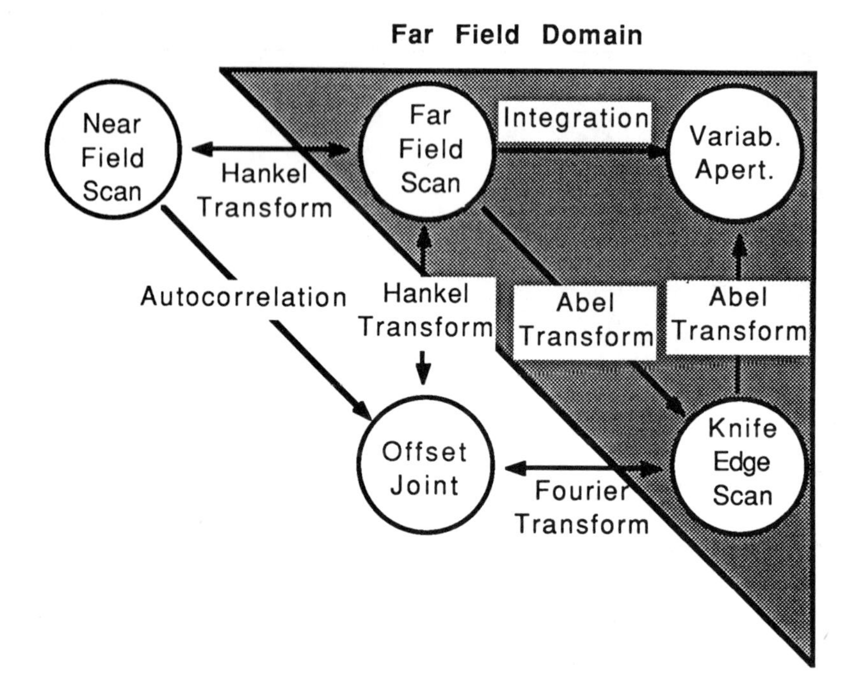

FIGURE 5.10 Relationships linking the five standard mode field diameter measurement methods.

The offset joint domain (or transverse offset) measurement technique involves measuring the power transmission coefficient between two identical fiber pieces, as a function of the transverse offset between the two core centers at the joint ([23,24]; Fig. 5.11). The joint between the two fiber pieces should be constructed so that the longitudinal separation between the two fiber ends is less than 5 μm. By offsetting the joint transversely in discrete steps of the order of 0.1 μm, the power transmission coefficient $T(\delta)$ is measured as a function of the lateral offset δ. The mode field diameter $2w_0$ is obtained from

$$w_0^2 = -2 \frac{T(0)}{\left(\frac{d^2T}{d\delta^2}\right)_{\delta=0}} \tag{5.4}$$

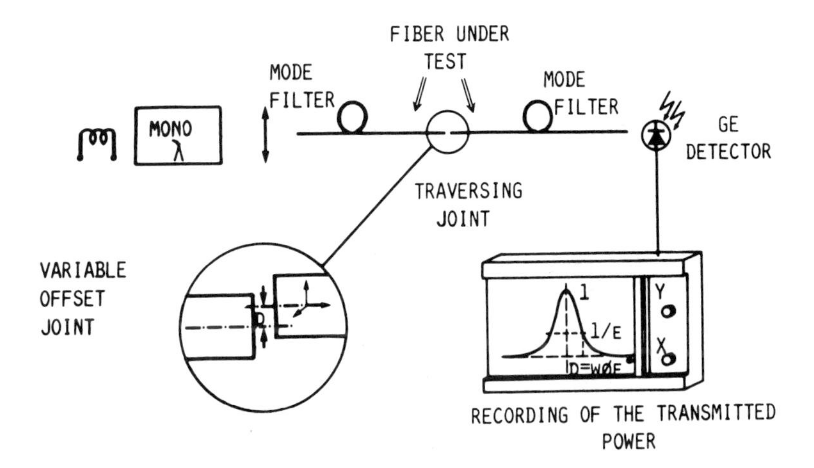

FIGURE 5.11 Experimental arrangement for the offset joint measurement technique of the mode field diameter. (After Ref. 20.)

One advantage of this method is that it allows us to measure w_0 as a function of wavelength, as a simple halogen lamp with a monochromator may be used as a light source. On the other hand, the initial preparation of the fiber ends and adjustment of the joint may be very time-consuming.

The far-field scanning technique is the easiest method to implement if one is interested in measurements at only one or two wavelengths. The experimental arrangement is shown in Fig. 5.12. Again a laser source shall be used because of the very small fraction of the total intensity received by the detector. The light beam escapes freely from the fiber output end, and the resulting beam is angularly scanned by a detector with a small pinhole aperture. The detector should be at a distance D at least 20 mm from the fiber end, and the pinhole diameter p should be smaller than $D \times \lambda/(20\ w)$, where 2w is the expected mode field diameter. This allows us to record the intensity distribution $F^2(q)$ as a function of the angle θ, and the mode field diameter $2w_0$ is obtained from

$$w_0^{\ 2} = 2\,\frac{\int_0^\infty qF^2(q)\ dq}{\int_0^\infty q^3F^2(q)\ dq} \qquad (5.5)$$

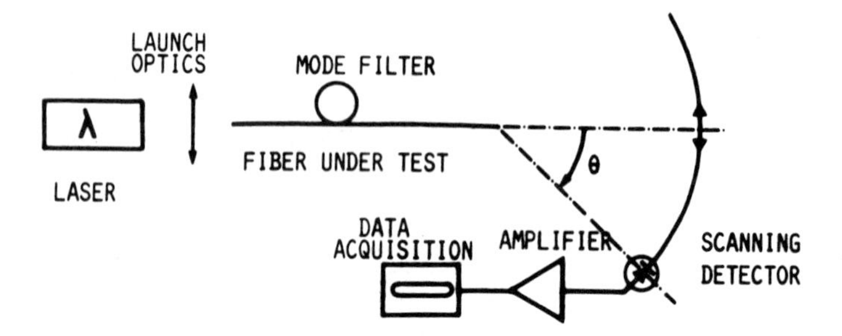

FIGURE 5.12 Experimental arrangement for the far-field scanning measurement technique of the mode field diameter.

$$q = \frac{2\pi}{\lambda} \sin \theta \tag{5.6}$$

This shows that w_0 has the physical significance of being the inverse of the rms half-width of the far-field intensity distribution. Except for the limitation arising from the requirement for a laser source, this method is the easiest one to use from both an experimental and a mathematical point of view.

The variable-aperture technique was first described as the variable-aperture launching technique [25]. Starting from a halogen lamp and a monochromator, the half-apex angle of the input beam focused onto the input end of the fiber was varied through a variable iris diaphragm. The transmitted power was then recorded as a function of the aperture of the input beam. A simpler approach consists in implementing the variable-aperture filter in the output beam emerging from the fiber ([26]; Fig. 5.13). A mechanism containing at least 12 apertures spanning the half-angle range of numerical apertures from 0.02 to 0.25 should be used. The light transmitted by the aperture of radius x is collected and focused onto the detector, yielding the power P(x). The maximum power transmitted by the largest aperture is P_{max}, and the complementary aperture transmission function a(x) is found as

$$a(x) = 1 - \frac{P(x)}{P_{max}} \tag{5.7}$$

Knowing the distance D between the fiber output end and the plane of the aperture allows us to convert x into θ or q through Eq. (5.6) and

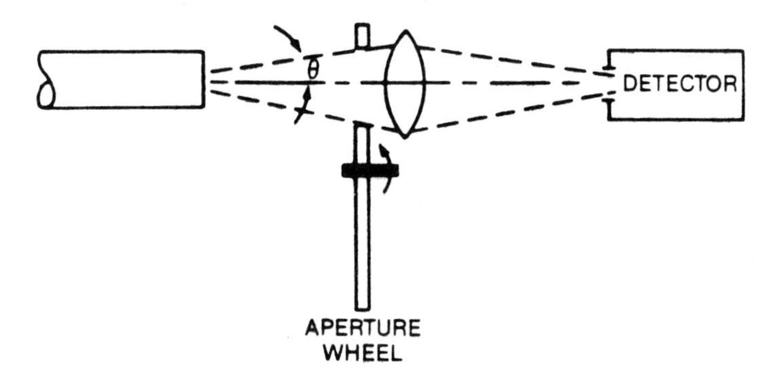

FIGURE 5.13 Experimental arrangement for the variable-aperture far-field measurement technique of the mode field diameter.

$$x = D \tan \theta \tag{5.8}$$

Finally, the mode field diameter $2w_0$ is obtained from

$$w_0^2 = \frac{1}{\int_0^\infty a(q)q\ dq} \tag{5.9}$$

This method can be easily implemented using the same setup as the one for spectral loss measurements. This may prove interesting for routine characterization as this eliminates one step of fiber sample preparation and manipulation. Also, this method has the capability of yielding the mode field diameter as a function of wavelength.

The knife-edge scanning technique consists in scanning a knife edge linearly in a direction orthogonal to the fiber axis and to the edge of the blade, in the beam emerging from the fiber ([27]; Fig. 5.14). The light transmitted by the knife edge is collected and focused onto the detector, yielding the power K(x) as a function of knife-edge lateral offset x. Equations (5.8) and (5.6), with D as the distance between the fiber output end and the plane of the knife edge, allow us to convert x into q. Finally, the mode field diameter $2w_0$ is obtained from

$$w_0^2 = \frac{\int_0^\infty [dK(q)/dq]\ dq}{\int_0^\infty q^2\ [dK(q)/dq]\ dq} \tag{5.10}$$

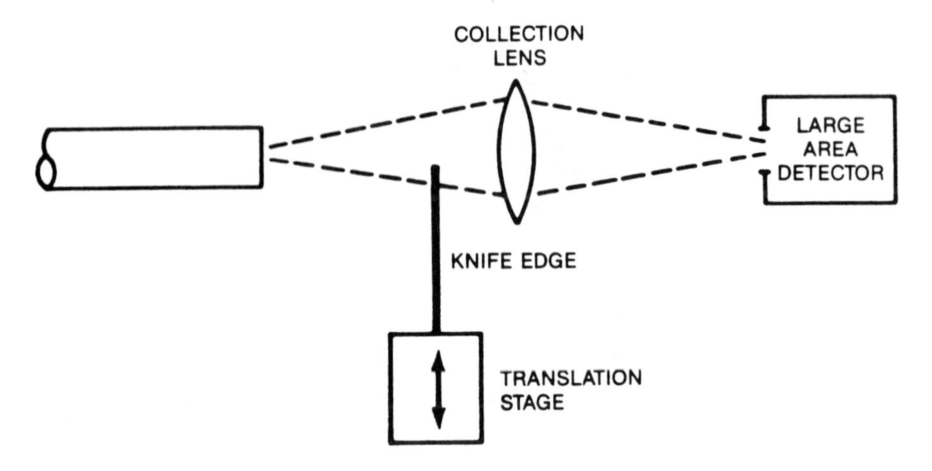

FIGURE 5.14 Experimental arrangement for the knife-edge scanning measurement technique of the mode field diameter.

5.3.3 Attenuation

The attenuation of a single-mode fiber can be measured by three methods: the cutback (two-point) method is the most widely used, and the most precise and accurate (Reference Test Method in CCITT-G.652), but it is destructive (1 to 2 m of fiber lost with each measurement). The insertion loss method is an alternative test method for CCITT-G.652 and it is nondestructive, allowing its application to connectorized cables. Backscattering measurements also permit determination of fiber attenuation in a nondestructive way, requiring access to only one fiber end. They are thus well suited to field measurements and are recognized as an alternative test method in CCITT-G.652.

Cutback Technique. We describe here the spectral loss measurement, but the restriction to single-wavelength measurements is straightforward. The general technique is well known and uses the measurement setup of Fig. 5.15. The measurement is carried out as for multimode fibers, by first measuring the power emerging from a long fiber sample, and then from a short fiber piece (a few meters) taken at the input end of the whole fiber, without changing the launching conditions. Several points are peculiar to single-mode fibers and should not be overlooked.

1. The absence of multiple modes means no transient losses (except for a small wavelength region corresponding to the range

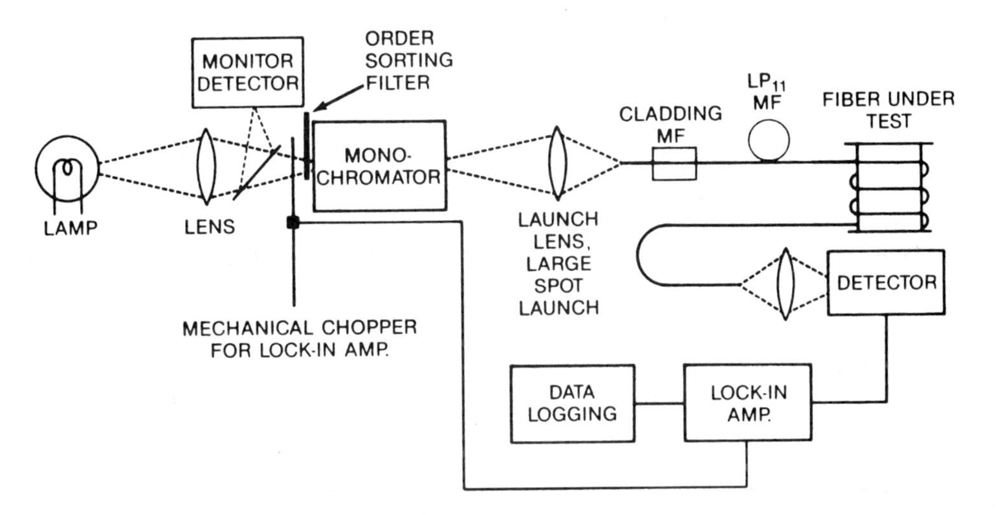

FIGURE 5.15 Experimental arrangement for the measurement of fiber attenuation with the cutback method. Total power received for the fiber of length L: $P_1(\lambda)$. Total power received after cutting the fiber 2 m after the input end: $P_2(\lambda)$. Average attenuation per unit length $\alpha(\lambda) = (1/L)\ 10 \log[P_2(\lambda)/P_1(\lambda)]$.

where the second order LP_{11} mode propagates in the short reference sample but is not transmitted through the long fiber). This means that there is no concern over a "steady-state" distribution. Also the measurement becomes insensitive to launching conditions. This results in better measurement precision as compared to multimode fibers—interlaboratory comparisons have shown standard deviations of about 0.01 dB/km for single-mode fibers, compared to 0.05 dB/km for multimode.

2. The very small mode field diameter means that the amount of light coupled into the fiber is much smaller than for multimode fibers. It is thus generally necessary to use cooled detectors.

3. The effective cutoff wavelength of the whole long fiber is generally shorter than that of the short reference piece, and this means that in the wavelength range where the LP_{11} mode is present in either of the fiber samples, the cutback technique does not permit evaluation of a loss per unit length. For the same reason the losses found in making the measurement with the two opposite launching ends will be identical only if the fiber is still single moded at both ends.

4. The cladding mode stripper (CMS) becomes much more critical than with multimode fibers. It is easy to show that the

measured loss will exceed the actual loss by the quantity $\delta\alpha$, given in decibels by

$$\delta\alpha = 10 \log \left(1 + T_r \frac{P_c}{P_g}\right) \tag{5.11}$$

where T_r is the residual transmission coefficient of the cladding modes through the CMS, P_c is the power launched into the cladding modes, and P_g is the power launched into the guided modes. Compared to standard multimode fibers (50-μm-core diameter, 125-μm outside diameter, 0.2 numerical aperture), in single-mode fibers P_c is multiplied by about 1.2 where P_g is divided by about 40, which means that T_r must be divided by 50 to get the same measurement inaccuracy.

To illustrate this effect, the power P_0 transmitted by a 1-m-long single-mode fiber sample with its PVDF coating and no CMS was first measured. Then the PVDF coating was removed for 5 cm midway along the fiber (which was held straight), and different liquids were used to immerse this part of the fiber as CMS, leading to a transmitted power P. Although P/P_0 is not really a measure of T_r, it is related to it, and the results are shown in Fig. 5.16 as a function of temperature for eucalyptol and paraffin oil. When the temperature increases, the liquid's refractive index decreases much more rapidly than that of silica. For eucalyptol, whose refractive index crosses that of silica at about 20°C for $\lambda = 1100$ nm, the rather good CMS efficiency ($P/P_0 \simeq 1\%$) drops abruptly above 20°C, as the refractive index of the liquid becomes smaller than that of silica. In the case of paraffin oil, the refractive index is higher than that of eucalyptol and hence the CMS efficiency is poorer than with eucalyptol at low temperatures, but it improves slowly with temperature, increasing as the index match becomes better. Similar phenomena could be observed as a function of wavelength, as the liquid's index dispersion curve is different from that of silica.

It can be concluded that the CMS for single-mode fiber measurements should receive close attention, especially in the choice of the liquid, and requires the additional use of bends to improve its efficiency; this can best be done using empirical trials. Finally, the specification and assessment of the CMS efficiency should be done prior to fiber loss measurements; the experiment described above can provide a simple evaluation of this efficiency.

The *insertion loss* technique uses a setup similar to the cutback technique. However, instead of cutting a short piece of the fiber at the input end to obtain the reference level $P_2(\lambda)$, one directly measures the power available at the launching point. This can be done by using a piece of fiber similar to the one to be measured

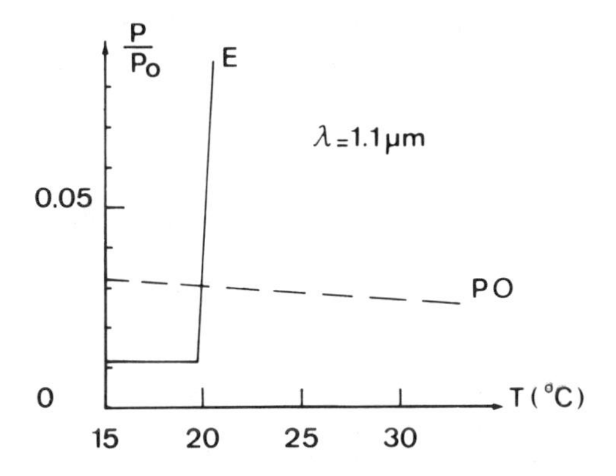

FIGURE 5.16 Cladding mode stripper (5 cm straight fiber + liquid) residual transmission as a function of temperature. E, eucalyptol; PO, paraffin oil; measurement wavelength, 1100 nm.

(except for a shorter length) as "launching optics" between the monochromator output and the input of the fiber under test. The total attenuation measured thus includes the loss of the fiber under test and the connection loss to the launching fiber. The typical value of this connection loss can be evaluated independently, and a correction for it can be implemented in the measurement procedure. Obviously the measurement will be more accurate for longer fibers.

Backscattering Techniques. These methods of loss measurement are based on optical time-domain reflectometry (OTDR) [28] or the more recent optical frequency-domain reflectometry (OFDR) [29,30], which can provide a higher dynamic range because of the use of narrow-band detection of modulated signals. As the problems associated with using these techniques in single-mode fibers are almost the same, we consider only OTDR, for which a detailed study applied to SMFs can be found in Ref. 31.

For a light pulse of peak power P and width S launched into the LP_{01} mode of a single-mode fiber at time t = 0, the backscattered signal at the input end has an optical power given by

$$P_B(t) = \left(\frac{\lambda}{2\pi n_1 w_0}\right)^2 P\alpha_s(z)S \frac{c}{2N_2} \exp\left[-\alpha(z)\frac{ct}{N_2}\right] \qquad (5.12)$$

the correspondence between time t and abscissa z along the fiber axis being $z = ct/2N_2$. α_s is the Rayleigh scattering loss in nepers per meter, N_2 is the group refractive index seen by the mode, w_0 is the mode field radius, and α is the total fiber loss per unit length.

One problem is associated with the very low level of the backscattered signal. We have typically

$$10 \log \left[\frac{P_B(0)}{P} \right] \simeq -54 + 10 \log S \quad \text{(dB)} \tag{5.13}$$

with S in microseconds. This corresponds to signal levels almost 10 to 20 dB smaller than in multimode fibers, for the same pulse duration (and hence spatial resolution) and the same peak power effectively coupled into the fiber. Simultaneously, for the same laser-emitted peak power, the power effectively coupled into the fiber is about 10 dB smaller in single-mode fibers than in multimode fibers. A possibility for increasing P (and thus P_B) is to use a 1320-nm Q-switched YAG laser, which could launch into the fiber 100 times more power than a laser diode while remaining below the power thresholds for nonlinear effects (see Chap. 9). Similar lasers are being developed for field applications of laser range finding and could be adapted to the requirements of OTDR. For links operating at 1550 nm, where the loss is about 0.3 dB/km smaller than at 1320 nm, 1320-nm YAG lasers would provide a signal P_B stronger than 1550-nm laser diodes only for lengths shorter than 30 km [31]. With typically 1 mW of power launched into the LP_{01} mode, averaging techniques yield a dynamic range of about 30 dB (two ways) [31].

Another problem is associated with the coupler, which should minimize the Fresnel reflection from the fiber's input end, which is about 40 to 50 dB higher than the backscattered power, according to Eq. (5.13). A basic arrangement that is usable almost exclusively in the laboratory is shown in Fig. 5.17a. The beam splitter (BS) should present transmission and reflection characteristics as insensitive as possible to the state of polarization of the light, or a polarization scrambler should be inserted between the laser and the coupler. As the laser emission is polarized, the state of polarization (SOP) of the light would evolve along the fiber and thus the backscattered SOP would vary with time. Then if the spatial resolution of the measurement system (given by $cS/2N_2$) is smaller than the SOP beat length (see Chap. 2), and if the beam splitter is SOP dependent, these SOP variations will be transformed into apparent loss fluctuations [32]. Figure 5.17b shows the detected signal with a nonpolarizing beam splitter, and Fig. 5.17c shows the detected signal from the same fiber but with an anlyzer inserted in front of the photodetector. This illustrates clearly that any polarization-sensitive coupler would be inadequate for loss measurements by backscattering techniques, whereas it is an interesting tool for characterizing the

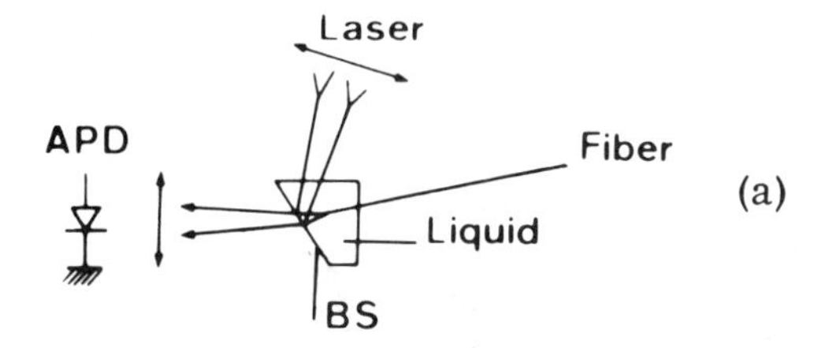

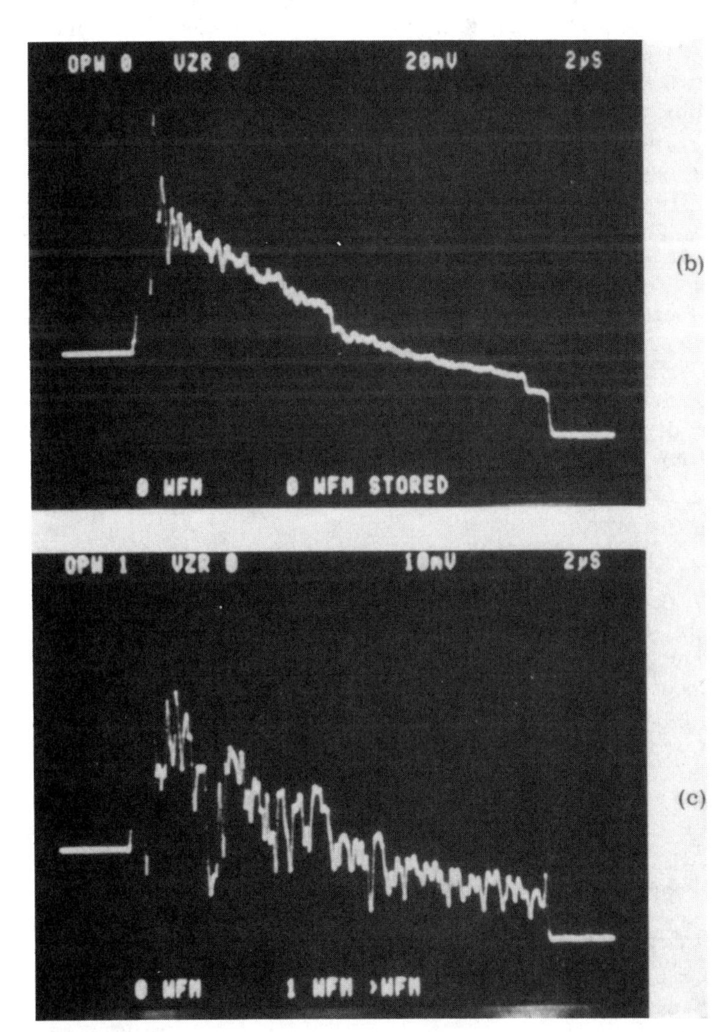

FIGURE 5.17 (a) Basic laboratory arrangement for OTDR. (b) OTDR signal with a nonpolarizing beam splitter. (c) Same signal with an analyzer inserted in front of the photodetector.

fiber birefringence (Sec. 5.3.4). Other possible couplers for field applications have used a launching fiber with a tilted end face, which led to a −67 dB residual reflection with a 12° tilt angle [33], or X-couplers such as those described in Chap. 6.

Finally, the residual Fresnel reflection can be cancelled completely by using two avalanche photodetectors (APDs), one of them illuminated with the backscattered signal, and the other by a small fraction of the probe pulse [31]. When the level of this probe pulse fraction is adjusted to provide the same photocurrent as that of the residual Fresnel reflection, the APD's circuitry is arranged such that the two current pulses cancel each other out.

A third problem may arise with interpretation of the measurement results. Equation (5.12) clearly shows that if the fiber is inhomogeneous [variation of $\alpha_s(z)$ through a variation of the index difference (see Chap. 3) or variation of w_0 along the fiber], the backscattered signal will vary in a way that may appear surprising. As an illustration, Fig. 5.18 shows the calculated variation of P_B when traversing a splice between two fibers with the same index difference, perfectly jointed but with mode field radii w_1 and w_2 such that $w_1 = 1.2w_2$. Equation (3.12) shows that the splice loss (independent of the propagation direction) is 0.14 dB, and we would thus expect a decrease of 0.28 dB for P_B, independently of the propagation direction. However, Eq. (5.12) shows that there will be a variation of ±1.58 dB in the backscattering coefficient, because of the variation in the mode field radius, the sign + corresponding to the decrease in the mode field radius. We thus obtain the two curves of Fig. 5.18, which are similar to experimental results already observed [34].

Similarly, if the mode field radius continuously decreases from the launching end toward the far fiber end, the backscattered signal may appear with an "amplification" due to the increase in the backscattering coefficient. As in the case of multimode fibers, the only way to separate both effects is to measure the backscattered signal for both propagation directions (launching from end A, and then from end B). For a given point of the fiber (abscissa z from A, and abscissa L-z from B) the actual attenuation (two ways) with respect to end A is the half sum of both measured attenuations (with respect to A), whereas the half difference gives the variation in the backscattering coefficient with respect to that of end A [35].

5.3.4 Birefringence

As shown in Chap. 2, the only type of intrinsic birefringence in a SMF is linear birefringence, and accurate measurements of this parameter require that the fiber be kept free of parasitic-induced birefringences. The fiber should thus be straight, without lateral

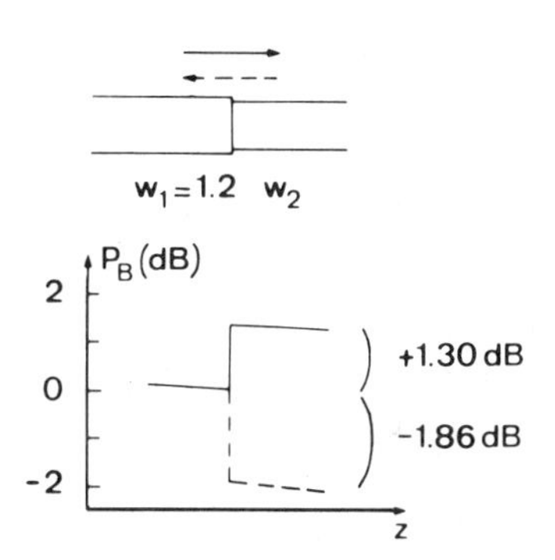

FIGURE 5.18 Calculated backscattering effect of a perfect splice between two fibers with the same index difference but with mode spot radii such that $w_1 = 1.2w_2$. The solid line is for launching from fiber with w_1, and the dashed line is for launching from fiber with w_2.

forces or twists (except in one measurement method, see below), and the fiber ends should be clamped into a 60° V-groove (Sec. 2.3.2) or into a hypodermic needle with wax. We first discuss various methods applicable to the determination of the intrinsic fiber birefringence, and then a measurement technique for the polarization-mode coupling. Many measurements require determination of the principal birefringence axes of the fiber, and a general method for this is described in Appendix 5.1.

Beat Length Measurement. One of the earliest methods is to measure the beat length L_b, from which the birefringence $\delta\beta$ is deduced through

$$\delta\beta = \frac{2\pi}{L_b} \tag{5.14}$$

The method consists of exciting equally the two orthogonally polarized LP_{01} modes with a circularly polarized input wave. As the scattering due to the vibration of a dipole is null along its vibration axis, each time the light becomes linearly polarized (i.e., at abscissas $L_b/4$, $5L_b/4$, $9L_b/4$, etc., or $3L_b/4$, $7L_b/4$, etc.), there will be no scattering detectable when observing the fiber transversely to its axis and at 45° of its principal birefringence axes [36] (see Fig. 2.5b for an example). This therefore allows direct measurement of L_b, but it is limited to fibers that are single moded in the visible range and requires a powerful laser source (10 to 100 mW), as the scattering is weak and becomes even weaker when the wavelength increases. Additionally, L_b should be between 1 mm and 10 cm for practical measurements.

Cutback Technique. This straightforward method consists of launching a given SOP in a fiber and determining the output SOP as a function of fiber length by cutting back small fiber pieces. The output SOP shows a periodicity L_b with fiber length, from which the birefringence can be deduced as above if the fiber length is greater than L_b. If this is not the case, simple considerations on classical linear retarders permit deduction of $\delta\beta$ from the SOP variations with fiber length. However, this theoretically simple method suffers several practical drawbacks. As the whole fiber influences the output SOP, it should be kept free of stresses on its whole length, and the best configuration is to hold the fiber vertically, which additionally avoids any twist. Second, because the fiber has to be cut several times, the fiber clamp (especially that of the output end, which is cut back) should induce the smallest stresses possible. From a practical point of view this makes it difficult to measure fibers with $L_b > 1$ m. Finally, as the length of fiber taken off between two measurements should be less than $L_b/2$ (in order to be able to determine a period equal to L_b), it is difficult to use this method with fibers such that $L_b < 1$ cm.

External Fields. Applying a transverse electric field or a longitudinal magnetic field to the fiber induces an externally controlled linear or circular birefringence, as shown in Chap. 2. With oscillating fields it is then possible to deduce $\delta\beta$ from the corresponding modification of the output SOP modulation when the external device is moved along the fiber [37]. The dimensions of the inducing device in practice require that L_b be greater than 1 to 2 cm, whereas the maximum value of L_b is limited only by the practical ability to keep the fiber straight in the laboratory.

Fiber Twisting. A circularly polarized wave is launched into the fiber, and the output SOP is determined as a function of the fiber twist angle [38]. The fiber is twisted through a quickly rotating analyzer between the fiber output end and the photodetector. This allows automatic recording of the maximum intensity and the minimum intensity of each SOP as a function of the twist (see Fig. 5.19). Comparisons of the behavior of $\varepsilon = (P_{max} - P_{min})/(P_{max} + P_{min})$ as a function of the twist N (turns per meter) with the theoretical model allows simultaneous determination of the linear intrinsic birefringence $\delta\beta$ and the circular elastooptic coefficient g defined in Chap. 2:

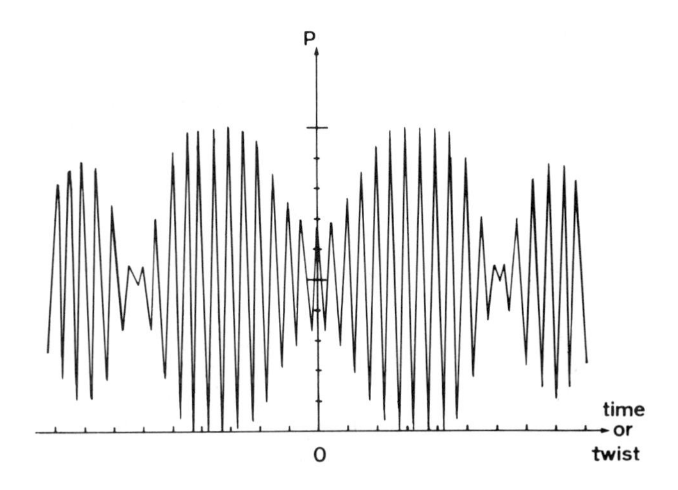

FIGURE 5.19 Detected power P passing through a rapidly rotating analyzer placed at the output of a slowly twisted fiber excited by a circularly polarized light. The horizontal coordinate is time, which is proportional to the twist angle.

$$\varepsilon = \frac{\delta\beta L}{Q}\left[1 - \left(\frac{\delta\beta L}{2Q}\right)^2 \sin^2 Q\right]^{\frac{1}{2}} |\sin Q| \tag{5.15}$$

where L is the fiber length and Q is given by

$$Q = \frac{1}{2}\{(\delta\beta L)^2 + [(2 + g)\ 2\pi N L]^2\}^{\frac{1}{2}} \tag{5.16}$$

This method is limited to L_b values greater than 1 cm, because at each end the fiber fixations impose a "dead zone" a few millimeters in length where the fiber is not twisted. On the other hand, the maximum value of L_b is limited only by the space available in the laboratory, as L should be greater than about $L_b/4$ for accurate measurements.

Polarization OTDR. The setup shown in Fig. 5.17a can be completed by an analyzer in front of the detector, allowing direct display of the evolution of the SOP along the fiber as shown in Fig. 5.17c (see Ref. 39). As shown in Fig. 5.17c, the oscillations in the signal are characteristic of the SOP evolution and also permit testing of the birefringence uniformity along the fiber.

Where a constant linear birefringence $\delta\beta$ exists, the oscillations would follow a $\cos(2\delta\beta z)$ law, where z is the single-way abscissa along the fiber (note that reciprocal circular birefringence will have no effect). For better data collection the signal can be fed into a spectrum analyzer [40], which will display a spike at frequency f related to L_b through

$$L_b = \frac{c}{fN_2} \tag{5.17}$$

where c is the light velocity in vacuum and N_2 is the group index seen by the wave. Equation (5.17) corresponds approximately to $L_b = 200$ m for $f = 1$ MHz (in Ref. 40 the beat length differs by a factor of 2 from our definition). The limitations of this technique are due to the practical impossibility of detecting backscattered signals with pulses shorter than about 10 ns, and thus L_b should be greater than a few meters.

Polarization Mode Coupling. As discussed in Sec. 2.4.3, the output state of polarization (SOP) from a fiber with random polarization mode coupling is unpredictable from a theoretical point of view and varies strongly with the exact wavelength or environmental conditions. For fibers aimed at maintaining a given SOP but presenting some residual mode coupling, the cross-polarization (power coupled to the undesired SOP) will thus be unpredictable and insignificant for a particular fiber, except for a given wavelength and in fixed environmental conditions (which has no practical interest). Such a parameter has practical significance only as an average value of the cross-polarizations in an ensemble of similar fibers. It is then given by Eq. (2.34) as long as the cross-polarization (average) remains small, or more generally by [41]

$$\frac{\langle P_2 \rangle}{\langle P_1 \rangle} = \tanh hz \tag{5.18}$$

where $\langle P_1 \rangle$ is the average power emerging from the excited eigenstate of polarization p_1, $\langle P_2 \rangle$ is the average power emerging from the fiber on the undesired eigenstate p_2 (p_2 is orthogonal to p_1; if p_1 is linear, p_2 is linear rotated by 90°; if p_1 is circular right, p_2 is circular left; etc.), z is the abscissa along the fiber, and h is given by $\langle |X_2(L)|^2 \rangle / L$ in Eq. (2.34).

For small values of $\langle P_2 \rangle / \langle P_1 \rangle$, h is a direct measure of the rate of power transfer from the excited eigenstate to the undesired

SOP, per unit fiber length. Measuring h requires averaging the values of P_2 and P_1 over a large ensemble of similar fibers, which obviously is difficult in practice. It can be done by averaging these quantities for one fiber for many different temperatures, or more easily for many different wavelengths. In the latter case, the fiber behaves differently at the various wavelengths because of birefringence dispersion, although it suffers the same perturbations.

It thus appears that a practical and direct averaging process is obtained by exciting the principal eigenstate p_1 with a broadband light source [42]. Using a light source of rms spectral width $\Delta\lambda$ (in terms of wavelength, or $\Delta\nu = c\Delta\lambda/\lambda^2$ in terms of frequency) with a fiber of length L and birefringence dispersion T (group delay difference per unit length) is equivalent to averaging a monochromatic measurement over N similar fibers, with $N = TL\,\Delta\nu$. This is due to the fact that a light source of spectral width $\Delta\nu$ can be considered as emitting a random sequence of pulses with a duration $1/\Delta\nu$; the part of these pulses carried by one polarization is completely separated from the part of the pulses carried by the other polarization after a length of fiber given by $1/(T\,\Delta\nu)$, which can thus be considered as the minimum length above which the powers can be added, instead of the fields. The ratio between the actual fiber length and this elementary depolarizing length yields the number of equivalent fibers taking part to the averaging process, which should be much greater than 1.

For the usual highly linear birefringent fibers, the source spectral width should be on the order of 10 to 100 nm. However, when a fiber is unknown, it may appear difficult to determine the required minimum spectral width. This can be estimated from the relative fluctuations on $\langle P_2 \rangle$ as a function of external perturbations (temperature variations, etc.), which are on the order of $N^{-\frac{1}{2}}$, and a high value for N thus provides a very low optical noise for varying ambient conditions. Thus heating the fiber should hardly affect the measured value of $\langle P_2 \rangle$ if the spectral width is sufficiently high [42].

Finally, it should be recalled that a low value of $\langle P_2 \rangle$ does not necessarily mean that under monochromatic illumination, the cross-polarization will always remain low. In some cases (temperature, etc.), P_2/P_1 may increase to very high values [42].

5.4 DYNAMIC TRANSMISSION PARAMETERS

These parameters comprise the two basic signal distortion sources, birefringence dispersion and chromatic dispersion, and overall signal distortion by the fiber or the fiber link.

5.4.1 Birefringence Dispersion

It was shown in Chap. 2 that the birefringence dispersion (i.e., the group delay time difference between the two eigenstates of polarization of the LP_{01} mode) may vary from less than 0.05 ps/km to several nanoseconds per kilometer. Although the local birefringence and hence the local birefringence dispersion at a given abscissa z along the fiber can always be defined theoretically, measurement of these parameters is meaningful only if the influence of polarization-mode coupling is negligible. Practically, this means that birefringence dispersion can be measured only on fiber pieces that are almost polarization maintaining, that is, either short fiber pieces or long fibers with a high birefringence (either linear or circular). As a consequence, the measurement of birefringence dispersion smaller than a few picoseconds per kilometer is difficult, as the corresponding birefringence is small and hence the fiber must be short to maintain polarization, leading to very small absolute time differences. Thus one approach is based on interferometric effects.

Interferometric Methods. These methods are based on the fact that a given input SOP can always be decomposed as a linear combination of the two eigenstates of polarization of the LP_{01} mode, and the experimental arrangement usually ensures that both eigenstates are excited equally. After traveling through the fiber, both eigenstates are separated by a delay of $T \times L$, where T is the birefringence dispersion and L is the fiber length (here we consider the power carried by the two modes, and thus the group delay difference). The source should have a nonzero spectral width and thus a finite coherence time, and when $T \times L$ is much greater than the coherence time, the output light is totally depolarized, as the two eigenstates are no longer able to interfere with each other. It should be noted that a simple rotating analyzer does not permit us to distinguish between a totally (or partially) depolarized light and a circularly (or elliptically) polarized light. For this, a compensator such as a Babinet–Soleil compensator should be used and one can conclude that the light is depolarized if it is impossible to recover a linearly polarized light by adjusting the compensator. Practically, the measurement of the degree of polarization requires searching through all possible compensator adjustments and all analyzer orientations for the maximum contrast between the maximum detected intensity and the minimum detected intensity.

In one application of this general method, described in Ref. 43, a short fiber is excited with a linear SOP at 45° of its principal axes and a variable light spectral width. It can be shown that if the spectral power density is constant over width $2\Delta\omega$ (in terms of pulsation), the degree of polarization determined is given by

$|\sin(\Delta\omega TL)|/(\Delta\omega TL)$, and hence the variation of $\Delta\omega$ permits measurement of T. As it is difficult to ensure that the spectrum is flat over very wide spectral ranges, this measurement technique is limited to values of T × L greater than about 0.1 ps.

An alternative technique consists of separating both eigenstates of polarization before entering the fiber (or after transmission through the fiber) and using an optical delay line for delaying one of the eigenstates with respect to the other ([44]; Fig. 5.20). As stated earlier, the degree of polarization at the output is at a maximum when the total delay between both eigenstates is zero. Thus, measuring the necessary optical delay line adjustment (3.33 ps for a 1-mm adjustment) for recovering a totally polarized wave at the output of the fiber, we are able to measure the product T × L introduced by the fiber. Again, the resolution is limited to about 0.1 ps.

Global Method. It is often desirable to measure average parameters on a long fiber length instead of local parameters on short fiber pieces which may not be representative of the whole fiber. A very simple method can be used for birefringence dispersion. A laser diode is sinusoidally modulated with a high-frequency-stabilized generator (typically 200 to 500 MHz), and the linearly polarized emitted wave is launched into the fiber through a rotating half-wave plate, in order to vary the direction of the launched linear SOP and thus avoid the need to search the principal axes of the fiber. Then a vector voltmeter measures the phase variation of the transmitted wave as a function of the input polarization direction, from which the birefringence dispersion is deduced as the maximum phase variation divided by $2\pi fL$ (f is the modulating frequency and L is the fiber length [45]). The limiting resolution of this method is about 0.5 ps + 0.5 ps/km for T.

5.4.2 Chromatic Dispersion

Although there are no problems associated with mode coupling (or its equivalent, as long as nonlinear effects are not generated inside the fiber), we again find a distinction between measurement techniques for short and long fibers. Measurements of short fiber pieces are more interesting for comparisons with theory, as it is likely that the structural parameters of the fiber are constant in the sample, but measurements of long fibers are more realistic from the user's point of view.

CCITT-G.652 has selected the interferometric measurement technique performed on short fiber samples as an alternative test method. The reference test method for CCITT-G.652 and EIA-455 consists in measuring the relative group delay experienced by various wavelengths during propagation through a long fiber sample.

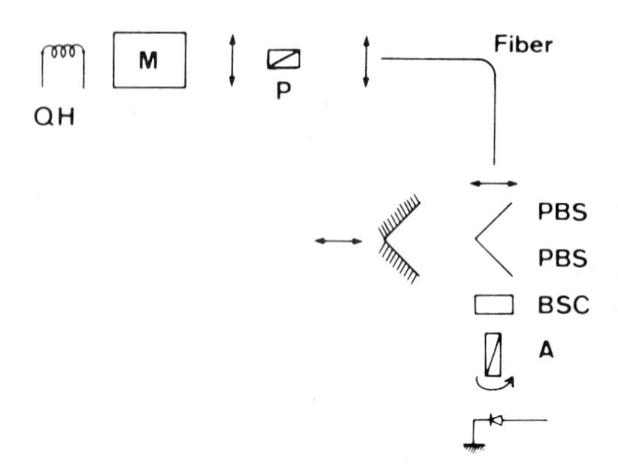

FIGURE 5.20 Experimental setup for the measurement of small birefringence dispersion in a linearly birefringent fiber. QH, quartz–halogen lamp; M, monochromator with variable slits, which permit varying the spectral width; P, polarizer oriented at 45° from the fiber's birefringence axes; PBS, polarization beam splitters oriented at 45° from the fiber's birefringence axes, and constituting a variable optical delay line for one polarization (with the movable mirror); BSC, Babinet–Soleil compensator; A, rotating analyzer for determining the degree of polarization.

The group delay can be measured in either the time domain or the frequency domain (phase shift). We first discuss the measurement methods applicable to short fiber samples.

Mode Field Radius Variation. It was shown in Sec. 4.3.3 that the chromatic dispersion is given approximately by the sum of a pure material dispersion term [first term on the right-hand side of Eq. (4.7)] and a pure waveguide dispersion term [second term on the right-hand side of Eq. (4.7)]. The corrections to this simplified equation are usually less than ±2 ps/nm × km for $C(\lambda)$, or equivalently less than ±40 nm for λ_D [$C(\lambda_D) = 0$]. Simultaneously, for high-silica-content fibers, the material dispersion term is well known and it thus appears that measurement of the waveguide dispersion allows approximate evaluation of the chromatic dispersion. It has been shown [46] that the waveguide dispersion $C_W(\lambda)$ can be deduced from a knowledge of the mode field radius spectral behavior, with an approximation better than 10% in most cases of interest:

$$C_w(\lambda) = -\frac{\Delta n}{c\lambda} V(Vb)'' \simeq -\frac{1}{\pi^2 n_2 c}\frac{\lambda}{w_0^2}\left(\frac{\lambda}{w_0}\frac{dw_0}{d\lambda} - \frac{1}{2}\right) \qquad (5.19)$$

This result is very useful for several reasons. First, it provides good accuracy for predicting the total chromatic dispersion $C(\lambda)$ once the material dispersion has been added to it (Sec. 4.3.3); and second, it involves only the measurement of a static parameters (w_0) as a function of wavelength and does not require knowledge of any other fiber parameter. Very simple methods for measuring $w_0(\lambda)$, possibly using the same bench as that used for spectral loss measurements, have been described in Sec. 5.3.2.

It thus appears that the (simple) measurement of $w_0(\lambda)$ makes it possible to evaluate the chromatic dispersion within a few ps/nm × km, without the need of any other fiber parameter. Obviously, this would not replace accurate measurements of the chromatic dispersion as described below, but could provide a simple and inexpensive routine test for quality control purposes. Finally, for the purpose of comparisons with theory, it should be noted that it is the only measurement method that provides a *localized* evaluation of dispersion, as the spot size measurements described in Sec. 5.3.2 involve only the spot size on the output end of the fiber.

Interferometric Methods (ATM for CCITT-G.652). These methods, again, are concerned with short fiber pieces, and basically they all use relatively wide spectrum sources and some kind of Fourier transform spectroscopy [47–51]. A possible experimental arrangement is shown in Fig. 5.21a. A broadband light source [such as a light-emitting diode (LED), a quartz–halogen lamp, or a fiber Raman source (see Chap. 9)] is filtered through a monochromator to a spectral width $\Delta\lambda$, yielding a coherence time $t_c = \lambda^2/(c\Delta\lambda)$ on the order of 3 ps for a spectral width of 2 nm at 1300 nm (or a coherence length $L_c = \lambda^2/\Delta\lambda$, on the order of 1 mm in the example chosen). The light beam is split into a reference beam with an optical path length L_0 (which can also be a reference fiber), and the test beam which traverses the fiber under test of length L and also a variable optical delay line of length $2(\ell_0 + \Delta\ell)$, before recombining with the reference beam on the photodetector.

When both optical path lengths are exactly equal,

$$L_0 = c\tau L + 2\ell_0 + 2\Delta\ell \qquad (5.20)$$

(where τ is the group delay time per unit length at the peak wavelength of measurement), interferences are produced, and either the

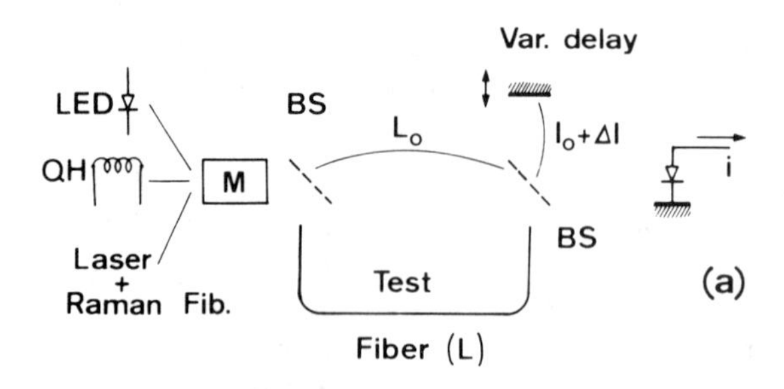

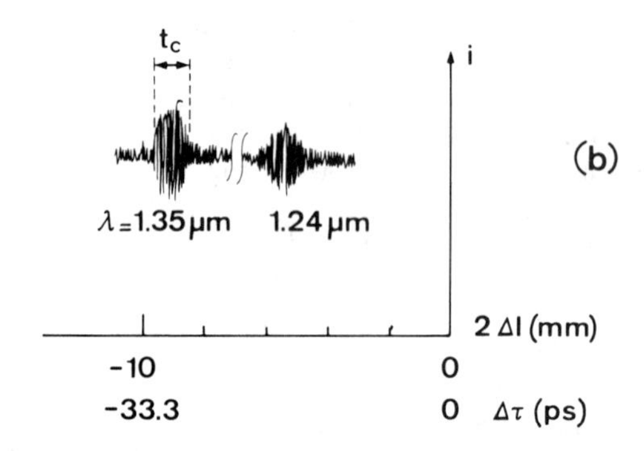

FIGURE 5.21 Interferometric measurement of chromatic dispersion. (a) Experimental setup. QH, quartz–halogen lamp; M, monochromator; BS, beam splitter. The movable mirror makes it possible to build an optical variable delay line. (b) Detected photocurrent as a function of optical delay line adjustment for two measurement wavelengths. (After L. G. Cohen and J. Stone, in *Optical Fiber Communication 1982 Digest of Technical Papers*, Optical Society of America, Washington, D.C., Apr. 1982, paper PD2.)

random-phase fluctuations due to air turbulences in the optical path or an intentional modulation (such as a vibrating mirror) induce strong fluctuations in the photocurrent [51]. When both optical path lengths differ by much more than the coherence length, the fields are totally uncorrelated and do not interfere. This aspect of the detected photocurrent is illustrated in Fig. 5.21b, where it is seen that the discussion above clearly applies.

Measuring $\Delta \ell$ as a function of wavelength λ yields $\tau(\lambda)$ directly through Eq. (5.20), from which the chromatic dispersion is easily deduced through $C(\lambda) = d\tau/d\lambda$. The resolution and accuracy of the measurement result from a trade-off between the detected power and the temporal resolution (which both require a high spectral width) and the spectral resolution [as $C(\lambda)$ is deduced from $d\tau/d\lambda$, a small spectral width improves the accuracy on $C(\lambda)$]. Typically, the source spectral width shall be selected to lie between 2 and 10 nm. The linear positioning device used for achieving the variable optical delay line should cover the range from 20 to 100 mm with an accuracy of about 2 μm. It is estimated in Ref. 51 that an accuracy of $\pm 0.1 t_c$ is readily achievable, which typically corresponds to ± 0.3ps on the measurement of $L\, d\tau$. Getting a very good accuracy in terms of ps/km for $d\tau$ thus requires the use of relatively long fiber lengths, which can be achieved practically only if the reference beam also traverses a fiber whose chromatic dispersion is known and serves as a reference [51].

Spectral Time Delay. This method is considered as RTM by CCITT-G.652 and EIA-455-168. It is based on the measurement of the relative group delay time experienced by light pulses transmitted through the fiber at various wavelengths. The light source may be either a fiber Raman laser with a monochromator ([52]; Fig. 5.22), or a set of several laser diodes emitting short pulses at various discrete wavelengths. In the case of the fiber Raman laser a broadband spectrum is generated inside a single-mode fiber through a stimulated Raman effect (see Chap. 9), and is then filtered by the monochromator. In both cases, the wavelength-adjustable 100-ps wide pulses are launched into the fiber under test. At the output a high-speed photodetector and a sampling oscilloscope are used to record the variation of group delay time as a function of wavelength λ with a resolution of about 10 to 20 ps, throughout the range 1100 to 1700 nm. Dividing by the fiber length yields the group delay time per unit fiber length $\tau(\lambda)$ which shall be fitted by a three-term Sellmeier expression:

$$\tau(\lambda) = \tau_0 + \frac{C'(\lambda_D)}{8}\left(\lambda - \frac{\lambda_D^2}{\lambda}\right)^2 \tag{5.21}$$

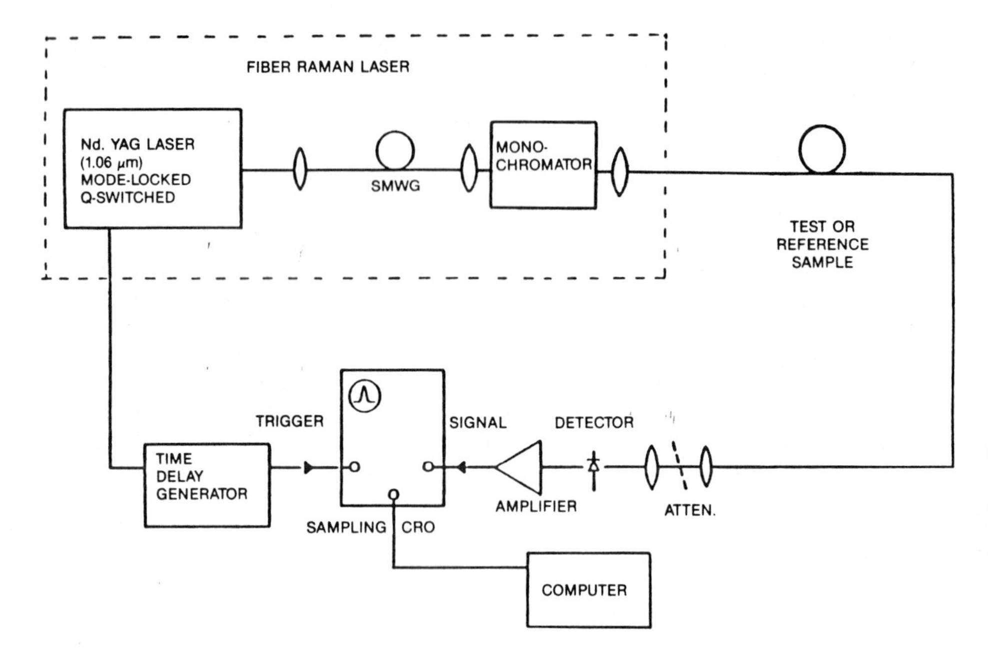

FIGURE 5.22 Experimental setup for measuring the chromatic dispersion with a fiber Raman laser source in the time domain.

The chromatic dispersion $C(\lambda) = d\tau/d\lambda$ can be determined from the differentiated Sellmeier expression:

$$C(\lambda) = \frac{C'(\lambda_D)}{4}\left(\lambda - \frac{\lambda_D^4}{\lambda^3}\right) \tag{5.22}$$

The fiber Raman laser source yields many measurement data because of its broadband spectrum, and this in turn provides better accuracy than using a few (3 to 7) laser diodes at discrete wavelengths. Differences between the two measurement techniques are about 0.1 ps/nm × km around 1300 mm for unshifted fibers and 0.4 ps/nm × km for dispersion-flattened fibers.

Spectral Phase Shift. This method is considered as RTM by CCITT-G.652 and EIA-455-169. It is based on the measurement of the relative phase of a sinusoidal modulating signal transmitted through the fiber at various wavelengths. The intensity of the light source is sinusoidally modulated at a frequency f with very good stability

($\Delta f/f < 10^{-7}$), and the phase of the signal detected after transmission through a fiber of length L is compared to the phase of the reference modulating signal [53]. If $\phi(\lambda)$ is the measured relative phase at wavelength λ, we obtain

$$\tau(\lambda) = \tau_0 + \frac{\phi(\lambda)}{2\pi f L} \tag{5.23}$$

The fit corresponding to Eqs. (5.21) and (5.22) is then applied to the measured data. The resolution is determined by both the frequency stability of the modulation and the phase resolution of the phasemeter (or vector voltmeter) used for the measurement. With a modulation frequency of 500 MHz, a frequency stability of 10^{-7}, and a phase resolution of 0.1°, we obtain a resolution of 0.5 ps/km + 0.6 ps for $\tau(\lambda)$ [53]. It is always very important to correct the measurements for possible spectrally varying delays in the emission of the light source. This may be done easily by making a second measurement on a short fiber sample or by taking a fraction of the emitted light as reference signal (instead of the modulating signal). It is also usual that varying delays are encountered on photodetectors, depending on the received light level. A variable attenuator should thus be used for constant light level detection.

This method has been implemented with a wide variety of light sources:

An *optical parametric oscillator* combined with an electrooptical modulator at 840 MHz has been used, together with the observation of Lissajous figures on an oscilloscope for the determination of the relative phase [54].

A *light-emitting diode* directly modulated at 30 MHz and filtered through a monochromator has been used to build a simple setup but with a very limited dynamic range (because of the very low launched power) and a limited time resolution (because of the low frequency) [55]. The accuracy has been reported as ±1 ps/(nm × km).

A *set of several laser diodes* operating at discrete wavelengths and sinusoidally modulated at frequencies up to 800 MHz, may be used [56–59]. The test fiber is then optomechanically switched sequentially to the various laser sources (Fig. 5.23). The accuracy here reaches ±0.1 ps/(nm × km).

Differential Phase Shift. This method is considered distinct from the preceding one by EIA-455-175, but only as a simple variation by CCITT-G.652. It uses basically the same principle as the spectral phase shift, but the light source is tunable on a small wavelength range and the measurement directly yields $d\tau(\lambda)/d\lambda$. The light

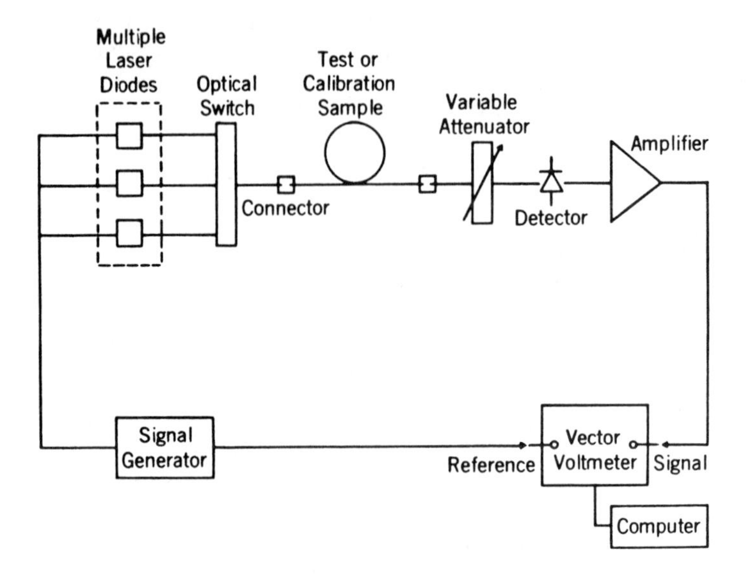

FIGURE 5.23 Experimental setup for measuring the chromatic dispersion with multiple laser diodes set in the frequency domain (phase shift).

source can be a laser diode whose wavelength is controlled by either temperature ([60]; Fig. 5.24) or an external cavity [61,62]. Using temperature as a means for controlling the laser emission wavelength makes it possible to cover a spectral range of about 10 nm at a rate of 0.8 nm/°C for InGaAsP lasers. An external cavity achieved with a 600 l/mm grating coupled to the rear facet of the laser through a collimating lens yielded a 30-nm tuning range, controlled by rotating the grating [61]. A much simpler device was achieved with a miniature concave spherical mirror (about 0.5 mm radius) positioned very close (200 μm) to the laser rear facet [62]. The mirror was supported by a piezoelectric translator which made it possible to control the wavelength by varying the distance between the mirror and the rear facet of the laser; 0.3 μm of variation in the external cavity length was sufficient to switch the laser wavelength from one longitudinal mode to the next (about 1 nm separation). The light intensity was directly modulated at 800 MHz and the wavelength was simultaneously modulated back and forth at 1 Hz, thereby allowing direct measurement of the differential phase variation $d\phi(\lambda)/d\lambda$, thus directly yielding $C(\lambda)$. The complete setup has been packaged into a compact, field portable, test instrument, having a dynamic

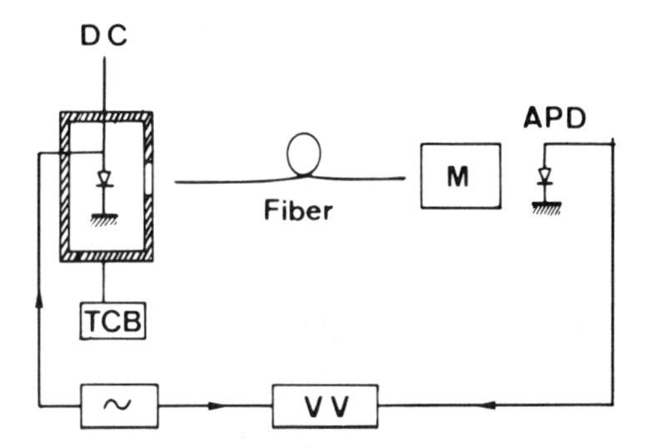

FIGURE 5.24 Chromatic dispersion measurement with a semiconductor laser, placed in a temperature-controlled chamber. TCB, temperature control box; M, monochromator; APD, avalanche photodetector; VV, vector voltmeter.

range of more than 20 dB (50 km of cable) and a precision of ±0.07 ps/(nm × km).

5.4.3 Signal Distortion Measurements

The combined effect of birefringence dispersion and chromatic dispersion on a long SMF link will lead to signal distortion, as described in Chap. 4. It should be recalled that it is not always possible to define a (power or field) frequency response for the fiber, and that when it is possible, it is not an intrinsic characteristic of the fiber but depends on the source spectral width (see Chap. 4). We can thus say that if signal distortion measurements are required, they should be carried out with the same source and the same modulation format as those used in the system. Otherwise, the only significant intrinsic parameters are the birefringence dispersion and the chromatic dispersion.

However, we have been that birefringence dispersion measurements can be made impossible by polarization mode coupling. In this case, the only way to separate signal distortions due to chromatic dispersion from those due to birefringence dispersion is to use a mode-locked laser, and measure the field spectrum amplitudes as a function of frequency before and after propagation through the fiber, using a scanning Fabry–Perot interferometer. It was shown in Sec. 4.3.3 that chromatic dispersion does not affect the amplitude of the spectrum because it is a purely nonlinear phase term, whereas birefringence dispersion will affect it. Thus the ratio of the two amplitude spectra will directly yield signal distortion due to birefringence dispersion. A schematic experimental setup is shown in Fig. 4.12.

5.5 SUMMARY

As evidenced by the discussion here and the long list of references that follows, the topic of single-mode fiber characterization has received considerable attention in the past and is likely to see further interest in the near future, together with increasing applications. The reader interested only in the use and application of single-mode fibers may thus be somewhat confused about the relevant parameters to be measured and how this should be accomplished. Restricting the discussion to the most common applications of single-mode fibers allows us to state that the measurements can be limited to (1) approximate index profiles and geometry through the near-field scanning technique (Sec. 5.2.2); and (2) spectral loss and spectral variation of mode spot size, with the same measurement bench discussed in Sec. 5.3.2. Let us recall that the effective cutoff wavelength is determined from the spectral loss curves, and that the spectral variation of mode spot size makes it possible to deduce the equivalent step-index fiber and the splice, bending, and microbending losses and to evaluate the chromatic dispersion. These very basic measurements can be completed in some cases by optical time-domain reflectometry and direct chromatic dispersion measurements. Obviously, when detailed investigations of some propagation characteristics are desired, other measurements and analyses should be carried out using the appropriate methods described in this chapter.

APPENDIX 5.1 Principal Birefringence Axis Determination in a Linearly Birefringent Single-Mode Fiber

A convenient measurement setup is shown in Fig. 5.25a, and a common procedure is as follows. A collimated light beam is linearly polarized by a polarizer whose direction can be varied (this holds for a nonpolarized light source; if the light source is a laser, a rotating half-wave plate allows us to vary the polarization direction) and launched into the fiber. The emerging beam is collimated before passing through a rotating analyzer and being detected (the analyzer rotation can be manually actuated, provided that the analyzer rotates much more quickly than the input polarizer). Figure 5.25b shows aspects of the detected photocurrent. For the direction θ_1 of the input polarization, the output beam appears elliptically polarized, but for the direction θ_2, it is linearly polarized ($i_{min} = 0$). In practice, i_{min} remains slightly above zero because neither the

coll.
light
P
A
i
(a)

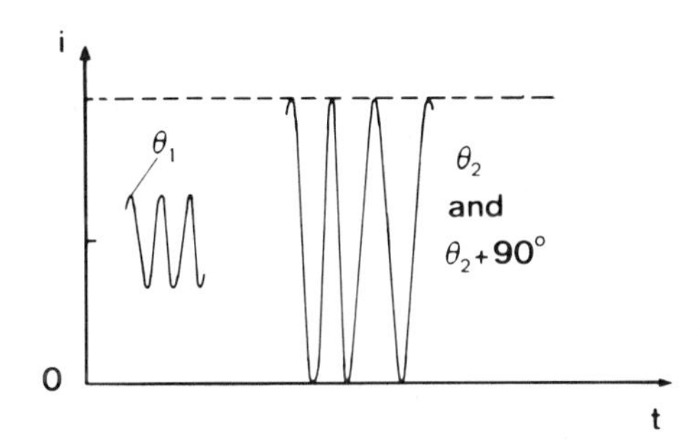

(b)

FIGURE 5.25 Principal birefringence axis determination. (a) Measurement bench, with a slowly rotating polarizer P and a rapidly rotating analyzer A. (b) Photocurrent i as a function of time (or analyzer orientation) for two orientations of the polarizer.

polarizer nor the analyzer is perfect. However, careful optical alignment and component selection make it relatively easy to get $i_{min} < 0.01 \times i_{max}$.

As stated in Sec. 2.4.1, one can conclude that the corresponding directions θ_2 and $\theta_2 + 90°$ determine the principal axes of linear birefringence in the fiber *only if* θ_2 *is stable against large temperature variations* for monochromatic light, or if a very wide spectrum light source has been used for the measurements. In the later case, if the value of i_{min} is significantly higher than the value expected from the measurement bench performances, the injection of i_{min}/i_{max} in Eq. (5.18) also permits evaluation of the degree of polarization-mode coupling.

REFERENCES

1. H. M. Presby, *Appl. Opt. 20*(3):446–450 (1981).
2. D. L. Philen and W. T. Anderson, in *OFC'82 Digest of Technical Papers*, Optical Society of America, Washington, D.C., Apr. 1982, paper THEE7.

3. H. M. Presby and D. Marcuse, *Appl. Opt.* *18*(5):671–677 (1979).
4. P. L. Chu and T. Whitbread, *Electron. Lett.* *15*(10):295–296 (1979).
5. I. Sasaki, D. N. Payne, and M. J. Adams, *Electron. Lett.* *16*(6):219–221 (1980).
6. T. Okoshi and M. Nishimura, *Appl. Opt.* *20*(14):2407–2411 (1981).
7. I. Sasaki, P. L. François, and D. N. Payne, *Proc. 7th Eur. Conf. Opt. Commun.*, Copenhagen, Sept. 8–11, 1981, paper 6-4.
8. T. Okoshi and M. Nishimura, *J. Opt. Commun.* *1*(1):18–21 (1980).
9. I. Sasaki, D. N. Payne, and R. J. Mansfield, in *OFC'82 Digest of Technical Papers*, Optical Society of America, Washington, D.C., Apr. 1982, paper WAA5.
10. W. J. Stewart, *Proc. IOOC'77*, Tokyo, June 1977, paper C2-2.
11. M. J. Saunders, *Appl. Opt.* *20*(9):1645–1651 (1981).
12. W. J. Stewart, in *IOOC'81 Digest of Technical Papers*, Optical Society of America, Washington, D.C., Apr. 1981, paper TuG6.
13. P. V. H. Sabine, F. Donaghy, and D. Irving, *Electron. Lett.* *16*(23):882–883 (1980); D. H. Irving, *Electron. Lett.* *19*(6): 190–191 (1983).
14. F. M. E. Sladen, D. N. Payne, and M. J. Adams, *Appl. Phys. Lett.* *28*(5):255–258 (1976).
15. W. T. Anderson and T. A. Lenahan, *IEEE J. Lightwave Tech.* *LT-2*(3):238–242 (1984).
16. Y. Murakami, A. Kawana, and H. Tsuchiya, *Appl. Opt.* *18*(7): 1101–1105 (1979).
17. V. A. Bhagavatula, W. F. Love, D. B Keck, and R. A. Westwig, *Electron. Lett.* *16*(18):695–696 (1980).
18. D. L. Franzen, *IEEE J. of Lightwave Tech.* *LT-3*(1):128–134 (1985).
19. W. T. Anderson and D. L. Philen, *IEEE J. of Lightwave Tech.* *LT-1*(1):20–26 (1983).
20. J. C. Augé and L. B. Jeunhomme, *Fiber Optics: Short-Haul and Long-Haul Measurements and Applications II* SPIE, Vol. 559, pp. 6–12 (1985).
21. K. Petermann, *Electron. Lett.* *19*(18):712–714 (1983).
22. C. Pask, *Electron. Lett.* *20*(3):144–145 (1984).
23. J. Streckert, *Opt. Lett.* 5 505–507 (1980).
24. C. A. Millar, *Electron. Lett.* *17*(13):458–460 (1981).
25. F. Alard, L. Jeunhomme, and P. Sansonetti, *Electron. Lett.* *17*(25/26):958–960 (1981).
26. J. M. Dick, R. A. Modavis, J. G. Racki, and R. A. Westwig, *Tech. Dig. OFC'84*, New Orleans, Optical Society of America, pp. 90–91 (1984).

27. J. P. Pocholle and J. Auge, *Electron. Lett.* *19*(6):191–193 (1983).
28. M. K. Barnoski, M. D. Rourke, and S. M. Jensen, *Proc. 2nd Eur. Conf. Opt. Commun.*, Paris, Sept. 1976, pp. 75–79.
29. R. I. MacDonald, *Appl. Opt.* *20*(10):1840–1844 (1981).
30. F. P. Kapron, D. G. Kneller, and P. M. Garel-Jones, in *IOOC'81 Digest of Technical Papers*, Optical Society of America, Washington, D.C., Apr. 1981, paper WF2.
31. K. I. Aoyama, K. Nakagawa, and T. Itoh, *IEEE J. Quantum Electron.* *QE-17*(6):862–868 (1981).
32. A. H. Hartog, D. N. Payne, and A. J. Conduit, *Proc. 6th Eur. Conf. Opt. Commun.*, York, England, Sept. 1980, post-deadline paper.
33. R. Ulrich and S. C. Rashleigh, *Appl. Opt.* *19*(14):2453–2456 (1980).
34. S. Heckmann, E. Brinkmeyer, and J. Streckert, *Opt. Lett.* *6*(12):634–635 (1981).
35. P. Di Vita and U. Rossi, *Opt. Quantum Electron.* *11*:17–22 (1980).
36. I. P. Kaminow, *IEEE J. Quantum Electron.* *QE-17*(1):15–22 (1981).
37. A. Simon and R. Ulrich, *Appl. Phys. Lett.* *31*(8):517–520 (1977).
38. M. Monerie and P. Lamouler, *Electron. Lett.* *17*(7):252–253 (1981).
39. B. Y. Kim and S. S. Choi, *Electron. Lett.* *17*(5):193–194 (1981).
40. M. Nakazawa, T. Horiguchi, M. Tokuda, and N. Uchida, *Electron. Lett.* *17*(15):513–515 (1981).
41. I. P. Kaminow, *IEEE J. Quantum Electron.* *QE-17*(1):15–22 (1981).
42. S. C. Rashleigh, W. K. Burns, R. P. Moeller, and R. Ulrich, *Opt. Lett.* *7*(1):40–42 (1982).
43. S. C. Rashleigh and R. Ulrich, *Opt. Lett.* *3*(2):60–62 (1978).
44. K. Mochizuki, Y. Namihira, and H. Wakabayashi, *Electron. Lett.* *17*(4):153–154 (1981).
45. M. Monerie, P. Lamouler, and L. Jeunhomme, *Electron. Lett.* *16*(24):907–908 (1980).
46. P. Sansonetti, *Electron. Lett.* *18*(15):647–648 (1982).
47. J. Piasecki, *Electron. Lett.* *16*(13):498–500 (1980).
48. J. Piasecki, B. Colombeau, M. Vampouille, C. Froehly, and J. A. Arnaud, *Appl. Opt.* *19*:3749–3755 (1980).
49. W. D. Bomberger and J. J. Burke, *Electron. Lett.* *17*(14): 495–496 (1981).
50. Hent-Tai Shang, *Electron. Lett.* *17*(17):603–605 (1981).

51. L. G. Cohen and J. Stone, in *OFC'82 Digest of Technical Papers*, Optical Society of America, Washington, D.C., Apr. 1982, paper PD2.
52. L. G. Cohen and C. Lin, *Appl. Opt. 16*:3136–3139 (1977).
53. L. Jeunhomme and P. Lamouler, *Opt. Quantum Electron. 12*: 57–64 (1980).
54. A. Sugimura and K. Daikoku, *Rev. Sci. Instrum. 50*(3):343–346 (1979).
55. B. Costa, M. Puleo, and E. Vezzoni, *Electron. Lett. 19*(25–26): 1074–1076 (1983).
56. P. J. Vella, P. M. Garel-Jones, and R. S. Lowe, *Electron. Lett. 20*(4):167–168 (1984).
57. S. Tanaka and Y. Kitayama, *IEEE J. Lightwave Tech. LT-2*(4): 1040–1044 (1984).
58. T. Horiguchi, M. Tokuda, and Y. Negishi, *IEEE J. Lightwave Tech. LT-3*(1):51–54 (1985).
59. J. E. Thomson, *Optical Fiber Characteristics and Standards*, SPIE, Vol. 584, p. 33 (1985).
60. T. Miyashita, M. Horiguchi, and A. Kawana, *Electron. Lett. 13*(8):227–228 (1977).
61. F. Mengel and N. Gade, *Proc. ECOC'84*, Stuttgart, pp. 126–127 (1984).
62. J. J. Bernard and B. Visseaux, *Optical Fiber Characteristics and Standards*, SPIE, Vol. 584, p. 32 (1985).

6

Passive Components

6.1 INTRODUCTION

The development of single-mode fiber technology gives rise to two important categories of passive components. The first category comprises the components that are required to implement SMF-based systems, such as demountable connectors, permanent cable splices, T-couplers and fiber switches, laser-to-fiber couplers, and couplers between fibers and integrated optical components. In many cases, the technologies used for manufacturing these passive components are simply improved versions of technologies developed previously for multimode-fiber systems.

The second category comprises fiber-based components which make it possible to perform, in the guided optics regime and in a compact and rugged form, some functions usually performed in bulk optics or even with high-frequency electronic devices: polarizers, isolators, modulation frequency filters, spectral filters, and others.

The first category is discussed in Secs. 6.2 and 6.3, and Sec. 6.4 is devoted to fiber-based components.

6.2 FIBER-TO-FIBER COUPLERS

In this section we examine connectors, splices, X- and star-couplers, and switches. References 1 and 2 provide an excellent and detailed review of fiber connectors and splices.

6.2.1 Connectors

Because of the very small core diameter of single-mode fibers, SMF connectors were originally seen as bogeys by both the system designer and the connector manufacturer. However, most single-mode fiber applications require only one truly demountable SMF connector, or even none. For sensor as well as for telecommunication applications, it is possible in most cases to connect single-mode fibers to multimode fibers at the detector side (receiving end), which obviously greatly simplifies the tolerance requirements and mechanical accuracy for this connector. This requires that the SMF connector be able to accommodate a multimode fiber without modification of the coupling hardware. At the source side (transmitting end), however, it is necessary to use only single-mode fibers.

Positioning Accuracy. Theoretical formulas for predicting joint losses as a function of various imperfections were given in Sec. 3.3.4 and illustrated in Figs. 3.12 and 3.14–3.16. From this discussion we observe that we can neglect the influence of fiber tolerances in practice, as the practical fiber-to-fiber reproducibility of core diameter and index differences corresponds to losses smaller than 0.1 dB. On the other hand, Figs. 3.12 and 3.15 show that the attainment of a loss smaller than 1 dB requires that the fiber end-to-end separation be maintained below about 30 μm, the core-to-core lateral offset below 1 to 1.5 μm, and the axis-to-axis tilt angle below 1° (these figures are for an operating wavelength of 1300 nm and vary linearly with wavelength for the dimensions and remain constant for the tilt angle, for a fixed index difference and λ/λ_c). Let us recall from Sec. 3.4.3 that the intrinsic core–cladding eccentricity varies between 0.2 and 0.5 μm on average, depending on the fiber manufacturing technique. We describe below the most usual single-mode fiber connectors: The biconic connector initially developed by AT&T represented about 60% of the single-mode connectors in use as of 1987; the FC connector specified by NTT represents, together with its variations, FC/PC and D4, about 30% of the market; the OPTABALL connector introduced by RADIALL offers very specific characteristics which make it well suited for high-performance applications; the ST connector introduced by AT&T represents a small number of single-mode connectors, but may become widely used, especially if single-mode fibers are chosen for subscriber loop, loop feeder and high-speed LAN applications.

Biconic Connector. This connector, described in Refs. 3 and 4, was originally designed by AT&T for multimode fibers. The technology involves the mating of two ground conical ferrules in an adapter to achieve repeatable low losses of less than 1 dB (Fig. 6.1). As in the case of multimode fibers, the conical plug is transfer-molded

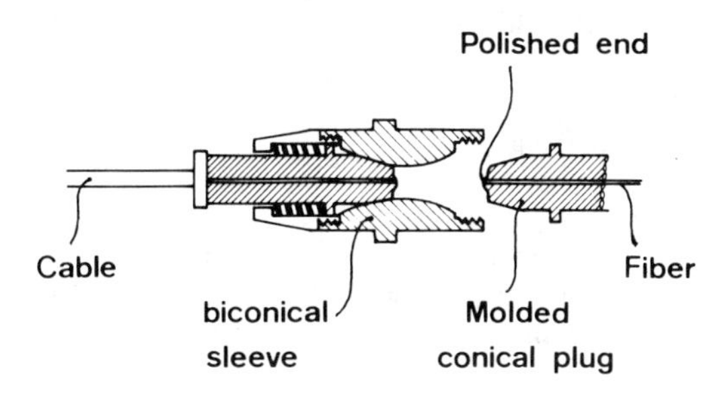

FIGURE 6.1 Schematic drawing of a biconical connector. [After A. H. Cherin and J. F. Dalgleish, *ITU Telecommunication J.* *48*(11): 657–665 (1981).]

around the single-mode fiber without stripping the fiber coating. Thereafter, the plug is precision-machined while monitoring the concentricity between the fiber core and the external conical surface, and finally the end face is polished to get a good optical surface and to adjust the plug length. An average eccentricity of 0.33 μm and an average tilt angle of 0.35° have been obtained for about 100 plugs [4]. The corresponding optical losses were 0.28 dB on average and 0.7 dB maximum. This type of connector is manufactured by several U.S. companies [2] and is widely used by telephone companies.

FC and FC/PC Connectors. This connector type is specified by NTT and has been designed directly for single-mode long-distance telecom applications. It is a field-installable unit based on the use of stainless steel ferrules with a ceramic interior [5]. The ferrules are precision-machined for the outer diameter, the inner diameter of the hole, and the concentricity between the hole and the outside surface.

The assembly steps are as follows:

1. Strip the coating from the fiber to the exact ferrule length.
2. Select a ferrule with a central hole diameter that exceeds the fiber outside diameter by less than 1 μm and insert the fiber into the hole (a complete set of ferrules with central hole diameters increasing by 1-μm steps is used).
3. Glue the fiber in the hole and polish the ferrule end.
4. Assemble the ferrule into its coupling nut to get a plug. Two plugs are then coupled by an adapter to complete the connection.

The high-accuracy capillary ferrules are manufactured according to the following specifications [5]:

Outside diameter of 2.499 mm ± 0.5 μm
Length of 10 mm
Capillary hole diameter with nominal values by 1-μm steps
Concentricity of capillary hole and outer surface better than 1 μm
Parallelism between cylinder axis and capillary axis better than 0.1°
External roundness better than 0.3 μm, external cylindriform better than 0.5 μm, and surface roughness better than 0.5 μm

After mounting in the plug, the goal for concentricity is 0.7 μm, and for tilt angle, 0.3° (average values). Once the fibers are connected, the two opposite ferrules are pushed against each other by springs located in the plugs, with a compressive force of 1200 g; the central adapter's holding force is 500 g.

Practical performances for 100 FC connector samples show average eccentricities (after mounting the fiber) of 0.73 μm, with 1.5 μm as a maximum, together with a tilt angle of less than 0.25°, both figures applying only to the fiber in the ferrule before plug assembly. For the optical characteristics, about 650 FC connectors yielded an average loss of 0.65 dB with a maximum of 1 dB (at 1300 nm [5]). The optical loss remains very stable against varying environmental conditions, showing less than a 0.2-dB loss variation in the −25°C to +80°C temperature range, and the connector has good aging properties (less than 0.2 dB variation after 500 mating–unmating cycles). Connectors of the FC type are manufactured by several Japanese companies and by some U.S. companies [2].

There are some variations around the original FC type. The NEC/D4 connector is very similar to the FC type in that it is based on a ferrule assembled in a coupling nut incorporating a spring, and on a collar-mating technique (Fig. 6.2). The main difference from the FC-type connector arises from a reportedly easier manufacturing technique yielding higher performances, called external trimming (Fig. 6.3). The connector ferrule's cylindrical surface is trimmed into a circle of the desired diameter, whose center is the same as that of the optical fiber core. This technique compensates for any eccentricity of the fiber core relative to the cladding, and for the clearance between the fiber and inner surface of the ferrule.

The FC/PC type of connector is a completely intermateable offspring of the FC-style connector. It features an all-ceramic ferrule instead of a stainless steel and ceramic one. Combined with a high-precision polishing process, this produces a convex spherical end face which ensures intimate physical contact between optical fiber end faces. This minimizes connection loss and, more importantly, reduces reflections to a low level of about −35 dB. This is

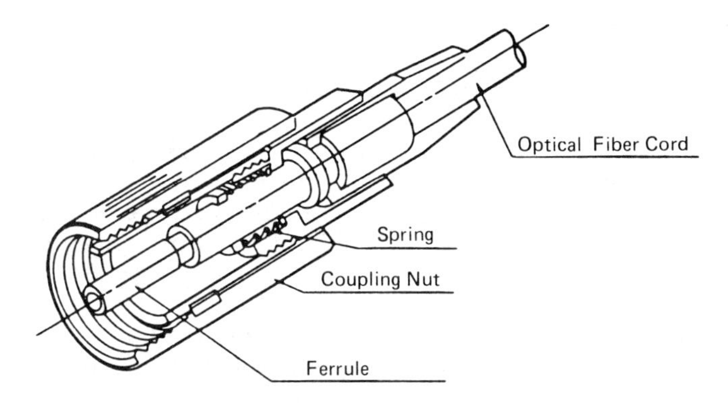

FIGURE 6.2 Schematic drawing of a D4 connector (courtesy of NEC Corporation, Yokohama, Japan).

necessary for high-performance systems using single-mode lasers, as the spectral purity of these lasers can be strongly affected by light reflections.

OPTABALL Connector. This connector, designed by RADIALL, uses a high-precision ferrule in a coupling nut. The two ferrules

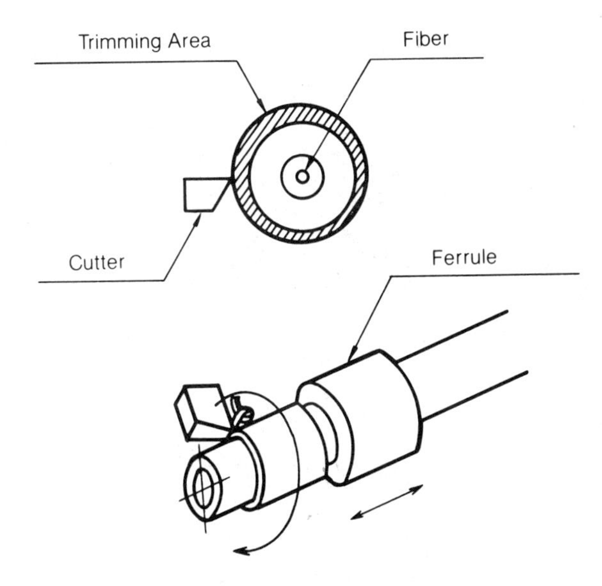

FIGURE 6.3 External trimming of the D4 ferrule for compensation of fiber core eccentricity (courtesy of NEC Corporation, Yokohama, Japan).

are centered inside a radial hole in a sphere, which is self-centered in the two cones of the plugs (Fig. 6.4). The insertion loss is less than 0.7 dB. The DF version of this connector type allows polishing of the fiber end faces with a tilt angle, yielding an ultralow residual reflection of −55 dB. There is also another version of the connector specifically developed for polarization-maintaining fiber systems. The plugs contain an angular orientation reference which provides an alignment of better than 3° of the fiber's birefringence axes.

ST Connector. The ST connector was originally designed by AT&T for loop feeder and LAN applications, with multimode fibers. It features a bayonet connection and keying to prevent variations in loss. The fiber is inserted into a ceramic capillary which is spring loaded and requires no polishing and no gaging. This connector is relatively inexpensive and provides acceptable losses, around 1 dB, with single-mode fibers. It is manufactured by various companies around the world.

Influence of Connectors. In addition to optical loss, connectors are responsible for other system performance degradations. Because of the small separation between the fiber end faces, it has been

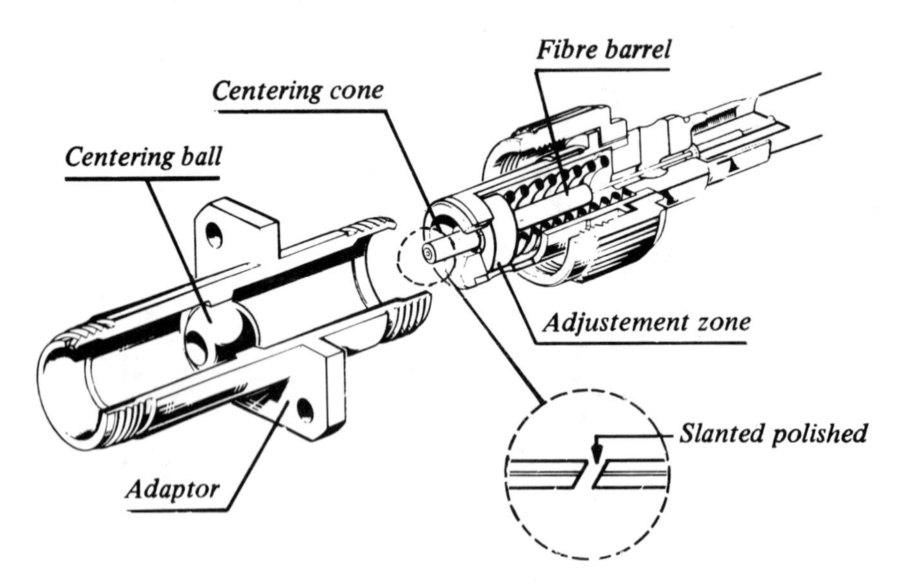

FIGURE 6.4 Schematic drawing of an OPTABALL/DF connector (courtesy of RADIALL, Rosny sous bois, France).

shown that connectors without index matching liquids behave like Fabry–Perot interferometers and induce harmonic distortion of analog signals [6,7]. This can be avoided by the use of index matching liquids (– 30 dB reflection), by convex polishing as in the FC/PC connector (– 35 dB reflection) or by polishing the fiber–ferrule end faces with a (small) tilt angle as in the OPTABALL/DF connector (– 55 dB reflection). This also allows a reduction in the backward-reflected power, which can induce some intensity and/or spectral fluctuations in the laser source.

6.2.2 Splices

As for multimode fibers, the splice loss, the assembly time, and the cost are parameters of primary importance for SMF-based systems, but the emergence of undersea systems has also focused attention on the mechanical strength and the lifetime of splices.

As in the case of demountable connectors, loss is governed primarily by positioning errors, as the influence of fiber parameter mismatch is negligible, as we saw in Chap. 3. Again, most of the methods used for single-mode fibers are sophisticated versions of methods developed for multimode fibers.

The question of the time actually needed to perform a splice and the total cost involved has been very controversial. Some say fusion splicing is the most economical and time-saving method; others say mechanical splicing is best. In fact, the various methods are difficult to compare, as their respective performances may be strongly influenced by environmental parameters. A detailed review of the various aspects of fiber splicing can be found in Ref. 1.

Fusion Splices. The fusion splice method generally uses an electric arc discharge (or sometimes a gas flame or CO_2 laser) for heating and melting the silica fibers at the splicing point (Fig. 6.5). This process offers two important advantages which make it the choice for undersea single-mode cables: splices can be recoated to the size of the original fiber and are then environmentally stable. In terrestrial applications these advantages are not so critical and are counterbalanced by the cost of the equipment, the average time needed to perform a splice, and the sensitivity of the result to the environment provided during splicing. Practically, this means that a splicing van should be used to obtain an average loss of 0.2 dB with an acceptable yield (10 to 15 min per splice).

One problem here is associated with the perpendicularity of the fiber end faces with respect to the fiber axis. As shown in Chap. 3, a fiber axis tilt angle of less than 1° is necessary for low losses, which necessitates cleaving the fiber end faces with a deviation from perpendicularity of well below 1°. In addition to that, the cleaved end faces should exhibit no protruding part, as this would

FIGURE 6.5 Arc fusion splicing of single-mode fibers. (Courtesy of L. d'Auria, Thomson-CSF, Orsay, France.)

introduce core offsets. To attain these performances, an improved fiber-cleaving machine has been fabricated, in which the fiber is held straight under prefixed tension but without torsion or twist, and a diamond tip "saws" the cladding until the fiber breaks [8]. For the arc fusion itself, the steps are the same as those for multimode fibers, but with somewhat different optimum parameters [9,10]. For 125-μm-outside-diameter high-silica-content fibers, the optimum electrode gap is 0.7 mm, the prefusion time is 0.2 s, the total electric discharge duration is 1 s under an intensity of 18 mA, and the fibers should be pushed one against the other with a stroke of 20 μm (the stuffing length of one fiber end face after the two fibers make contact). Figure 6.6 shows a histogram of arc fusion splicing losses over 20 splices, where the shaded area corresponds to nonperpendicular end faces.

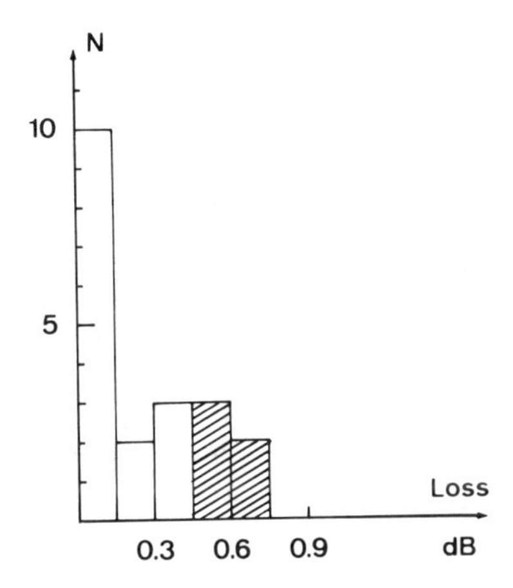

FIGURE 6.6 Histogram of arc fusion splice losses for 20 splices. The shaded area corresponds to nonperpendicular fiber end faces. (From Ref. 11.)

Finally, it has been suggested that applying tension to the fibers during fusion will result in a biconical taper, which could decrease the effect of lateral offsets (and increase that of tilts) because of the large increase in mode field radius [12]. In Fig. 1.10 we observe that starting from a fiber operated at its nominal cutoff wavelength and stretching the fiber to decrease its core radius by a factor of 2 at the splice (thus decreasing its cutoff wavelength by a factor of 2) will result in a 1.5-fold increase in the mode field radius. This will increase the lateral offset tolerance by the same factor.

Mechanical Strength. Single-mode fibers are very attractive for applications in undersea telecommunication systems, and this use requires very high mechanical performances for the fiber line, including splices, because of cable laying and recovery conditions and lifetime requirements. From this point of view it seems that CO_2-laser fusion splices are very attractive, as will be discussed below. Let us first point out that CO_2 lasers could be used in the field, as there exist compact air-cooled lasers in a rugged field design with the appropriate power output (less than 5 W [13]). It is well known that in fusion splices, fiber breakage under load occurs most often near the splicing point itself, and it has been shown that when the fusion heating time is increased from 3 to 20 s, the mechanical strength is reduced by 50% [14]. On the other hand, it has also been shown that the critical zone where most breaks originate corresponds to the zone that is heated to around 700 to 1500°C during fusion. Where the temperature is higher, all cracks are closed because of revitrification, and where it is lower, the crack propagation is not significantly stimulated. With CO_2-laser heating, the heating time is only 1/8 to 1/4 s, compared to a few seconds with arc fusion, and at the same time the directly heated zone is much more concentrated. This explains why CO_2-laser fusion splices yield better mechanical performance than electric arc fusion splices, as has been observed experimentally [15].

Obtaining high-strength splices requires, in addition to the proper choice of heating source, that the fiber coating be chemically etched and not mechanically stripped, and that the section where the fiber is unprotected be free of physical contact [15]. Also, after splicing, the use of hydrofluoric acid etching of the outside surface of the fiber (for approximately 3 min, to remove about 0.6 μm of the glass where cracks are present), and the molding of a 0.4-mm-diameter silicone buffer for protecting the splice, make it possible to get very high strengths, such as 2.6% elongation [16].

Dynamic Alignment. Because of the very small mode field radius involved and the high positioning accuracy required, it is common practice to actively align the two fiber cores with respect to each other. A first method consists in transmitting a stable light level from one far end of the fiber link and measuring the received light level at the other far end. The equipment involved is standard (a stable light source and a power meter) but some means is required for transmitting the information back to the splicing operator. This can be done with a power meter able to retransmit its readings onto one fiber or a copper wire if one is available in the cable. The splice operator can then adjust the micrometers holding the fiber so as to maximize the received light level. The advantage of this method is that it provides a real measure of the actual splice loss by comparing the light level received after the splice has been completed to the reference level obtained before. The drawback is that it requires three people to operate the various instruments.

A second method consists in locally detecting the light level, a few centimeters after the splice point along the receiving fiber. This greatly simplifies the process, as the splicing operator can directly read the power meter without the need for a remote transmission line. By strongly bending the fiber to a radius of a few millimeters, the fundamental mode experiences some leakage losses. The leaking light is detected with a large-area photodetector. Obviously, a very good cladding mode stripper should be used between the splice point and the local detection device. The sensitivity of the device can be strongly improved by the method described in Ref. 17 and illustrated in Fig. 6.7. The receiving fiber is inserted into a slotted glass tube, and the light launched in the unguided modes by the incoming fiber when the two cores are not perfectly centered is collected by the slotted glass tube and the photodetector. Figure 6.7b shows that the detected signal is very sensitive to the fiber offset, thus permitting accurate alignment. Another advantage is that the fiber is not bent sharply, and this helps to maintain its mechanical strength. On the other hand it becomes almost impossible to make a true measurement of the splice loss.

Finally, a further simplification can be obtained which reduces the personnel required to only the splice operator. By bringing a laser diode or a high brightness LED in contact with a sharp bend in the incoming fiber placed a few centimeters before the splicing point, one is able to launch a small amount of light in the fiber core. This local injection device allows the splice operator to operate autonomously. Many splicing machines are equiped with a local injection and detection (LID) system.

Mechanical Splices. Mechanical methods are generally based on a mechanical alignment of the fibers into some sort of V-groove as illustrated in Fig. 6.8a. The fibers may be retained in place either by mechanical clamping due to the use of somewhat elastic materials or by epoxy bonding. A variety of products are available [2].

A first typical example involves an elastomeric cylinder with a V-groove etched on the outer surface. After the fibers have been pushed in place inside the V-groove, the elastomeric cylinder is positively strained in radial compression by an external cylindrical sleeve, thus providing retention of fibers. In another product, the V-groove is etched on the outer surface of a solid cylinder, and the elasticity of the external sleeve is used as the retention mechanism.

Another widely used design is the rotary mechanical splice [17] illustrated in Fig. 6.8b,e. The fibers to be spliced are first cemented with a UV-curable adhesive in precision glass capillary tubes designed to make use of the small eccentricity that is present. The capillaries are then inserted in a three-glass-rod alignment sleeve which has built-in offset so that as each ferrule is rotated within the sleeve, the two circular paths of the center of each core cross each other. A local scattering detector is used for optimizing the alignment of the cores. An average loss of 0.03 dB has been obtained (with index-matching gel) over more than 4000 splices performed in the field.

6.2.3 Couplers

The X-coupler is a two-fiber four-port coupler, which is the guided-wave equivalent of the bulk-optics beam splitter. All these couplers

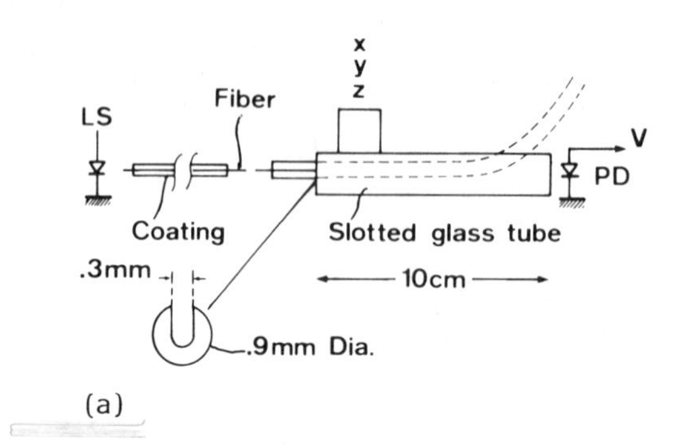

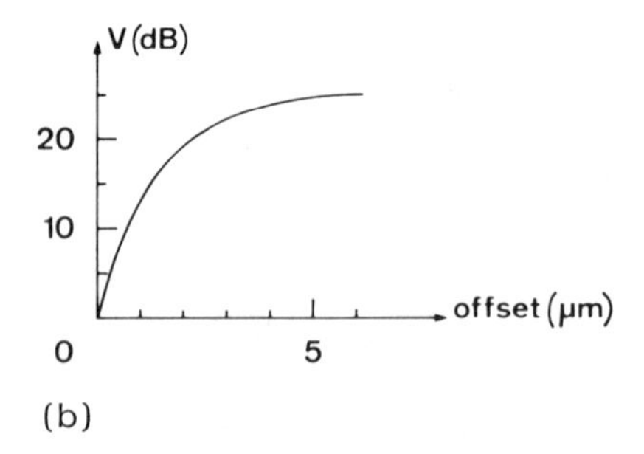

FIGURE 6.7 Apparatus for fiber alignment before splicing (a), and detected signal as a function of the offset (b). (From C. M. Miller, in *Optical Fiber Communication 1982 Digest of Technical Papers*, Optical Society of America, Washington, D.C., Apr. 1982, paper THAA2.)

are based on the fact that when two fiber cores are brought sufficiently close to each other, the two LP_{01} modes become coupled through their evanescent fields. This results in a reciprocal power transfer from one fiber to the other, with a power transfer ratio depending on the core spacing and interaction length. A typical example of such couplers, obtained by partial abrasion of the cladding of both fibers, is shown in Fig. 6.9.

Principles. One general method of obtaining such couplers consists of polishing the cladding on one side of the core of both fibers, and then bringing the cores in close proximity with an index matching liquid interface. A technique for polishing the cladding away consists in fixing the fibers inside slots grooved in quartz blocks, and polishing the whole block in order to remove

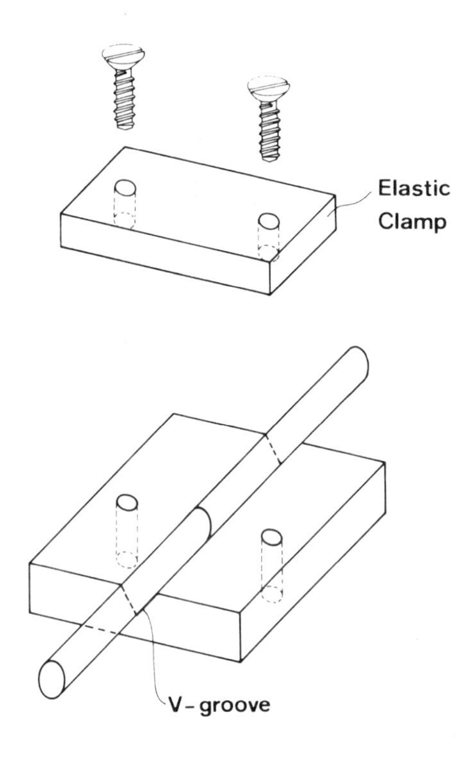

(a)

FIGURE 6.8 Schematic drawing of a basic mechanical splice.

nearly all the cladding on one side of the cores. The two quartz blocks are then brought in contact, with careful superposition of the cores [18]. A slightly different technique, which consists of fixing the fibers on semiconvex lenses prior to polishing, allows easy monitoring of the polishing depth by launching light into the fiber during polishing, or by measuring the dimensions of the axes of the elliptical polished surface [19]. A more recent technique allows direct monitoring of the remaining cladding thickness and yields a polished fiber of very good mechanical strength, which allows it to be manipulated without the need for support [20]. The polishing apparatus is depicted in Fig. 6.10, where the fiber is pressed between two rollers, and one of them covered with

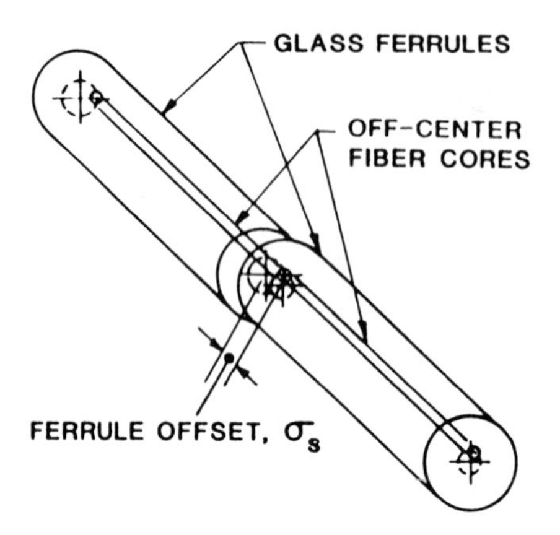

(b)

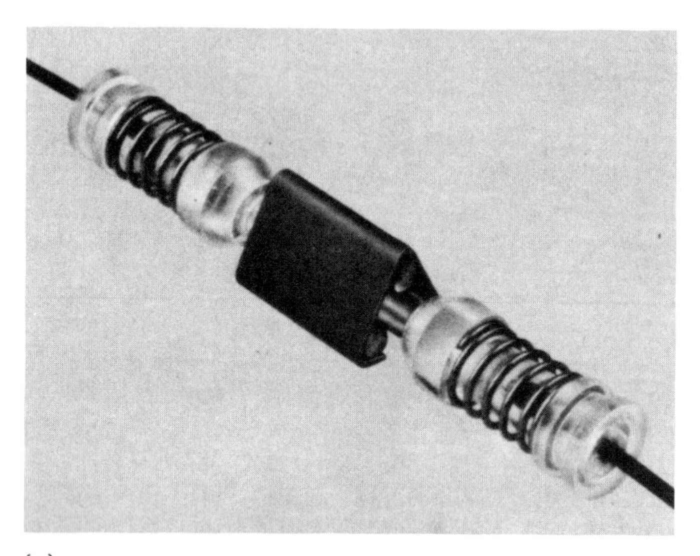

(c)

FIGURE 6.8 (Continued)

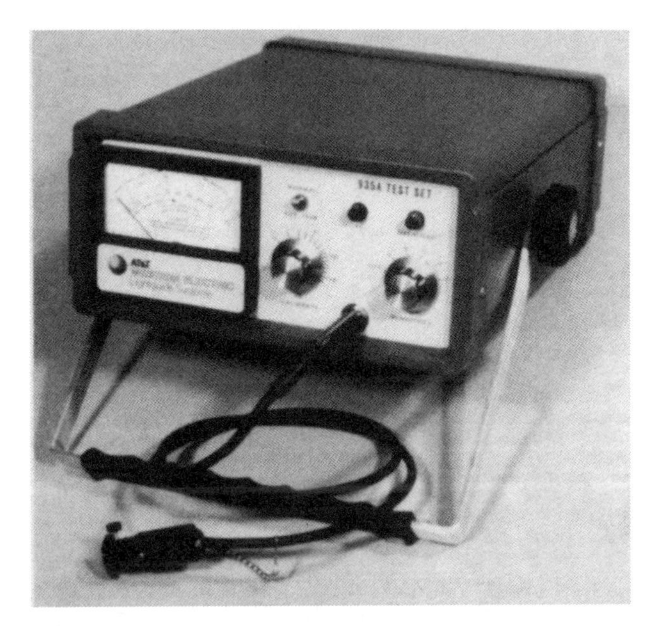

(d)

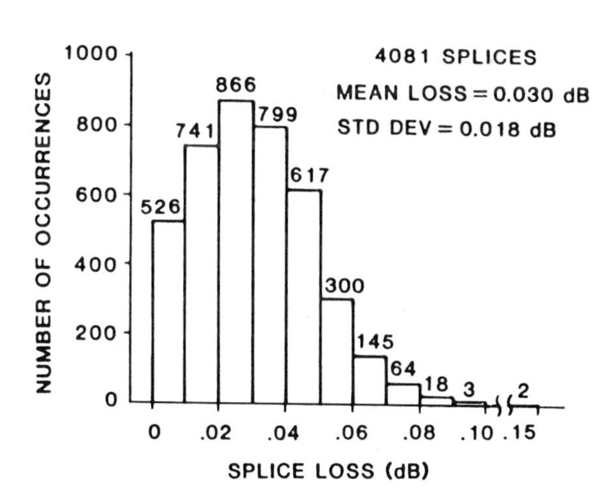

(e)

FIGURE 6.8 (Continued)

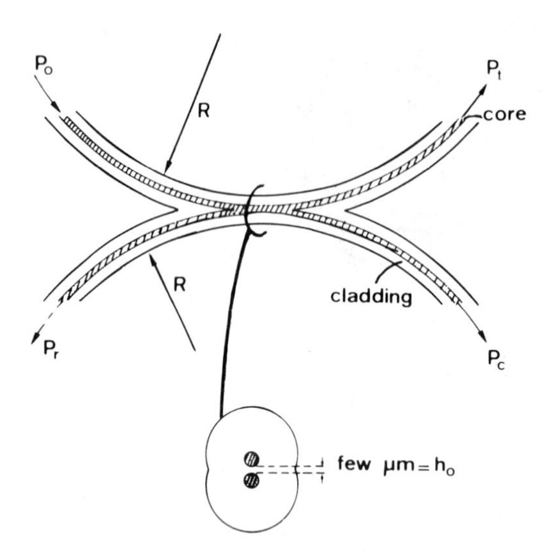

FIGURE 6.9 Polished fibers X-coupler. P_0, input power; P_t, transmitted power; P_c, coupled power; P_r, (weak) backward-coupled power.

polishing paste, and rotated. The basic reason the initial fiber strength (per unit cross-sectional area) is preserved seems to be that all the polishing-induced cracks are parallel to the fiber's axis.

A detailed theoretical and experimental analysis of this type of coupler can be found in Ref. 21, where it is shown that the

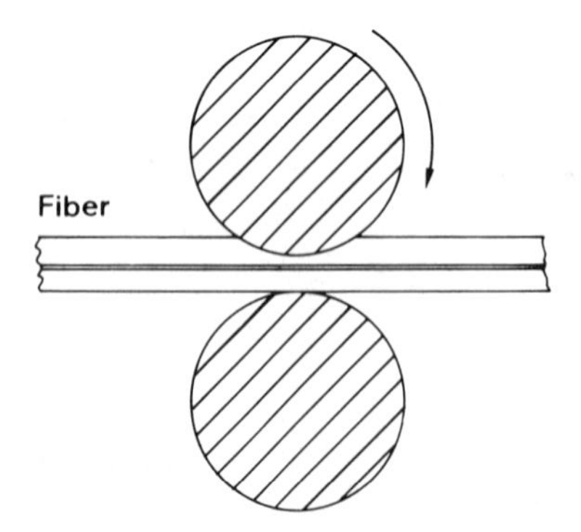

FIGURE 6.10 Cladding abrasion apparatus with polishing rollers.

transmitted power P_t and the coupled power P_c (see Fig. 6.9) are given by

$$P_t = P_0 \cos^2 C_0 L_c$$
$$P_c = P_0 \sin^2 C_0 L_c \tag{6.1}$$

where C_0 represents the coupling strength and L_c the effective interaction length, these parameters being separately controlled by the minimum fiber core spacing h_0 and the radius of curvature R, respectively.

$$C_0 = \frac{\lambda}{2\pi n_1} \frac{u^2}{a^2 V^2} \frac{K_0(vh_0/a)}{K_1^2(v)} \tag{6.2}$$

$$L_c = \left(\frac{\pi R a}{v}\right)^{\frac{1}{2}} \tag{6.3}$$

where u, v, and V are the guide parameters defined in Chap. 1, a is the core radius, and n_1 is the core refractive index. Useful approximations are given in Chap. 1 for most of these parameters and functions. For a given core spacing, the coupling strength increases with the light wavelength (this is due to the fact that the field extends deeper and deeper into the cladding), and typical values are in the range 0.1 to 2 mm^{-1} for core spacings in the range 4 to 15 μm. It also shows a strong decrease when the fiber core spacing increases, as expected from Eq. (6.2). Similarly, the interaction length increases with the radius of curvature and with the light wavelength [21], with typical values in the range of a few millimeters. It is thus obvious that the core spacing adjustment permits precise control of the power transfer ratio, and also that these couplers are wavelength selective. The power transfer ratio can be controlled from nearly 0% (actually, less than 0.01% has been obtained) up to nearly 100% (less than a 0.5 dB loss). Also notable is the fact that these couplers appeared insensitive to the state of polarization of the launched light (less than 0.5% coupling efficiency variation as a function of the SOP, for 100% and 50% couplers) and exhibited good directivity (less than −70 dB for P_r/P_0). However, these couplers are found to be temperature dependent, even after mechanical offsets due to differential expansion have been eliminated, because of the temperature dependence of the liquid's refractive index [21]. For a given coupler, the coupling efficiency drops from 100% to about 10% between 15 and 40°C.

Another convenient method for manufacturing X-couplers in a rugged form without fiber material removal has been described in Ref. 22. The experimental manufacturing bench, which also allows the fabrication of star couplers, is shown in Fig. 6.11a. The fabrication process makes use of the fused biconical taper technology, in which two (or more) fibers twisted around each other are flame-heated and fused while two stages of the fusion station are moved apart. During the operation, the power output values from the output ports are monitored, and the process can be stopped at any desired coupling ratio. Figure 6.11b shows an example of the measured detected powers as a function of taper length. The basic phenomenon which permits coupling in this process is that the biconical taper brings the two fiber cores close to each other, but simultaneously decreases the core diameters (and thus the cutoff wavelength) considerably, which increases the mode field radius (see Fig. 1.10). Both effects contribute to strong coupling through the evanescent tails of the guided modes.

Practical Couplers. This fused biconical taper technology is the most widely used for manufacturing standard single-mode fiber couplers. It provides low-cost, high-performance couplers in large quantities. Several manufacturers offer couplers with a specified excess loss as low as 0.1 dB and good temperature stability.

The polishing process is now almost exclusively used for manufacturing special single-mode fiber couplers: polarization-maintaining couplers or polarization-splitting couplers. Polarization-maintaining couplers are made with polarization-maintaining fibers with the aim of preserving the state of polarization of the light launched into either one of the two possible input fiber ends. In addition to what is required for a standard coupler, their manufacture requires good orientation of each fiber's birefringence axes. The orientation of the birefringence axes controls the amount of unwanted parasitic light at the output of the coupler, which is not polarized in the same direction, as the original light launched with a linear polarization oriented parallel with one of the fiber's birefringence axes. Polarization-splitting couplers split the light between the two output fiber ends, depending on the direction of light polarization. If linearly polarized light is launched in either one of the two input fiber ends with a polarization direction parallel to one of the fiber's birefringence axes, all the light (except for the excess loss) is either transmitted or coupled (and vice versa for the other polarization direction). The manufacture of these couplers again require a very good control of the fiber's axes orientation [23]. Such couplers are manufactured with an excess loss of about 0.2 dB and a typical polarization selectivity of −20 dB.

Although such couplers have been produced by the fusion technique [24–27], the polishing process yields more reproducible results

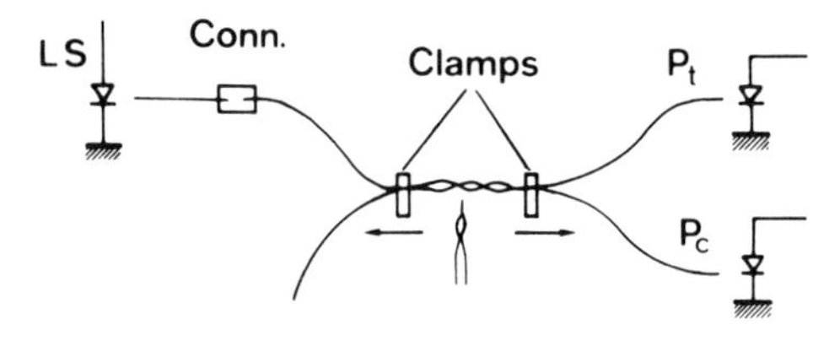

(a)

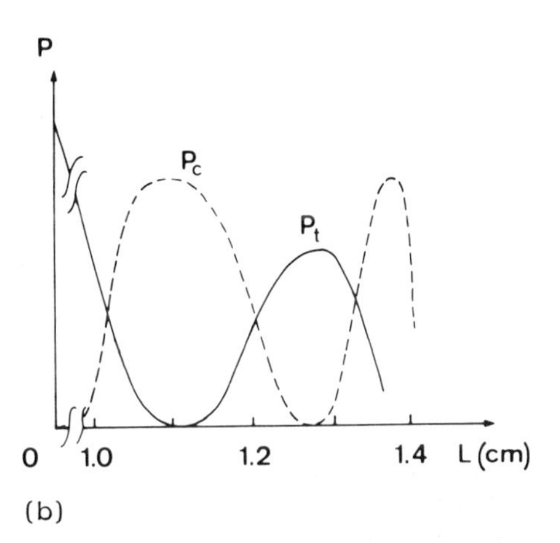

(b)

FIGURE 6.11 Single-mode-fiber fused biconical taper coupler. (a) Fabrication apparatus for an X-coupler (two output ports). (b) Example of detected power outputs as a function of taper length L. (From M. H. Slonecker, in *Optical Fiber Communication 1982 Digest of Technical Papers*, Optical Society of America, Washington, D.C., Apr. 1982, paper WBB7.)

because of better control of the relative orientation of the fiber's birefringence axes. Also, polished couplers can be manufactured with a very small core separation requiring only a small coupling length. This is favorable to a large wavelength tolerance, allowing these couplers to be used in a broader wavelength range than the fused couplers. Finally, a very high polishing quality allows direct optical contact between the two polished surfaces, thereby reducing the temperature sensitivity induced by the index-matching gel initially used.

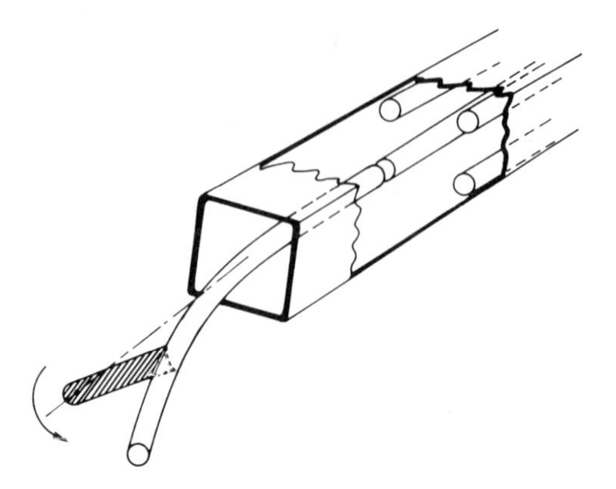

FIGURE 6.12 A 1 × 4 mechanical single-mode-fiber switch. (After Ref. 29.)

6.2.4 Switches

Single-mode fiber switches are very important components, especially in undersea systems, because of the long MTBF required for the repeaters and the comparatively short lifetime of laser diodes, especially at 1300 nm. It is thus currently envisaged to implement several lasers in the repeaters, together with a SMF switch allowing the selection of a new laser when one is running out-of-specifications.

A first, 1 × 4 switch initially developed for multimode fibers has been optimized for single-mode fibers [28]. Four fibers are fixed in the corners of a square glass tube with all end faces coplanar; a movable fiber enters the tube from the opposite side and has its end face separated from the others by a few micrometers (Fig. 6.12). The tube is filled with index matching liquid, and by using a fixed deviation of the movable fiber's axis outside the tube and a four-step rotation, the movable fiber is forced into any one of the four corners. The typical loss is about 0.5 dB with a repeatability of 0.02 dB. A second, 1 × 2 switch makes use of two fibers fixed in superposed V-groove arrays, sandwiched between two other V-groove arrays, with one fiber fixed in a V-groove array moving up and down between the two sandwiching arrays. Accurate alignment is provided by the sandwiching arrays, and cascading such elements makes it possible to construct a 1 × 4 switch with a 0.8 dB loss [29].

6.3 GUIDE-TO-FIBER COUPLERS

We examine here the techniques used for coupling single-mode fibers to single transverse-mode semiconductor lasers and to integrated optical elements.

6.3.1 Laser-to-Fiber Couplers

In addition to the problem of accurate positioning encountered in fiber-to-fiber couplers, we must adapt the laser's mode geometry to that of the LP_{01} fiber's mode. For some lasers using buried heterostructures, this reduces to a problem of isotropic magnification, as the laser's mode has a circular symmetry like that of the fiber, but for classical laser structures the problem is complicated by the asymmetry of the laser's mode, which has almost elliptical near-field and far-field patterns.

There are two broad classes of coupling techniques: those using integrated microlenses on the fiber end and those using external lenses. After discussing these two techniques, we present a practical laser diode module developed for a field trial.

Integrated Lenses. One technique consists of depositing glass on the fiber end face, taken from a silica rod 30 to 50 μm in diameter facing the fiber end, using electric arc heating [30]. Continuation of the arc heating melts the deposited glass into a hemispherical lens whose radius depends on the heating time (Fig. 6.13a). With a buried-heterostructure (BH) laser and an 8.5-μm-radius lens, a coupling efficiency of −3 dB has been obtained after optimization through micromanipulating; the tolerances for less than 1 dB degradation are 1.5 μm in the transverse direction and 8 μm in the axial direction.

A second technique consists of polishing the fiber end into a quadrangular pyramidal shape, and then obtaining hemiellipsoids through electric arc heating ([31]; Fig. 6.13b). With a classical laser, a coupling efficiency of −3 dB has been obtained (after optimization), with tolerances of 1 μm (transverse) and 7 μm (axial) for less than 1 dB degradation.

A third technique consists of hydrofluoric acid etching of the fiber end, which creates a lens-like structure because of the differential etching mechanism due to dopant differences [32]. The lens is obviously self-centered and its optical properties depend on the etching time. With a classical laser, a coupling efficiency of −4.5 dB has been reported (after optimization). Such a lens is shown in Fig. 6.14.

Another method uses the apparatus depicted in Fig. 6.10, where the two rollers should rotate. This makes it possible to transform the external surface of the fiber into a rectangular shape, leading to a screwdriver-like aspect for the fiber tail, as shown in Fig. 6.15 [20]. Using electric arc or CO_2-laser heating then permits the creation of a cylindrical lens with two different curvatures in both transverse directions, making this device suitable for coupling with a classical laser.

External Lenses. A typical solution of this kind is reported in Ref. 33 for use with a BH laser. A ruby ball lens (radius 0.5 mm,

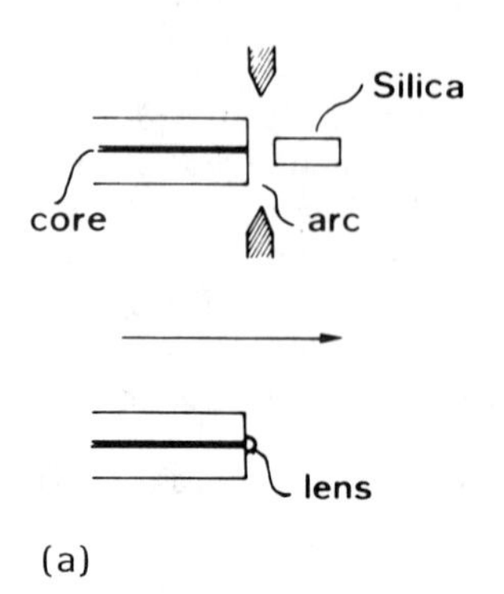

(a)

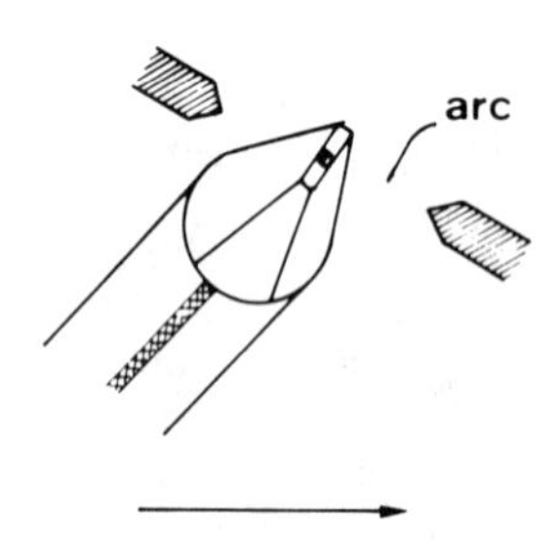

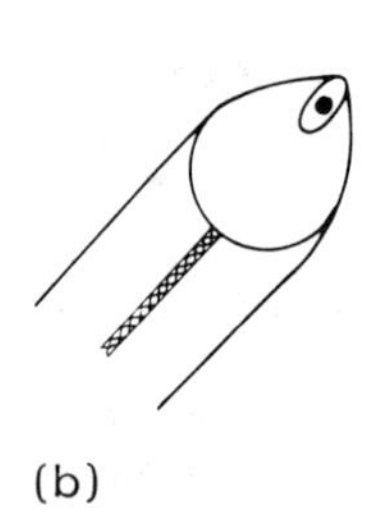

(b)

FIGURE 6.13 Integrated microlenses on fiber end faces for laser-to-fiber coupling. [(a) After Ref. 30; (b) after Ref. 31.]

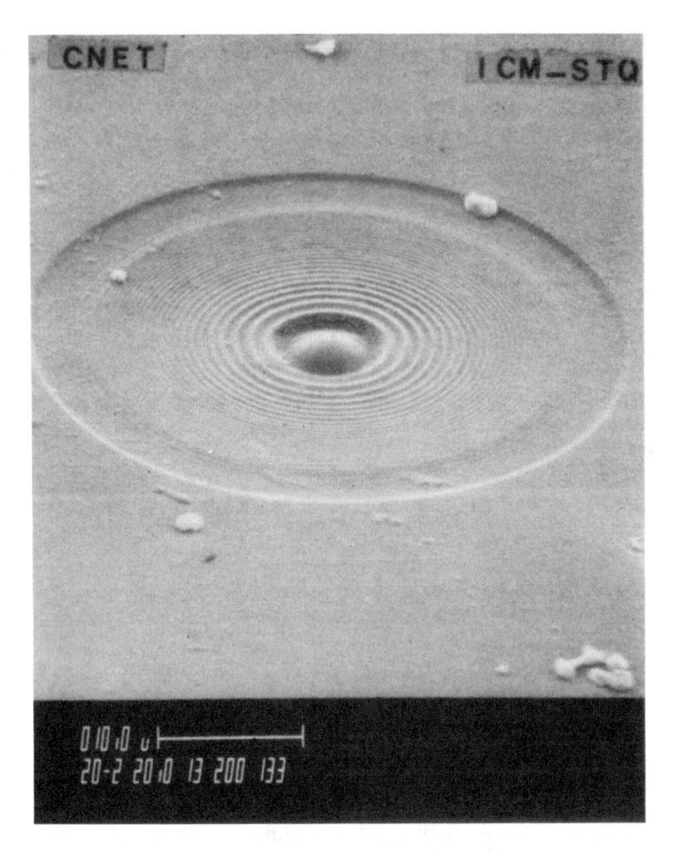

FIGURE 6.14 Chemically etched single-mode-fiber end face for creating a self-centered lens-like structure.

focal length 0.58 mm) is placed in front of the laser, followed by a GRIN rod lens (quarter pitch, focal length 1.8 mm) in a confocal arrangement. The coupling efficiency is -3.5 dB and the lens positioning tolerances for less than 1 dB degradation are 80 μm (transverse) and 100 μm (axial) for the ruby lens, and two to three times more for the GRIN lens. For classical lasers with a strong astigmatism, it is necessary to use a cylindrical lens in front of the laser. Calculations carried out for a 0.58-mm-thick half-cylinder of silica followed by two classical lenses indicate a loss of -3 dB (without Fresnel reflections) [34].

One advantage of the external lenses is that they make it possible to use an isolator between the laser and the fiber, which may be useful as it has been observed that fiber-induced optical

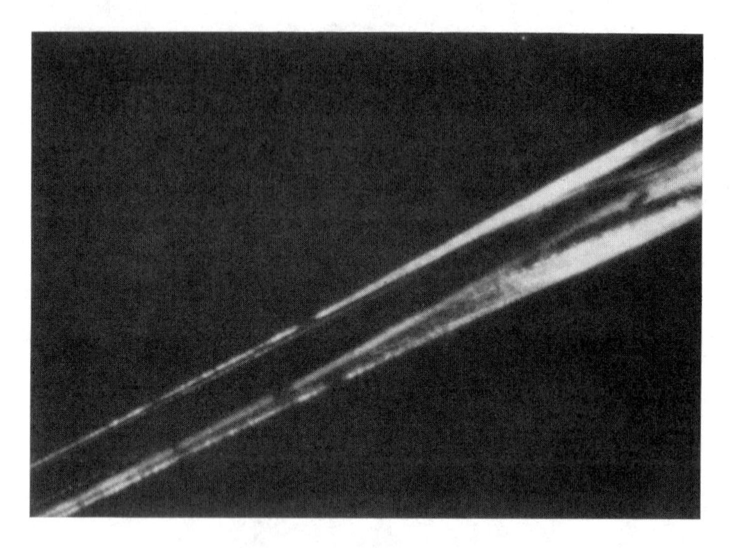

FIGURE 6.15 Screwdriver-like fiber tail obtained with the apparatus of Fig. 6.10, where the two rollers are rotating.

feedback into the laser may degrade its stability. A schematic description of a possible arrangement is shown in Fig. 6.16. The laser beam is collimated by the first lens with focal length f_1 and passes through a polarizer (aligned with the polarization direction of the laser beam), a nonreciprocal Faraday rotator (YIG crystal, transparent above 1100 nm, and permanent magnets) which rotates the polarization direction by 45°, an analyzer which is aligned with the emerging polarization direction, and a lens of focal length f_2, which focuses the beam onto the fiber end face with a magnification of f_2/f_1. Any light coming back from the fiber line is blocked by the polarizer, whatever its initial state of polarization, because of the nonreciprocity of the Faraday rotator. As will be discussed below, all the required individual components have been realized with all-fiber structures, and this function could thus also be realized with integrated lenses coupling. A prototype of a bulk optical laser–isolator–fiber coupler has been reported, with a total coupling loss of −5.2 dB at 1500 nm [35].

Experimental Laser Diode Module. A practical laser–diode–single-mode fiber module has been developed for a field trial of a long-haul system operating at 1300 nm [5]. First attempts to use integrated microlenses or very small external microlenses led to rather poor performances. The average coupling loss of −6 dB varied by

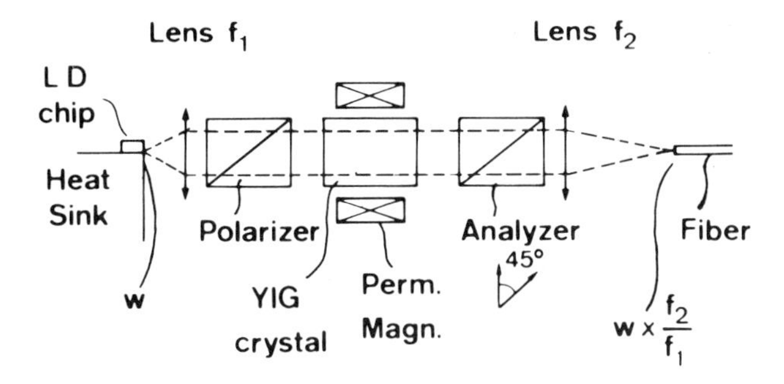

FIGURE 6.16 Schematic arrangement of a laser-to-fiber coupler, including an isolator for minimizing the laser instabilities induced by light reflected back from the fiber line.

more than 3 dB (and up to 5 dB) with temperature, primarily because of the very small positioning tolerances available with these microlenses. On the other hand, a module based on two confocal lenses, a YAG ball lens 0.8 mm in diameter and a quarter-pitch GRIN rod lens 1.8 mm in diameter, yielded average coupling efficiencies of −5.5 dB, with about a 1 dB variation with temperature [5].

6.3.2 Integrated Optics-to-Fiber Couplers

At first this problem seemed quite similar to that of laser-to-fiber coupling, and some solutions have been used for both (e.g., Ref. 32). However, more recent approaches seem to obtain very good results by simply butt-jointing the integrated optical chip with the fibers, after careful attention to the mode profile adaptation. Good performances have been obtained by adjusting the titanium in-diffusion process parameters in a Ti:$LiNbO_3$ waveguide, in order to match the modal fields of both the waveguide and the fiber. A fiber—waveguide—fiber insertion loss of 1 dB has been obtained [36]. Finally, a complete guided-wave Ti:$LiNbO_3$ amplitude modulator with two single-mode fiber pigtails has been reported [37]. At a wavelength of 1320 nm it provides a total fiber—modulator—fiber insertion loss of 1.5 dB, a 9 dB extinction ratio for the TM mode (which seems to indicate that linear polarization-maintaining fibers are required here), and a modulation bandwidth up to 1 GHz [37]. This modulator is shown in Fig. 6.17, where it is seen that besides the optimization of the waveguide manufacturing conditions, the optimization of the waveguide geometry also contributes to a decrease of insertion losses (in the on-state, no bend is encountered by the wave).

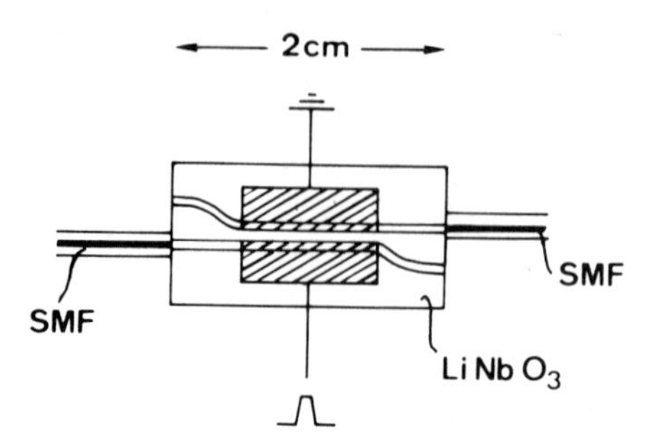

FIGURE 6.17 Low-loss fiber-coupled waveguide modulator. (After Ref. 37.)

6.4 FIBER DEVICES

As noted earlier, SMFs can be used to achieve devices with functional characteristics similar to some bulk-optic or electronic devices, and this makes it possible to perform directly on the guided wave some processes that would normally require extracting the light from the fiber.

6.4.1 Polarizers and Polarization Transformers

Polarizers. An in-line SMF polarizer has been described in Ref. 38, for which after the fiber's cladding has been laterally polished down to the core (with only a small remaining cladding buffer layer), a metallic film is deposited on the polished surface. In such a structure, the TM_0 mode has no cutoff, whereas the TE_0 mode has a low-frequency cutoff (TE_0 has its electric field parallel to the metallic film). Instead of using this method, which leads to very high insertion losses, it seems better to use the differential attenuation between both modes below cutoff (the TE_0 mode has the lowest loss). A −14 dB extinction ratio was obtained but at the expense of an 11 dB insertion loss for the transmitted polarization (TE).

Another possibility consists of polishing the fiber's cladding and bringing a birefringent crystal in contact or close to the core. If the core refractive index is intermediate between the ordinary and extraordinary indices of the crystal, one polarization is no longer guided and is thus lost, while the other (orthogonal) polarization is transmitted with small losses.

Polarization Transformers. The possibility of performing devices able to transform any input state of polarization into a given output

SOP is based on externally induced birefringences in the fiber, mainly through external forces or bends. The use of an electromagnetic relay, where the fiber is pressed between the armatures and the pole, makes it possible to induce a 360° phase shift between both linear eigenstates of polarization in 1 ms [39]. Combining three such relays with orientations at 45° angles to each other, and independently controlling the current in their coils, allows us to build a device equivalent to a Babinet–Soleil compensator.

The same function can be achieved by replacing the relay-induced birefringences by bending-induced birefringence. Here the induced linear birefringence is fixed in amplitude but its direction is varied. This is achieved by winding two independent fiber coils whose axes are perpendicular to the fiber's axis, and allowing them to rotate independently around the fiber line axis (Fig. 6.18). The coils should each induce a 90° phase shift. This has been achieved with an 80-μm-outside-diameter SMF wound with a radius of 8.5 mm (with one turn for the coils) operated at 633 nm wavelength [40].

6.4.2 Circulators and Isolators

The possibility of achieving all-fiber SMF optical isolators and circulators is based on the fiber polarizers described above and on externally induced nonreciprocal circular birefringence through an axial magnetic field (see Chap. 2). All the necessary functions for an isolator—a linear polarizer, a 45° nonreciprocal rotator, and an analyzer rotated by 45° with respect to the first polarizer—are illustrated (in bulk optics) in Fig. 6.16. A circulator would be obtained by replacing the input polarizer by a polarization beam splitter.

A straightforward all-fiber isolator would thus use two polarizers as described in Sec. 6.4.1, with an intermediate fiber length L with no linear birefringence, and submitted to a longitudinal magnetic field H such that (see Chap. 2)

$$\phi = V_f HL = \frac{\pi}{4} \tag{6.4}$$

where V_f is the Verdet constant of the fiber core material. For undoped silica, $V_f \simeq 3.03 \times 10^{-6}$ rad/G × cm (at 633 nm); thus a field of 10 kG should be used for a fiber length of about 25 cm. This obviously is impractical, and this solution can be used only if fibers with much higher Verdet constants (possibly with some special dopings) can be obtained, as stronger magnetic fields are difficult to obtain practically. Note finally that if some residual linear birefringence $\delta\beta$ is present in the fiber (due to bends or intrinsic), the isolator power extinction ratio will be given by the

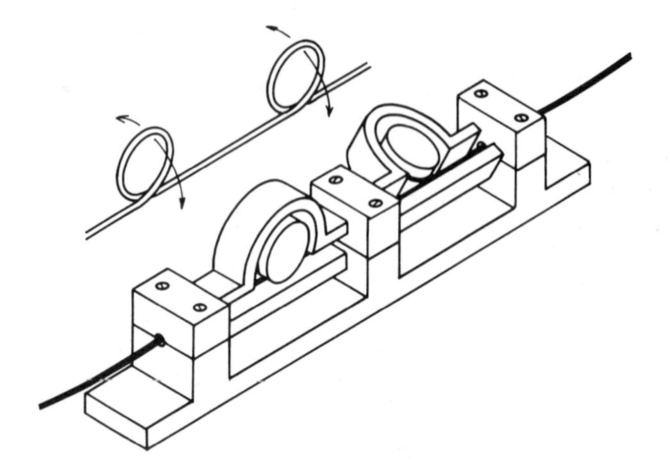

FIGURE 6.18 All-fiber polarization transformer (courtesy of H. C. Lefèvre, Photonetics, Marly le Roi, France).

cross-polarization, deduced from Chap. 2 as approximately $(\delta\beta/2V_fH)^2$, or equivalently here as $(4L/L_b)^2$, where L_b is the linear polarization beat length. With fused silica, a −40-dB power extinction ratio thus corresponds to $L_b \simeq 100$ m, which can be obtained with appropriate manufacturing conditions (see Chap. 2).

Another arrangement can be found for achieving an isolator (or circulator) with linearly birefringent fibers, by using alternate directions for the magnetic field [41]; a typical arrangement is shown in Fig. 6.19. Practically, the fiber had a polarization beat length of 3.3 cm and the 0.53-cm-thick magnets were separated by 1.65 cm. The slots where 0.03 cm wide, producing a field of approximately 5 kG. This device produced a −20-dB isolation, but has the drawback of being very temperature sensitive (a change of 2°C degrades the isolation effect drastically, because of changes in the fiber length) and wavelength dependent (as L_b varies with wavelength) [41].

6.4.3 Filters and Wavelength Multiplexers

Optical Transmission Filters. Such devices have been achieved by using the analog of a bulk-optic N-elements Solc filter, in which N identical birefringent crystal plates with different azimuths are placed between two polarizers (Fig. 6.20). With single-mode fibers, the polarizers are obtained as described in Sec. 6.4.1, and the birefringent crystal plates are replaced by SMF coils whose planes

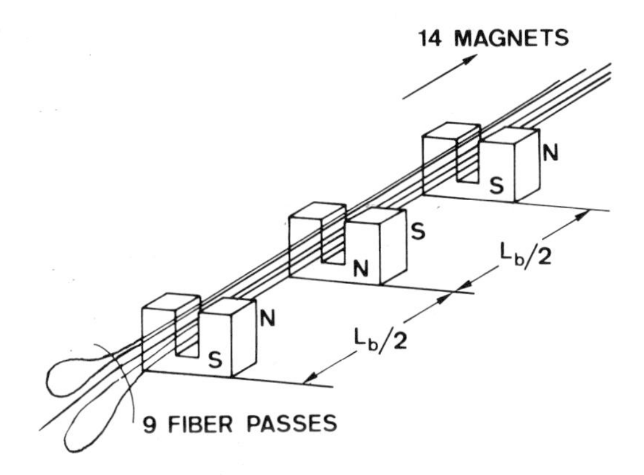

FIGURE 6.19 Arrangement for an all-fiber nonreciprocal 45° rotator, comprising 14 magnets and a linearly birefringent fiber (polarization beat length L_b) making nine passes inside the magnets. Note the alternate orientation of the magnets. (After Ref. 41.)

are successively rotated by an angle equal to 360°/4N. Only the wavelength for which the phase shift between the two linear eigenstates of polarization in each coil is a multiple integer of 360° are transmitted. As the normalized bending-induced birefringence is almost independent of wavelength (see Chap. 2), this results in

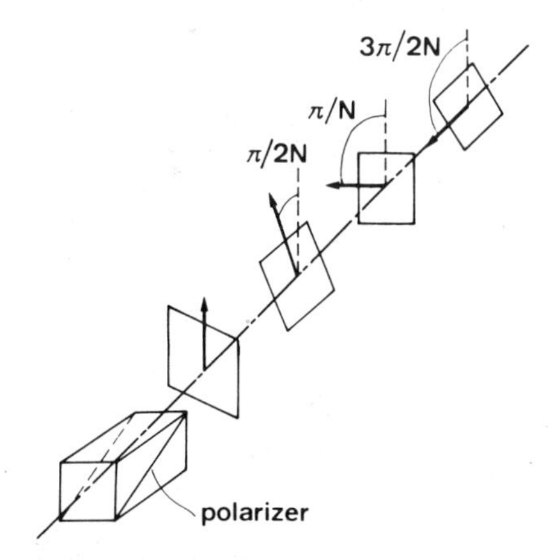

FIGURE 6.20 Schematic arrangement of a bulk-optics Solc filter.

regularly spaced transmitted optical frequencies with a Fabry–Perot-like transmission curve. The spectral width of the peaks is given approximately by their frequency separation divided by (N + 1), and the transmission is limited on the high-frequency (short-wavelength) side by multimode operation, and on the low-frequency (long-wavelength) side by bending losses. A four-element filter has been constructed with a free spectral range (transmission peaks separation) of 120 nm and an individual peak width of 24 nm at a wavelength of about 700 nm [42].

Optical Reflection Filters. These filters are based on the photo-induced refractive index change in Ge-doped silica. An argon-ion laser beam is launched by both fiber ends (or by one end, and a mirror is placed at the other end), and in a few minutes the standing-wave pattern created in the fiber causes a periodic index grating to be written in the fiber core [43]. This grating can then be used as a narrow-band Bragg reflection filter, and it has been used successfully as an external mirror for the writing laser. A detailed study of the performances of these filters can be found in Ref. 44. As an example, a writing power variation from 35 to 140 mW results in a spectral width variation from 0.001 to 0.03 nm, at a nominal wavelength of 500 nm. Apparently, thermal problems may arise from the high absorption of fibers at these wavelengths, and it has thus been recommended that the writing power be increased gradually before inserting the standing-wave mirror, to stabilize the fiber length before the grating is written [45].

Electrical Filters. Various devices resembling common microwave devices can be constructed with fibers (not necessarily single mode) and couplers such as those of Sec. 6.2.3. Examples of filters previously reported are shown in Fig. 6.21, which use lasers with a coherence length much shorter than the fiber loop length. Because of this lack of coherence, these multiple-order interferometers act only on the modulation frequency, not on the optical carrier frequency [46].

In the example shown in Fig. 6.21a, the coupling value of 38% provides a frequency response similar to that of a notch filter, with vanishing transmissions for frequencies of $c/2N_2L$, $3c/2N_2L$, and so on, and an overall frequency response given by $\cos(\pi f N_2 L/c)$, where N_2 is the group index seen by the mode and L is the fiber loop length. Notch frequencies between 0.5 and 400 MHz have been obtained, with possible operation up to 1 GHz or even more (even for a notch frequency of 0.5 MHz). The notch depth is in the range 45 to 55 dB [46].

Similarly, the example shown in Fig. 6.21b corresponds (for high coupling coefficients) to a passband filter with Q factors on the

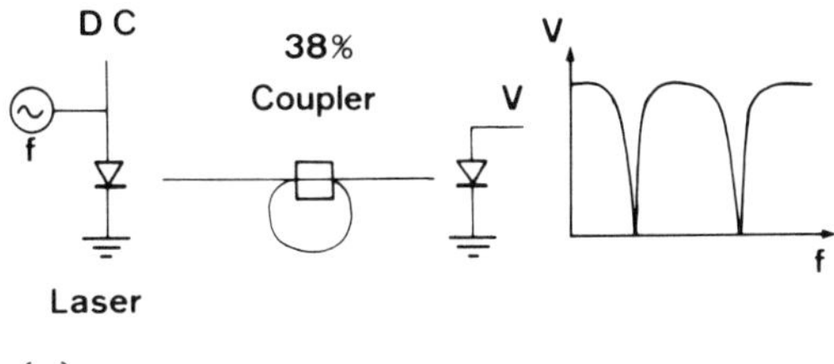

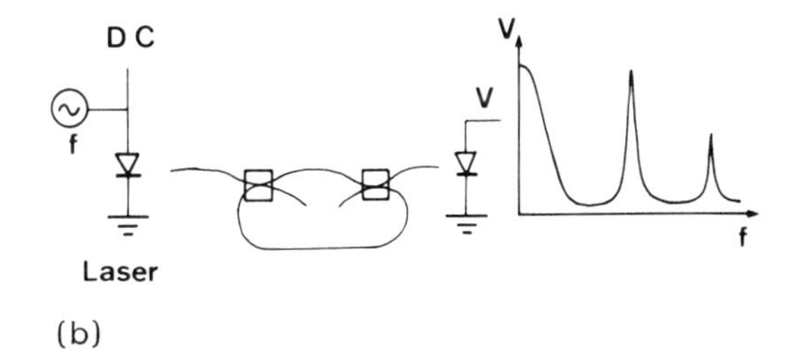

FIGURE 6.21 Examples of SMF modulation frequency filters. (After Ref. 46.)

order of 15. Higher Q's can be obtained by using signal regeneration in the loop, either direct optical amplification or optoelectronic amplification. In the latter case, a Q factor of 420 has been obtained [46].

Wavelength Multiplexers. As seen in Sec. 6.2.3, the fiber couplers are wavelength selective, and the wavelength of maximum coupling can be controlled by the core-to-core separation. This has been attempted experimentally to obtain rather large wavelength separation [47], and some results are shown in Fig. 6.22. The design of such multiplexers is first, to compute the radius of curvature for the desired wavelength separation, and second, to adjust the core-to-core spacing for maximizing the coupling at the desired wavelength. Simultaneously, this ensures that the coupling is minimized at the second wavelength. Such wavelength-selective couplers (or multiplexers) are now available for operation at 1300 nm and 1550 nm, with an excess loss of 0.5 dB and a wavelength isolation of −20 dB.

However, future very high data rate communication links could require a much narrower channel separation than is allowed by the coupler described above. Much better wavelength selectivity can

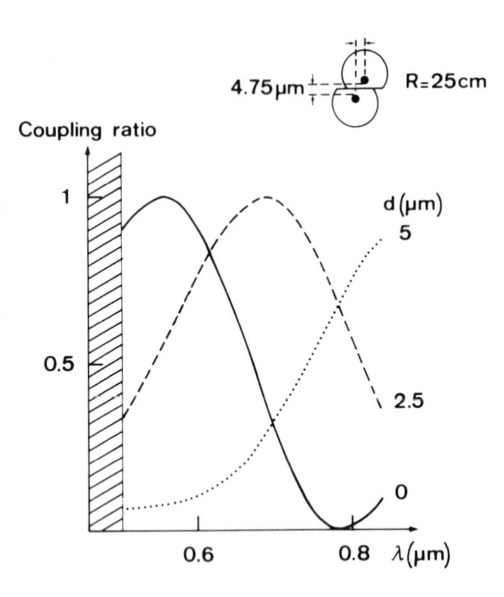

FIGURE 6.22 Polished cladding fiber wavelength multiplexer. (From M. J. F. Digonnet and H. J. Shaw, in *Optical Fiber Communication 1982 Digest of Technical Papers*, Optical Society of America, Washington, D.C., Apr. 1982, paper WBB8.)

be obtained by using two different fibers (here labeled 1 and 2) in the couplers, such that [48]

$$\Delta_2 > \Delta_1; \qquad a_2^2\Delta_2 < a_1^2\Delta_1 \tag{6.5}$$

where Δ_1 and Δ_2 are the relative refractive index differences of fibers 1 and 2, respectively, and a_1 and a_2 are their core radii. It can be shown under these conditions that curves $\beta_1(\lambda)$ and $\beta_2(\lambda)$ (where β_1 and β_2 are the propagation constants in fibers 1 and 2) cross at some wavelength (see Chap. 1). As a strong coupling and thus a significant power transfer between both fibers requires that $|\beta_1 - \beta_2|$ be small compared to the coupling strength, a proper design of the coupler can lead to a wavelength-selective coupling.

Even better wavelength selectivity can be obtained by using $n_2(2) < n_2(1)$ and $n_1(2) > n_1(1)$, where the n_2 are the cladding indices and the n_1 are the core indices, as indicated in Fig. 6.23a for the particular case

$$n_2(1) - n_2(2) = n_1(1) - n_2(1) = n_1(2) - n_1(1) \tag{6.6}$$

$\Delta_2 > \Delta_1$; $a_2^2\Delta_2 < a_1^2\Delta_1$

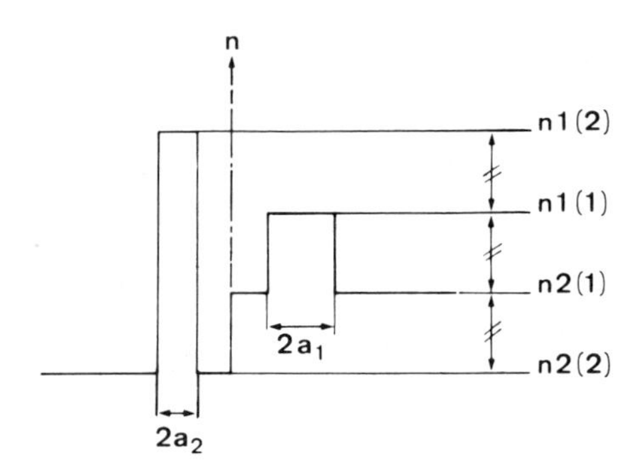

(a)

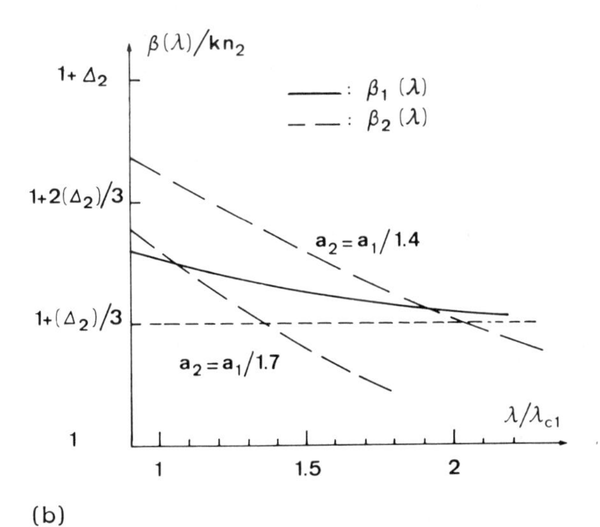

(b)

FIGURE 6.23 (a) Wavelength-selective coupler structure ($\Delta_2 = 3\Delta_1$). (b) Propagation constants as a function of wavelength in fibers 1 and 2, for $\Delta_2 = 3\Delta_1$ and $a_2 = a_1/1.7$ or $a_2 = a_1/1.4$.

This can be achieved by using pure-silica cladding for fiber 1, with fluorine or boron doping of the cladding of fiber 2 and a higher GeO_2 concentration in the core of fiber 2 than in fiber 1. This results in a greater difference between the slopes of the curves $\beta(\lambda)$ than in the previous case, and thus better wavelength selectivity. Examples are shown in Fig. 6.23b for $a_2 = a_1/1.7$ ($\lambda_{c2} = \lambda_{c1}$) and $a_2 = a_1/1.4$ ($\lambda_{c2} = 1.5\lambda_{c1}$).

REFERENCES

1. C. M. Miller, *Optical Fiber Splices and Connectors*, Marcel Dekker, New York (1986).
2. *Laser Focus/Electro Optics Magazine* *23*(6):130–156 (1987).
3. A. H. Cherin and J. F. Dalgleish, *ITU Telecommun. J.* *48*(11).657–665 (1981).
4. W. C. Young, L. Curtis, and P. Kaiser, in *OFC'82 Digest of Technical Papers*, Optical Society of America, Washington, D.C., Apr. 1982, paper PD5.
5. J. I. Minowa, M. Saruwatari, and N. Suzuki, *IEEE J. Quantum Electron.* *QE-18*(4):705–717 (1982).
6. H. Kuwahara and M. Goto, *Electron. Lett.* *17*(18):626–627 (1981).
7. R. E. Wagner and C. R. Sandahl, *Appl. Opt.* *21*(8):1381–1385 (1982).
8. L. Rivoallan, J. Y. Guilloux, and P. Lamouler, *Electron Lett.* *19*(2):54–55 (1983).
9. H. Murata and N. Inagaki, *IEEE J. Quantum Electron.* *QE-17*(6):835–849 (1981).
10. Y. Kato, S. Seikai, and M. Tateda, *Appl. Opt.* *21*(7):1332–1336 (1982).
11. A. Tardy and M. Jurczyszyn, CGE Research Laboratories, unpublished work.
12. K. Furuya, T. C. Chong, and Y. Suematsu, *Trans. IECE Jap.* *E61*(12):957–961 (1978).
13. K. Kinoshita and M. Kobayashi, *Appl. Opt.* *18*(19):3256–3260 (1979).
14. I. Hatakeyama, M. Tachikura, and H. Tsuchiya, *Electron. Lett.* *14*(19):613–614 (1978).
15. J. T. Krause, C. R. Kurkjian, and U. C. Paek, *Electron. Lett.* *17*(6):232–233 (1981).
16. Y. Miyajima, K. Ishihara, T. Kakii, Y. Toda, and S. Tanaka, *Electron Lett.* *17*(18):670–672 (1981).
17. C. M. Miller, G. F. Deveau, and M. Y. Smith, in *OFC'85 Digest of Technical Papers*, Optical Society of America, San Diego, CA, Apr. 1985, paper MI2.
18. R. A. Bergh, G. Kotler, and H. J. Shaw, *Electron Lett.* *16*(7) 260–261 (1981).

19. O. Parriaux, S. Gidon, and A. A. Kuznetsov, *Appl. Opt.* *20*(14):2420–2423 (1981).
20. L. Rivoallan, CNET, unpublished work.
21. M. J. F. Digonnet and H. J. Shaw, *IEEE J. Quantum Electron.* *QE-18*(4):746–754 (1982).
22. M. H. Slonecker, in *OFC'82 Digest of Technical Papers*, Optical Society of America, Washington, D.C., Apr. 1982, paper WBB7.
23. H. C. Lefèvre, P. Simonpiétri, and P. Graindorge, *Electron. Lett.* *21*(24):1304–1305 (1988).
24. M. Kawachi, B. S. Kawasaki, K. O. Hill, and T. Edahiro, *Electron. Lett.* *18*(22):962–964 (1982).
25. C. A. Villarruel, M. Abebe, and W. K. Burns, *Electron. Lett.* *19*(1):17–18 (1983).
26. M. S. Yataki and D. N. Payne, *Electron. Lett.* *21*:249–251 (1985).
27. T. Bricheno and V. Baker, *Electron. Lett.* *21*:251–252 (1985).
28. C. M. Miller, R. B. Kummer, S. C. Mettler, and D. N. Ridgway, *Electron. Lett.* *16*(20):783–784 (1980).
29. W. C. Young and L. Curtis, *Electron. Lett.* *18*(16):571–573 (1981).
30. Y. Murakami, J. I. Yamada, J. I. Sakai, and T. Kimura, *Electron. Lett.* *16*(19):321–322 (1980).
31. H. Sakaguchi, N. Seki, and S. Yamamoto, *Electron. Lett.* *17*(12):425–426 (1981).
32. P. Kayoun, C. Puech, M. Papuchon, and H. J. Arditty, *Electron. Lett.* *17*(12):400–402 (1981).
33. M. Saruwatari and T. Sugie, *Electron. Lett.* *16*(25):955–956 (1980).
34. O. Krumpholz and F. Westermann, *Proc. 7th Eur. Conf. Opt. Commun.*, Copenhagen, Sept. 8–11, 1981, paper 7-7.
35. Y. Odagiri, M. Seki, H. Nomura, M. Sugimoto, and K. Kobayashi, *Proc. 6yh Eur. Conf. Opt. Commun.*, York, England, Sept. 1980, *IEE Public. 190*, pp. 282–285.
36. V. Ramaswamy, R. C. Alferness, and M. Divino, *Electron. Lett.* *18*(1):30–31 (1982).
37. R. C. Alferness, L. L. Buhl, and M. Divino, *Electron. Lett.* *18*(12):490–491 (1982).
38. W. Eickhoff, *Electron. Lett.* *16*(20:762–764 (1980).
39. M. Johnson, *Appl. Opt.* *18*(9):1288–1289 (1979).
40. H. C. Lefèvre, *Electron. Lett.* *16*(20):778–780 (1980).
41. E. H. Turner and R. H. Stolen, *Opt. Lett.* *6*(7):322–323 (1981).
42. Y. Yen and R. Ulrich, *Opt. Lett.* *6*(6):278–280 (1981).
43. K. O. Hill, Y. Fujii, D. C. Johnson, and B. S. Kawasaki, *Appl. Phys. Lett.* *32*(10):647–649 (1978).
44. D. K. W. Lam and B. K. Garside, *Appl. Opt.* *20*(3):440–445 (1981).

45. J. Lapierre, J. Bures, and G. Chevalier, *Opt. Lett.* 7(1): 37–39 (1982).
46. J. E. Bowers, S. A. Newton, W. V. Sorin, and H. J. Shaw, *Electron. Lett.* 18(3):110–111 (1982).
47. M. J. F. Digonnet and H. J. Shaw, in *OFC'82 Digest of Technical Papers*, Optical Society of America, Washington, D.C., Apr. 1982, paper WBB8.
48. O. Parriaux, F. Bernoux, and G. Chartier, *J. Opt. Commun.* 2(3):105–109 (1981).

7

Telecommunication Applications

7.1 INTRODUCTION

One of the major applications of single-mode fibers concerns telecommunications, particularly for trunk networks, where long-haul high-data-rate links predominate. These links can take full advantage of the very high data rate offered by single-mode fibers on long lengths. The demand of system designers for practical SMF transmission systems exists for terrestrial as well as undersea applications. For terrestrial applications it appears that the emphasis is on maximizing the data rate rather than increasing the repeater span (above 25 to 35 km), whereas in undersea applications the data rate is likely to remain below 560 Mb/s while the repeater spacing is maximized. Figure 7.1 shows some of the experimental SMF transmission systems reported, on a scale of repeater span versus bit rate.

Designing such systems requires analysis of the possibilities and characteristics of all components (transmitters, fibers, cables, receivers, etc.). In light of Chaps. 3 and 4, fiber optimization will be extensively discussed in Sec. 7.2. Section 7.3 is devoted to the analysis of direct detection transmission systems, and Sec. 7.4 concentrates on systems using coherent heterodyne or homodyne detection. Direct optical amplifiers are also discussed in Sec. 7.4, as their best performance is obtained when they are associated with coherent detection, although they may also be used with direct detection [14].

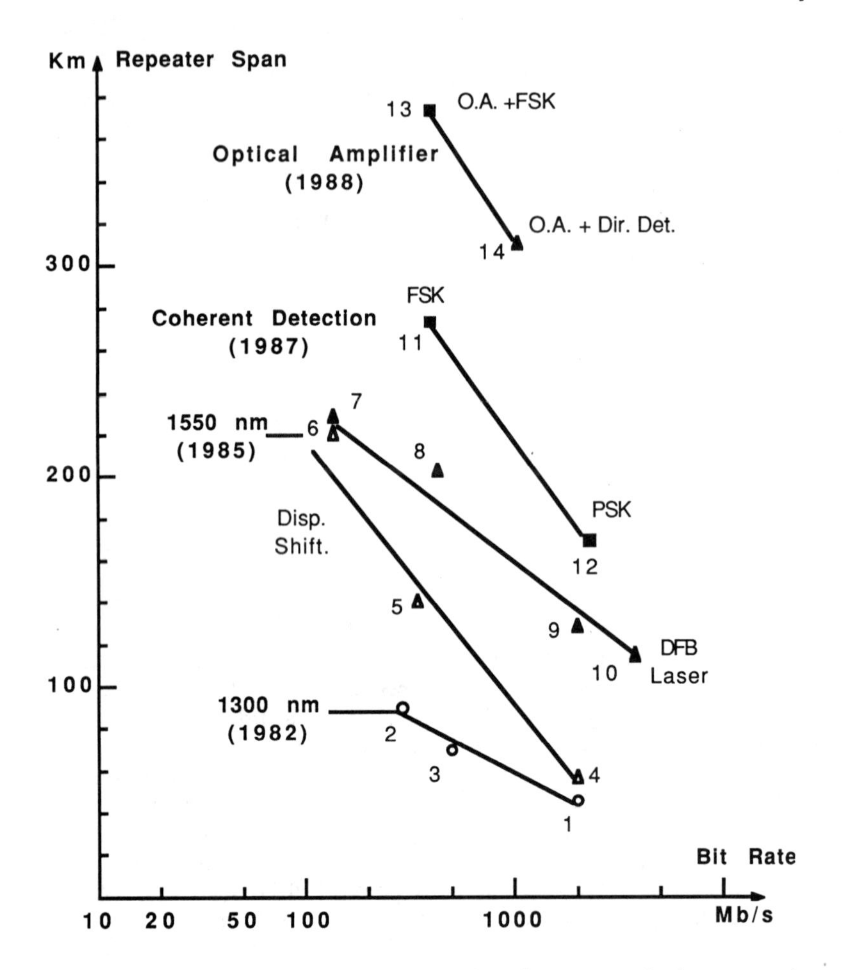

FIGURE 7.1 Best laboratory results for transmission experiments over single-mode fibers. Labels refer to the corresponding reference.

7.2 FIBER PARAMETERS OPTIMIZATION

As seen in Chaps. 3 and 4, the transmission characteristics of a single-mode fiber line (loss and signal distortion) depend on the structural parameters (index difference, core diameter, index profile) of the fibers. For example, the Rayleigh scattering loss would require a low index difference, whereas the bending and microbending losses require a high index difference. The first problem is thus to determine, for a given application and given operating conditions, the optimum SMF structural parameters that lead to the best trade-off for all loss and distortion sources. The second problem is to compare the relative performances of the 1300-nm and 1550-nm transmission windows in terms of repeater spacing, and to determine the conditions under which it is preferable to choose one or the other.

In this section we provide some insight into these problems by using the theoretical material assembled in Chaps. 1 to 4, especially the analytical approximations developed thus far, which allow us to discuss these difficult questions using a relatively simple formalism. A similar and more detailed approach is given in Ref. 15.

7.2.1 Optimization Procedure

Operating Conditions. To characterize the application and operating conditions, we need several parameters to enter into the various equations in Chaps. 3 and 4:

Transmission wavelength λ.

The microbending loss due to fiber drawing and cabling, which would be induced in a standard multimode fiber (step index, core diameter 50 μm, numerical aperture 0.2), α_{mm}. This parameter is easy to determine by using such multimode preforms and fibers in the same drawing and cabling assembly lines, and it is generally known when the same technology has been used previously to manufacture multimode-fiber lines.

The average lateral offset d and axes tilt angle θ at splices. These parameters result from both the fiber manufacturing imperfections (core-cladding eccentricity, tolerance on external diameter) and the splicing technology. They can thus be determined from an in-depth characterization of both processes.

Average distance between splices L_s. This parameter is likely to be representative of a particular kind of application (terrestrial trunk network, or undersea), and depends greatly on the cable manufacturing and laying technology.

The radius of curvature of the fiber in the cable, R.

The radius of curvature of the fiber in the splicing pots, R_s, together with the proportion of fiber length in these splicing pots. Generally, a few meters of fibers are wound under a small radius of curvature in boxes on each side of a splice, for practical convenience. As will be seen below, this puts a limitation on the usable index differences, and therefore some attention should be paid to this practical engineering detail for single-mode-fiber-based systems.

Transmission Characteristics Evaluation. We characterize the fiber by its index difference Δn and its LP_{11} cutoff wavelength λ_c, which should be understood as being the ESI parameters. This means that the present analysis is restricted to fibers having an index profile not too far from the step, although a similar philosophy can be applied for other fibers. The core diameter 2a of the fiber can be deduced through Eq. (5.2) as

$$2a = 0.447\lambda_c(\Delta n)^{-\frac{1}{2}} \tag{7.1}$$

where the index of refraction of the cladding has been taken as 1.45. Let us recall that these ESI parameters can be measured by various methods (see Chap. 5), but it seems preferable to rely on those methods using the mode field radius w_0 determination, as this parameter directly enters important loss factors (microbending and splices). From the basic structural parameters of the fiber we calculate the mode field radius w_0 through Eq. (1.29).

To evaluate the loss performance of a given fiber under given operating conditions, we then add various loss sources:

1. The intrinsic fiber loss is composed of the Rayleigh scattering loss [Eq. (3.1)] and of a residual absorption term which we take as 0.05 dB/km independently of wavelength (OH contribution around 1300 nm and IR absorption tail around 1550 nm). In the following, we consider the scattering loss as if the only dopant used for the index difference were GeO_2. This gives a worst-case estimate, and it should be remembered that fibers with depressed inner claddings (DIC) obtained by fluorine doping decrease the germanium contribution.

2. The microbending loss per unit length is computed through Eq. (3.10).

3. For the bending losses, we first have the continuous curvature in the cable whose contribution is evaluated through Eqs. (3.4)–(3.6). We then have the effect of the loss due to curvature in the splicing pots, which affects a small proportion of the fiber length, assumed here to be 1/1000 (which means that for splices separated by 1 km, 1 m of fiber is wound in the splicing pots). We can thus compute this loss through Eqs. (3.4)–(3.6), and divide it by 1000 to obtain an equivalent distributed loss in dB/km.

Note from Chap. 3 that the exact proportion of fiber length is not very important, because these bending losses vary so abruptly with the exact curvature that they never take a contribution of more than 0.001 dB/km in practical safe operation.

4. The splicing loss is composed of the effects of lateral offset and axes misalignment computed through Eq. (3.16), and the effect of fiber parameter tolerances, taken as 0.05 dB per splice (which corresponds to a tolerance of ±5% on the core diameter and ±10% on the index difference; see Chap. 3). The localized splicing losses are then transformed into an equivalent distributed loss per unit length by using the mean distance between splices L_s.

The total link loss per unit length α_t is then determined by adding the contributions of all these loss sources. Note that all calculations are within the capability of a pocket calculator, although they are tedious if carried out systematically.

For the chromatic dispersion C, we simply use Eqs. (4.7)–(4.10) for a first evaluation, ignoring dopant contributions and W-type fibers (only step-index fibers correspond to these equations). The effect of dopants is known to be small (less than 2 ps/(nm × km) in most cases, or less than a 30-nm zero-dispersion wavelength shift), and can be taken into account by Eqs. (4.18)–(4.22) for a more precise evaluation of the chromatic dispersion in the few selected fibers (remember, however, that manufacturing tolerances are likely to induce variations of the same order of magnitude). For W-type fibers, calculations have to be carried out separately, and some examples have been given in Sec. 4.3.1. Note again that except for W-type fibers, all calculations used here are within the capability of a pocket calculator.

Parameter Optimization. For a given set of operating conditions, we are able to draw a surface representing the loss α_t and another surface representing the chromatic dispersion C, as a function of Δn and λ_c. The loss surface has a minimum, and around 1300 nm it is generally possible to find a set of fiber parameters which lead to C smaller than 2 ps/(nm × km) and to a loss very close to the minimum (within 5%). For a given application at 1300 nm, this point can be selected as the optimum point. Around 1550 nm, it is generally not possible to combine C = 0 and the loss minimum, and the definition of the optimum point therefore depends on other considerations. If single longitudinal-mode lasers are available, the minimum loss is the only important point, whereas if the source has several modes, C becomes a stringent parameter.

In fact, the loss α_t and dispersion C should not be computed for a given set of parameters Δn and λ_c, but rather as the average of values obtained for slight variations around the nominal parameters because of manufacturing tolerances, taken here as ±10% for

Δn and ±5% for a. For this reason, λ_c values are acceptable only if they are at least 10% shorter than the operating wavelength, to ensure that spurious coupling with the second-order mode will not lead to an additional signal distortion.

7.2.2 Application to Long-Haul Systems

The general method described above is applied here to some typical operating conditions which are likely to be representative of many applications. However, it may happen that the relevant operating conditions for particular cases are very different from those assumed here and the fiber optimization procedure thus has to be carried out again.

Terrestrial Systems. For terrestrial systems in the trunk network, we consider the following operating conditions:

Nominal operating wavelength: 1300 nm or 1550 nm.

Multimode microbending loss: $\alpha_{mm} = 0.15$ dB/km (fiber drawing + fiber cabling). This parameter is strongly dependent on the cable design and may appear high for loose tube or cylindrical V-groove cables, or low for tight buffer cables. We deliberately chose here a figure intermediate between the extreme figures actually experienced on the first-generation multimode-fiber systems reported. However, it turns out that the fibers finally selected (see below) exhibit a microbending loss $\alpha_{sm} \simeq 0.4\alpha_{mm}$, and the exact value of α_{mm} is thus not a critical point for the purpose of the present comparison.

For splices, we consider an axis tilt angle of $\theta = 0.2°$ (on average), which seems to be easily obtained in most current splicing technologies. The average axes lateral offset d is varied between 1 and 2 μm (including the fiber's imperfections), the latter being considered typical of passive splicing techniques derived from multimode technology, and the former corresponding to dynamic fiber positioning (see Chaps. 3 and 6).

The average distance between splices L_s is varied between 1 and 2.5 km. These rather long spans should be obtained without major difficulty in the trunk network, with appropriate cable-laying techniques.

Finally, we consider a radius of curvature of the fiber in the cable of R = 0.2 m, which is in fact much smaller than the actual radii encountered in most practical cables. However, the effect of this curvature is completely hidden by that of splicing pots, for which we consider a radius $R_s = 0.05$ m, despite the fact that this curvature affects only 1/1000 of the total fiber length. This is due to the exponential curvature loss increase with 1/R (see Chap. 3), and again shows that some attention should be paid to the problem of splicing pot curvature (and also that of fiber pigtails, for which the critical radius of curvature defined in Chap. 3 should be considered).

Undersea Systems. For undersea systems, we consider the following operating conditions:

Nominal operating wavelength: 1300 nm or 1550 nm.

The multimode microbending loss α_{mm} is varied between 0.15 dB/km (as for terrestrial systems) and 0.3 dB/km. This is justified by the fact that the cable design for such applications (requirement for high tensile strength and protection against outside pressure and water intrusion; see Sec. 7.3.2) may increase the fiber's axis distortion.

For splices, we retain an axis tilt angle of 0.2° and the axis offset of 2 μm, because the manufacturing conditions for long cable spans may not allow the use of dynamic fiber positioning at splices.

The average distance between splices is taken as 20 km, this figure being made possible by the fact that the only practical limitation to this parameter is the maximum continuous fiber length available with high tensile strength. The present value of 20 km seems a reasonable figure in view of recent results concerning the fabrication of very long fiber pieces (e.g., Ref. 16).

Finally, we consider the same radius of curvature of 5 cm everywhere along the fiber, in the cable as well as in the repeater housings. Again, careful attention should be paid to the repeater housing mechanical design, in order not to increase the curvature imposed on the fiber over that of the figure given above, which should be considered as a limit. For the cable we took this small radius of curvature because some undersea cable designs (see Sec. 7.3.2) impose such a strong curvature to the fiber to support the considerable cable elongation during layout and recovery.

Results. A typical result for the total loss per unit length in a terrestrial system operating at 1300 nm with the parameters noted above and d = 2 μm, L_s = 2.5 km, is shown in Fig. 7.2. The sharp rise for short cutoff wavelengths and small index differences is due to curvature losses in the splicing pots and shows clearly how stringent this parameter is. The less pronounced rise for long cutoff wavelengths and small index differences is due to microbending losses. The general loss increase for high index differences is due to both Rayleigh scattering and splicing loss, with a relative maximum occurring (for a given Δn) at the λ_c value corresponding to the minimum mode field radius.

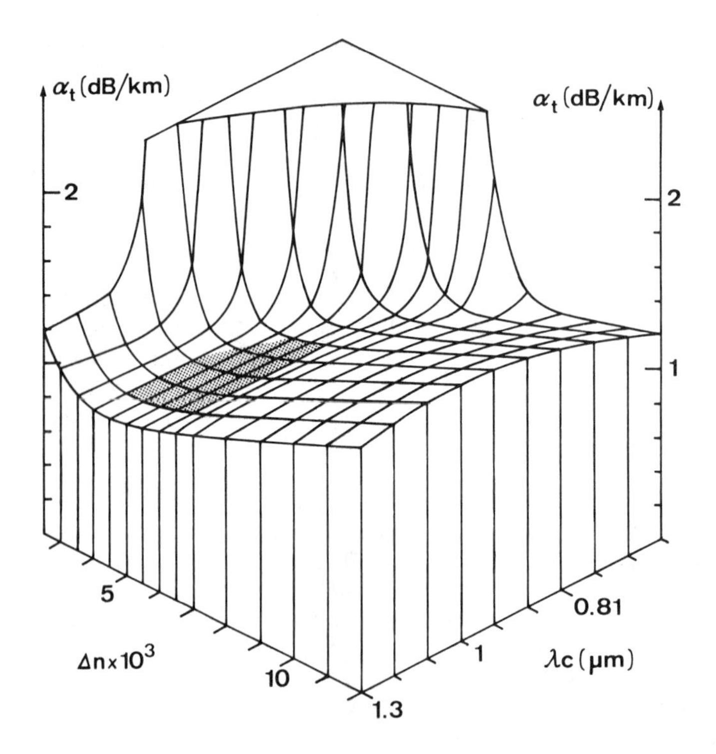

FIGURE 7.2 Total link loss per unit length for terrestrial systems at 1300 nm, with an average offset at splices d = 2 μm and an average distance between splices L_s = 2.5 km. Other operating parameters multimode microbending added loss, 0.15 dB/km; axes tilt at splices, 0.2°; radius of curvature in splicing pots, 5 cm (2.5 m of fiber in each splicing pot); radius of curvature in the cable, 20 cm. The shaded area corresponds to the "usable surface," in which the loss remains within 10% from its minimum value. (The λ_c scale is linear in terms of λ/λ_c.) (From Ref. 15, © IEEE 1982.)

A limited qualitative comparison with practical results is found by looking at Fig. 7.3, where the 1300-nm loss measured for fibers with various index differences and approximately constant cutoff wavelengths (between 1100 and 1200 nm) is shown [17]. As the fibers were wound on polystyrene drums, all loss factors are qualitatively present, except the splicing losses. Also shown are the theoretical intrinsic fiber loss (scattering + absorption) considered in this chapter and the sum of the intrinsic fiber loss and of the calculated microbending loss with α_{mm} adjusted at 0.25 dB/km. The only serious difference with our theoretical results occurs for high index differences (above 0.6%), as the actual loss increases more strongly than expected. This may be due to the absorption increase for high GeO_2 contents and should be improved by using T-DIC structures and better control of the drawing conditions together with triangular core index profiles (see Chap. 3).

For several reasons, such as standardization and fiber cost, it would be unrealistic to optimize fiber parameters for only one set of operating conditions, and we thus consider as acceptable in each case all the fiber parameters that yield a loss α_t exceeding the minimum loss by less than 10%, thus defining a "usable surface" in terms of α_t (Δn, λ_c).

Figure 7.4 shows the total link loss per unit length for undersea systems operating at 1550 nm, with α_{mm} = 0.3 dB/km. The general behavior is similar to that of Fig. 7.2, although the loss increase for high index differences is hardly distinguishable because of the smaller Rayleigh scattering loss influence (λ^{-4} dependence), and of the negligible splicing loss contribution (20 km between splices).

A detailed study of various optimum points obtained when combining different operating conditions can be found in Ref. 15, and here we retain only two fiber types (recall that all structural fiber parameters should be understood as the parameters of the equivalent step-index fiber, having the same mode field radius as the actual fiber).

1. Fiber type A is characterized by a relative index difference Δ = 0.37% and a core diameter of about 8.3 μm. From a practical measurement point of view, this fiber shall exhibit an LP_{11}-mode effective cutoff wavelength of 1280 nm (except for W-type structures), and a mode field radius w_0 of about 4.5 μm at 1300 nm, and 5 μm at 1550 nm (whatever the fiber structure is). The dispersion characteristics of this fiber are C (1300 nm) between −3 ps/(nm × km) for a step index GeO_2-doped core structure, and −0.6 ps/(nm × km) for a T-DIC structure with a contribution of 57% from the fluorine doping in the total index difference between core and inner cladding (see Chap. 4). The corresponding zero-dispersion wavelengths are 1330 nm and 1310 nm, respectively. These characteristics

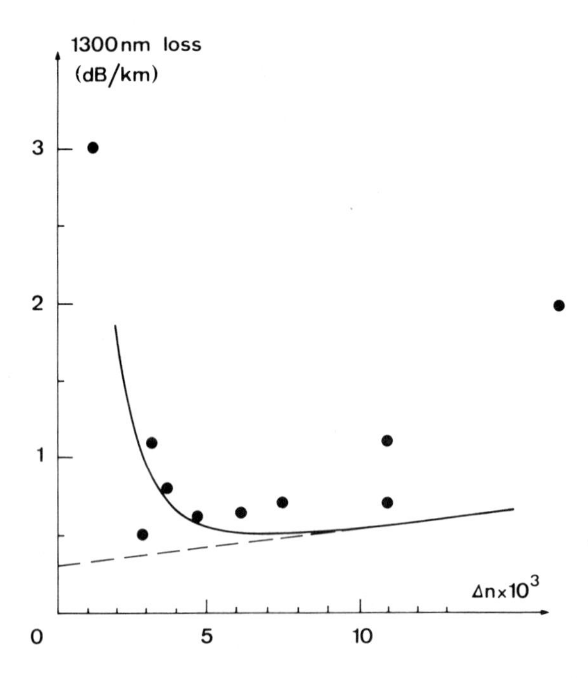

FIGURE 7.3 Dots are measured losses at 1300 nm on fibers wound on polystyrene drums, as a function of the index difference for cutoff wavelengths between 1100 and 1200 nm. The dashed curve corresponds to the theoretical intrinsic fiber loss (Rayleigh scattering + absorption) considered in this chapter, and the solid line corresponds to the intrinsic loss and the calculated microbending loss with α_{mm} adjusted empirically at 0.25 dB/km.

are summarized in Table 7.1. Finally, let us recall that W-type fibers with similar ESI parameters but very different dispersion characteristics have been presented in Sec. 4.3.1.

2. Fiber type B is characterized by an absolute index difference $\Delta n = 1.1\%$ and a core diameter of 4.2 μm. From a practical measurement point of view, this fiber shall exhibit an LP_{11}-mode cutoff of 970 nm (except for W-type structures) and a mode field radius of 3.7 μm at 1550 nm. The zero-dispersion wavelength of this fiber is about 1565 nm for a step-index GeO_2-doped core structure, or a T-DIC structure with a contribution of 57% from the fluorine doping in the total index difference (Table 7.1). This fiber requires more stringent control of the manufacturing tolerances to keep its dispersion within acceptable limits (Sec. 4.3.1).

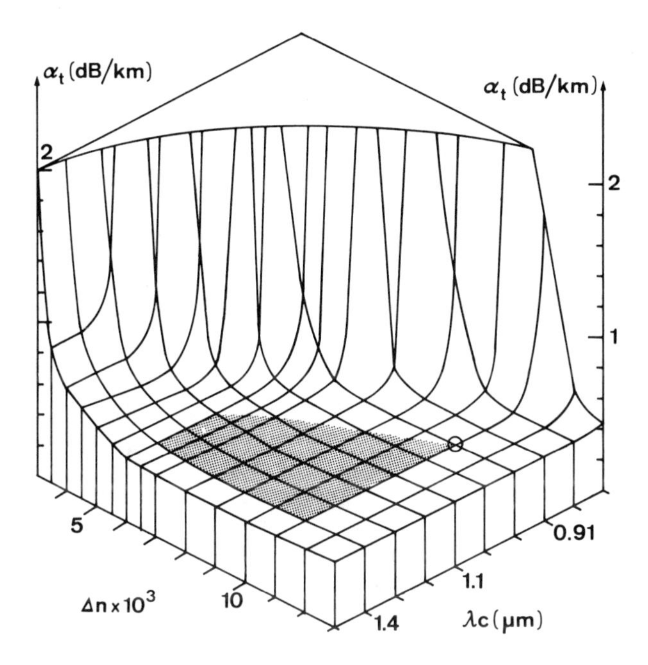

FIGURE 7.4 Total link loss per unit length for undersea systems at 1550 nm, with a multimode microbending loss α_{mm} = 0.3 dB/km. Other operating parameters: axes offset at splices, 2 μm; axes tilt angle at splices, 0.2°; distance between splices, 20 km; radius of curvature, 5 cm. The shaded area corresponds to the "usable surface," in which the loss remains within 10% from its minimum value; the circle corresponds to a fiber with zero dispersion. (The λ_c scale is linear in terms of λ/λ_c.) (From Ref. 15, © IEEE 1982.)

Fiber type A corresponds to the intersection of all usable surfaces for the loss α_t, for all terrestrial and undersea systems envisaged above, at either 1300 or 1550 nm. These fiber parameters (ESI) thus appear to be very interesting, as they provide the lowest loss over a wide spectral range and varied operating conditions. For the dispersion, we saw in Sec. 4.3.1 that its λ_D value can be adjusted either around 1300 nm with classical step-index structures, or anywhere between 1300 and 1600 nm (possibly also with a wide zero-dispersion spectral range) with W-type structures, at the expense of very stringent manufacturing tolerances (Fig. 4.6 and 4.12).

TABLE 7.1 Characteristics of two fiber types optimized for maximizing the repeater spacing under various operating conditions

Type	ESI index difference	ESI core diameter (μm)	LP_{11} cutoff (nm)	Mode radius (μm)	Zero dispersion λ_D (nm)
Fiber A	5.5×10^{-3}	8.3	1280	4.5 (at λ_D)	1300–1320
Fiber B	1.1×10^{-2}	4.2	970	3.7 (at λ_D)	1565

Actually, fiber type A is one of the fiber types that provide transmission characteristics in agreement with recommendation G.652 issued for a standard single-mode fiber by the International Telegraph and Telephone Consultative Committee (CCITT). As per CCITT-G.652, a standard single-mode fiber optimized for use in the 1300-nm wavelength region (which can also be used in the 1550-nm region although it is not optimized for this wavelength) shall exhibit the following characteristics:

Nominal mode field diameter between 9 and 10 μm, with a deviation smaller than ±10% of the nominal value.

Cladding diameter: 125 μm ± 3 μm.

Mode field concentricity error below 1 μm.

Effective cutoff wavelength λ_c below 1280 nm (and above 1100 nm) for a short primary-coated fiber piece, or below 1270 nm for a cabled fiber (see Sec. 5.3.1 for measurement methods).

The bending loss induced by winding the fiber loosely with a 75-mm diameter over 100 turns shall not exceed 1 dB at 1550 nm.

The chromatic dispersion is specified by several parameters. The zero-dispersion wavelength λ_D and the zero-dispersion slope $C'(\lambda_D)$ shall satisfy

$1295 \text{ nm} < \lambda_D < 1322 \text{ nm}$

$C'(\lambda_D) < 0.095 \text{ ps}/(\text{nm}^2 \times \text{km})$

The chromatic dispersion limits are then calculated through Eq. (5.22) with the maximum value measured for $C'(\lambda_D)$ and with the measured values of $\lambda_{D,min}$ and $\lambda_{D,max}$. The absolute value of the chromatic dispersion limits shall satisfy

for $1285 \text{ nm} < \lambda < 1330 \text{ nm}$, $|C(\lambda)| < 3.5 \text{ ps}/(\text{nm} \times \text{km})$

for $1270 \text{ nm} < \lambda < 1340 \text{ nm}$, $|C(\lambda)| < 6 \text{ ps}/(\text{nm} \times \text{km})$

for $\lambda = 1550 \text{ nm}$, $|C(\lambda)| < 20 \text{ ps}/(\text{nm} \times \text{km})$

Such a set of transmission characteristics can be achieved with a variety of single-mode fiber designs. However, two designs basically dominate the market: a matched-cladding design, which usually offers a 10-μm mode field diameter based on a step-index core with 0.3% relative index difference and 9 μm core diameter; and a depressed index cladding design, which corresponds to the present fiber type A, as listed in Table 7.1 [18]. A detailed comparison between these designs is presented in Table 4.1.

Fiber type B is the step-index core fiber structure which provides λ_D in the range 1500 to 1600 nm and the lowest loss at this wavelength in all operating conditions (among other 1550-nm zero-dispersion fibers with step-index structures). Practically, it should be manufactured with a T-DIC structure, which should minimize the problems related to GeO_2-associated fiber drawing induced loss and even decrease the Rayleigh scattering loss by about 0.05 dB/km (see Chap. 3). Compared to a W-type structure, it is less sensitive to manufacturing tolerances (Sec. 4.3.1). However, it has been shown that the most practical design for low-loss dispersion-shifted fibers is a triangular core index profile, with either a depressed index cladding or a raised index ring around the core (see Sec. 4.5.2). The application of equivalent step index fiber formulas to the fiber parameters listed in Table 4.2 shows that the fiber type B is very close to the ESI of these dispersion-shifted fibers.

Summary. The major results of the detailed loss comparisons are summarized in Table 7.2 and can be interpreted as follows:

Considering first that today's technologies limit the splicing offset to 2 μm and make it necessary to operate a fiber at its zero-dispersion wavelength (because of the lack of reproducible and reliable truly single-longitudinal-mode lasers, or optical pulse equalizers, or W-type structures), we observe that terrestrial systems are not significantly improved by using fiber type B at 1550 nm instead of fiber type A at 1300 nm. For undersea systems, fiber type B provides approximately a 40% loss reduction compared to fiber type A at 1300 nm.

Second, the improvement of splicing technology in terrestrial systems not only decreases the loss of both fibers, but also gives a significant advantage to fiber B at 1550 nm compared to fiber A at 1300 nm. The loss of fiber B at 1550 nm with a 1-μm splicing offset is half the loss of fiber A at 1300 nm with a 2-μm splicing offset.

However, if some reproducible and reliable solution for reducing the chromatic dispersion effects (industrially developed truly single-longitudinal-mode lasers, or optical pulse equalizers, or W-type structures with ESI parameters of fiber A) become available, then fiber type A appears as the best overall trade-off for a great variety of applications. This is clearly confirmed by the results of laboratory experiments where experimental systems based on

TABLE 7.2 Typical loss of cabled and spliced links based on fibers type A and B (Table 7.1) under various operating conditions

	Loss (dB/km)	
Type	1300 nm	1550 nm
Terrestrial systems, 2 km between splices		
Fiber A		
2 μm offset	1	0.7
1 μm offset	0.6	0.4
Fiber B[a]		
2 μm offset	—	0.9
1 μm offset	—	0.45
Submarine systems, multimode microbend loss 0.3 dB/km		
Fiber A	0.6	0.35
Fiber B[a]	—	0.35

[a]Loss figures for fiber B could be decreased by 0.05 dB/km by using T-DIC structures with a low GeO_2 content.

dispersion-unshifted fiber (type A) and DFB lasers at 1550 nm offer better performance than those based on type B dispersion-shifted fiber (see Fig. 7.1).

7.3 DIRECT DETECTION TRANSMISSION SYSTEMS

Although detailed discussions of appropriate transmitter, receiver, and cables design for long-haul high-bit-rate near-IR transmission systems can be found extensively in the literature, we recall here briefly the relevant characteristics of these devices, before discussing system performance.

7.3.1 Transmitters and Receivers

Transmitter Hardware. The transmitters basically involve a laser diode, together with its control and modulation circuitry, and optical coupling. The electrical control and modulation circuitry will not be discussed here as it is not different from that for multimode systems. Let us simply recall that these long-wavelength GaInAsP lasers are very temperature sensitive and require accurate temperature control. The optical coupling to the fiber was discussed in Chap. 6; practical coupling efficiencies of 50% have readily been obtained. Two other points are worth mentioning. A fiber pigtail is commonly used in the transmitters, and the fiber is then tightly wound under rather small radii of curvature. We can consider that the critical radius of curvature defined by Eq. (3.7) is an absolute lower limit for these pigtails, and we find limiting curvatures of radius 15 mm (at 1300 nm) and 25 mm (at 1550 nm) for fibers A and B defined above. This should not be overlooked in practical transmitter designs. Also, the projected lifetime of long-wavelength laser diodes may not be sufficient for the very high reliability required for undersea repeaters, and it is currently envisaged to implement several lasers per fiber, with an optical switch allowing the use of a new laser when the operational laser is running out-of-specifications. Such switches were described in Chap. 6.

Finally, a very important characteristic of the transmitter is its spectral behavior, as this dominates the maximum achievable bit rate on the system when fibers are operated at wavelengths different from λ_D (zero-dispersion wavelength). The fiber single-pulse distortion was discussed extensively in Chap. 4, but we encounter here another type of distortion due to laser-mode partition noise [19].

Transmitter Spectrum. It is generally observed that under high-speed modulation a laser diode exhibits multi-longitudinal-mode behavior, and more precisely that its spectrum varies randomly from pulse to pulse. As the typical longitudinal-mode spacing is in the range 1 to 2.5 nm (at 1550 nm nominal wavelength), this effect leads to severe intersymbol interference if the fiber is not operated at its zero-dispersion wavelength. This phenomenon, illustrated qualitatively in Fig. 7.5, is called laser-mode partition noise. Obviously, this plays a significant role only in fibers with $C(\lambda) \neq 0$, illuminated with a laser whose dynamic spectral distribution involves more than one mode. Calculations carried out for lasers having 1-nm mode spacing show that the product (bit rate × fiber length × chromatic dispersion × rms laser spectral width) is limited to about 0.08 by the phenomenon, instead of the value of 0.18 given in Sec. 4.4.1 for the limitation due to the fiber frequency response [20]. This phenomenon brings a more stringent limitation to the

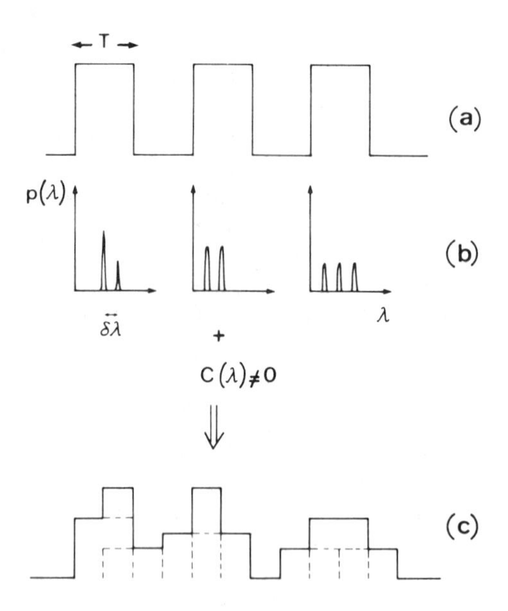

FIGURE 7.5 Qualitative effect of laser-mode partition noise. (a) is the pulse modulation, (b) represents the instantaneous laser spectrum, (c) shows the output pulse from a fiber with a product (chromatic dispersion × length × mode spacing = T/2).

use of dispersive fibers with multi-longitudinal-mode lasers. However, it is not yet clear how this phenomenon is influenced by various parameters such as the laser-mode spacing (for a given rms spectral width) or the laser structure (e.g., index or gain guiding). Note finally that for practical purposes, if more than five longitudinal modes are present, the rms spectral width can be approximated by 0.4 × the full width at half maximum of the spectrum envelope (dynamic spectrum, under modulation).

As soon as the fiber exhibits a nonzero chromatic dispersion (and when no optical pulse equalizer is used), the laser should oscillate on a single longitudinal mode, even under modulation, to overcome the drawbacks of bandwidth limitations and mode partition noise. Note, however, that the presence of a weak satellite mode may have dramatic effects, as illustrated in Fig. 4.15. For two satellite modes spaced $\delta\lambda$ around the central mode, with a relative intensity of $x << 1$ (on average), the rms spectral width to be taken into account is $\Delta\lambda = \delta\lambda(2x)^{\frac{1}{2}}$ [only $\delta\lambda(x)^{\frac{1}{2}}$ if only one satellite mode is present]. If lasers are to be used with such fibers, it is thus necessary to characterize their dynamic spectral behavior under modulation in great detail, and to ensure that the amplitude of satellite modes remains compatible with the desired system performance over the entire range of operating conditions (temperature, aging, etc.).

From a practical point of view, such truly single-longitudinal-mode transmitters can be obtained in two ways:

1. Use of a distributed feedback laser (DFB) structure as in Refs. 7 to 9, which provides truly single-mode operation under high-speed modulation.
2. Use of a continuous-wave (CW) single-longitudinal-mode laser oscillator and an external modulator, such as an integrated optical directional coupler (as in Ref. 10), or another laser which acts as a modulator after being injection-locked by the first oscillator (e.g., Ref. 21).

In both cases, an isolator (see Chap. 6) and very accurate temperature control are required. However, it is not yet clear how these spectral properties may be affected by aging, temperature fluctuations, laser degradation, and other factors.

Receivers. Receivers involve basically a photodetector (PIN or APD), together with its control and amplification circuitry, and optical coupling. The electrical control and amplification circuitry will not be discussed here.

The optical coupling can be achieved through a multimode optical fiber tail, which could reduce the loss on the receiving side. As for the receiving sensitivity, Ge APDs have been studied extensively. The best result has been achieved with a $p^{+}nn^{-}$ APD, with a quantum efficiency of 60% and an excess noise factor $x = 0.83$ at 1550-nm wavelength [4]. Tests carried out with a 1550-nm laser source modulated under an RZ (return to zero) pulse pattern showed a practical minimum detectable power P_{min} for a 10^{-9} BER (bit error rate) of [4]

$$P_{min} \simeq -44 + 10 \log\left(\frac{BR}{140}\right) \quad \text{dBm} \tag{7.2}$$

where BR is the bit rate in megabits per second. Other studies carried out at 1300 nm with a commercially available Ge APD (Optitron GA-1) followed by a transimpedance amplifier and an NRZ modulation format yielded [22]

$$P_{min} \simeq -40 + 10 \log\left(\frac{BR}{140}\right) \quad \text{dBm} \tag{7.3}$$

This figure was degraded by about 3 dB when the temperature increased from 20 to 60°C. For InGaAs PIN-FET receivers, the study in Ref. 22 indicates a sensitivity at 1300 nm of

$$P_{min} \simeq -46 + 15 \log\left(\frac{BR}{140}\right) \quad \text{dBm} \tag{7.4}$$

This figure is degraded by only 1 dB when the temperature increases from 20 to 60°C. Taking the average of Eqs. (7.2) and (7.3) for Ge APDs shows that InGaAs PIN-FETs are more sensitive up to a bit rate of about 800 Mb/s at 20°C, and 2.2 Gb/s at 60°C. However, these results concern top performance and may be affected by other phenomena when going to the industrial development stage.

Besides the signal detection and current amplification, the receivers usually contain some pulse equalization functions to overcome the slight intersymbol interference caused by pulse distortion due to the fibers. As in the case of single-mode fibers, the pulse distortion is usually dominated by chromatic dispersion (birefringence dispersion effects are negligible in most cases; see Chap. 4), optical pulse equalizers (OPEs) implemented in front of the photodetector can be envisaged to cancel a small residual chromatic dispersion of the fiber line, or even to compensate a large dispersion such as that of fiber type A operated at 1550 nm. This would then strongly decrease the requirements regarding source spectral purity in this kind of application.

Optical Pulse Equalizers. Basically, any medium capable of providing a strong chromatic dispersion in a rather compact form (at the desired wavelength) could serve as an optical pulse equalizer, provided that its chromatic dispersion is able to cancel that of the total fiber length.

A simple possibility is to use a single-mode fiber with a strong chromatic dispersion at the operating wavelength, but opposite in sign to that of the fiber line. As this is of interest only if the compensating fiber length is much smaller than the link length (a 1-km length can be considered as a maximum), it can be envisaged only for correcting a slight systematic difference between the zero-dispersion wavelength of the fiber link and the operating wavelength. For example, around 1550 nm an error of 10 nm in the operating wavelength corresponds roughly to a residual chromatic dispersion of 18 ps/nm for a 30 km type B fiber link [see Eq. (4.13)]. This can be compensated by a 1.2-km type A fiber, whose exact length can be adjusted to the exact correction required.

A more universal and versatile solution is shown schematically in Fig. 7.6. The principle is to use collimating lenses and a

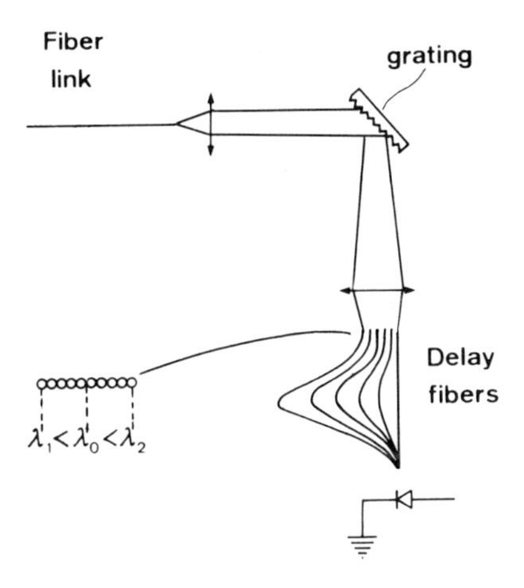

FIGURE 7.6 Schematic representation of an optical pulse equalizer. The various wavelengths with different arrival times emerging from the fiber link are coupled into different delaying fibers by a dispersive grating. If the delay variation among the delaying fibers (adjusted by their length difference) just compensates the difference in arrival times, the detected pulse shows no distortion.

diffraction grating to couple the various wavelengths emerging from the fiber link into different fibers (as in a monochromator where the input slit is replaced by the fiber link output end face, and the output slit is replaced by a linear array of adjacent core fibers). After the total chromatic dispersion of the link has been determined (in ps/nm), and knowing the spatial resolution of the dispersive system (in nm/mm of abscissa along the output fiber array end face), it is easy to calculate the appropriate length of each delaying fiber to compensate almost exactly for the total dispersion. The pulse detected by the photodetector will then show almost no distortion and should no longer be affected by laser-mode partition noise [23]. To obtain orders of magnitude, if we consider that we allow a maximum residual pulse distortion of 100 ps (which corresponds to a bit rate above 2 Gb/s), this means that the length difference between one delaying fiber and its neighbor should be about 2 cm. For a link 70 km long operating at 1550 nm with fiber type A (15 ps/nm × km), the total dispersion will be about 1 ns/nm, indicating that the spectral resolution of the OPE should be 0.1 nm (between each delaying fiber).

It is thus seen that all these orders of magnitude are still within the technological possibilities of currently available optical components.

For operation closer to the zero-dispersion wavelength of the link, and/or with shorter links, the requirements are even less stringent. Note that such an OPE can be fitted to each installed link, and that this principle can be used at any currently envisaged wavelength between 1200 and 1700 nm.

7.3.2 Cables

Terrestrial Cables. Cables for single-mode-fiber-based terrestrial transmission systems are generally identical to the cables developed previously for multimode-fiber systems, provided that they yield a small microbending-induced loss. The most appropriate structures seem to be cylindrical V-groove structures or loose tube cabling. Figure 3.11 shows that single-mode fibers cabled in such structures do not exhibit an increase in loss due to cabling. Other structures still used comprise tight buffering of the fiber.

Undersea Cables. The application of optical fibers to undersea telecommunications is more recent than for terrestrial links, and it is generally envisaged to use only single-mode fibers, to maximize the repeater spacing while maintaining a high bit rate (up to 560 Mb/s). For this application, appropriate cable structures have been developed and a discussion of these applications can be found in Ref. 24. Some of the cable structures are shown in Fig. 7.7. The objectives are to get a high tensile strength (75 to 100 kN) for withstanding tensile loading during laying and recovery, and to accommodate high hydrostatic pressures (up to 80 MPa) over 20 years. As a good silica fiber breaks at about 2% elongation, it is desired to minimize the fiber elongation during laying and recovery. Also, it is generally admitted that the fibers should be kept free from outside pressure and water intrusion, even in the case of cable rupture. Such cables are thus constituted of a central optical core hermetically closed, surrounded by steel wires which provide most of the tensile strength and pressure resistance, a metallic conductor for repeaters power feeding, and an outside insulating plastic jacket. The central optical core comprises a high-tensile-strength member surrounded by the fibers, which are either tightly buffered with plastic materials or embedded inside a grease coating which prevents the fibers from water intrusion even in the case of cable rupture. For shallow-water use, the cable is armored with additional steel wires.

The requirement for small fiber elongation, even under the high loads sustained by the cable during laying and recovery, tends toward designing a stranding of fibers around the central strength member with a small pitch, leading to a small radius of curvature for the fiber helical path. Also, the presence of a tight buffer or

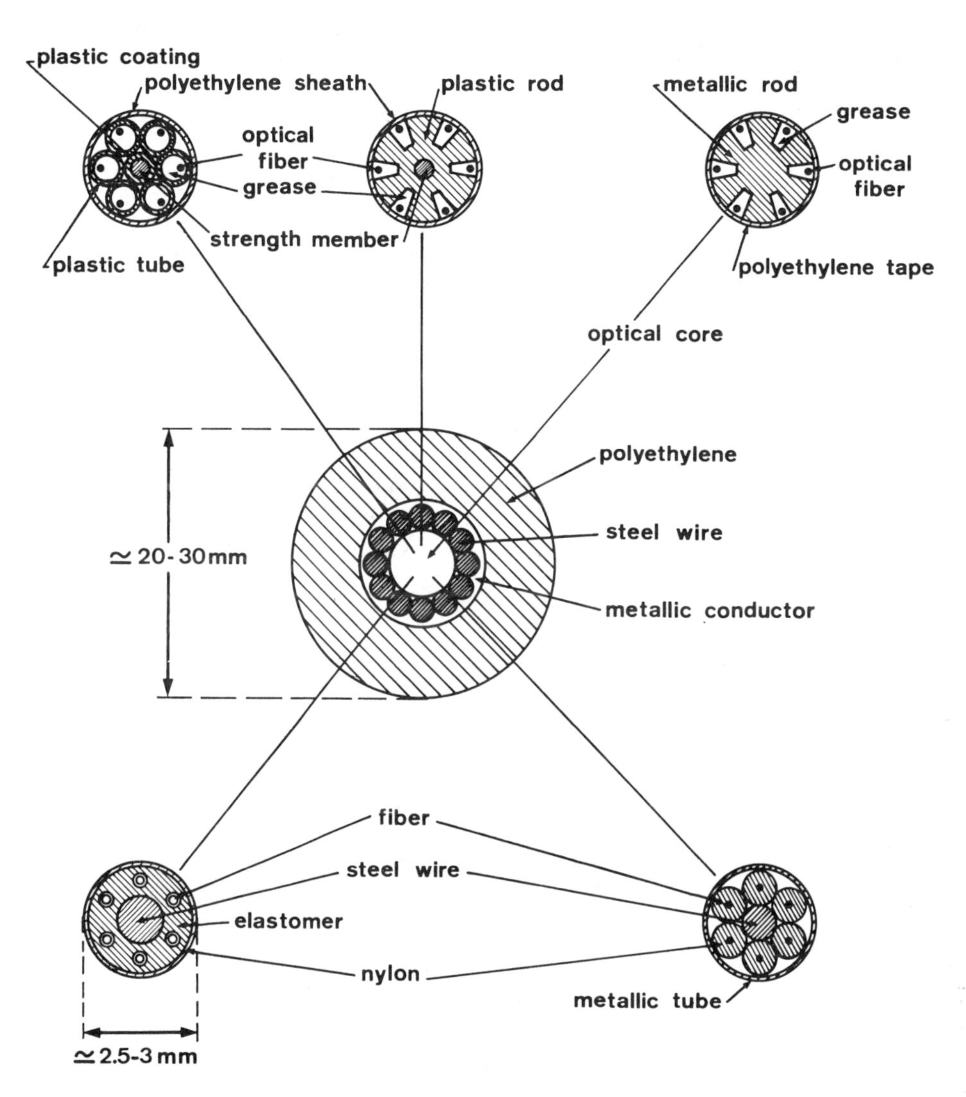

FIGURE 7.7 Typical examples of undersea optical cable structures. (From Ref. 24.)

grease on the fiber can increase the microbending losses to levels higher than those of terrestrial loose tube or cylindrical V-groove structures.

Finally, it should be mentioned that severe detrimental effects on fiber attenuation caused by hydrogen evolution from a variety of cable materials have been observed in cables exposed to sea water [25,26]. This has resulted in significant changes in the design and fabrication of optical cables, especially for undersea applications. A comprehensive overview of these problems can be found in Ref. 27.

Moisture in the cable can liberate hydrogen through electrolysis of dissimilar metals present in the cable. This effect is greatest when the cables contain iron and aluminum, or stainless steel and aluminum. Copper and stainless steel yield an insignificant amount of hydrogen. The other most troublesome material in a cable is silicone, and nylon or polyethylene should be preferred. It is also possible to protect the fiber from hydrogen permeation, even if significant quantities of hydrogen are generated. Silicon oxynitride has been found to be an effective coating acting as a barrier against hydrogen [27].

7.3.3 Digital System Performances

In this section we examine the expected repeater spacings for both fiber types A and B in terrestrial and undersea systems with various sources and discuss some examples of operational or experimental systems.

For evaluating the attenuation allowed for the link as a function of the bit rate BR, we assume that the minimum detectable power (for a 10^{-9} BER) is given by Eq. (7.2) at both 1300 nm and 1550 nm. We assume a coupled power of 0 dBm into the fiber, a fixed loss of 2 dB for connectors, and a system margin of 6 dB. The total allowed link loss is thus $36 - 10 \log (BR/140)$(dB, BR in Mb/s).

For the limitations due to dispersion, we use the results of Secs. 4.4.1 and 4.4.2 and do not specifically take the effects of laser-mode partition noise into account, because they give a limitation of the same order of magnitude as that given in Chap. 4. Fiber types A and B will be considered, and it should be remembered that W-type fibers with the same ESI parameters as that of fiber A should provide the same performances as that of type A for a zero-source line width, independently of the source line width, provided that their dispersion properties are well controlled.

Evaluation of Terrestrial Systems. For the sake of simplicity, we consider only a splicing offset of 2 μm and a splice spacing of

2.5 km. We consider the following sources: lasers with spectral width of less than 5 nm, emitting at the nominal wavelength of zero chromatic dispersion for fiber types A and B, respectively; lasers with spectral widths of 2.5 or 5 nm, and an emission wavelength shifted by 20 nm from the zero chromatic dispersion wavelength. Figure 7.8 shows the resulting repeater spacing L as a function of the bit rate for the various fiber-source combinations.

Several features can be observed:

Fiber type A at 1550 nm provides a very significant improvement when single-mode operation can be obtained (DFB laser), but any source spectral broadening has dramatic effects.

Fiber type A at 1300 nm requires good control of the laser's emission wavelength if very high bit rates are looked for, and fiber type B at 1550 nm provides reduced sensitivity to source spectral fluctuations.

As said above, it is likely that the bit rate of terrestrial systems will be increased progressively while the repeater spacing should stay in the range 25 to 35 km. Figure 7.9 shows the allowed maximum bit rate (voluntarily limited to 2 Gb/s) as a function of the source rms spectral width for various operating conditions of fiber type A in terrestrial systems with a splice offset of 2 μm and a splice spacing of 2.5 km. For the optical pulse equalizer, an insertion loss of 3 dB has been used in the calculations. The practical interest of the OPE is clearly observed in this figure and would even be reinforced by an improvement in the splicing technology, because all horizontal lines of Fig. 7.9 would be up-shifted by a factor of about 10 for an offset reduced to 1 μm, while the diagonal lines would remain unchanged.

Since 1983, a huge number of single-mode long-distance transmission systems have been installed in the United States. By the end of 1988, the total number of fiber-kilometers involved was around 3 million with an average count of 30 fibers per cable. All these systems use fiber type A or similar fibers operated around 1300 nm. In such systems the average link attenuation is about 0.5 dB/km including splices, and the repeater spacing is in the range 30–40 km. This performance is slightly better than that given in Table 7.2 for fiber type A and splices with 1 μm offset every 2.5 km, because of improvements in the actual fiber intrinsic loss and splicing technology.

More recently, British Telecom has installed the first long-distance link based on dispersion-shifted fiber to be operated around 1550 nm [28]. The average cable loss was 0.29 dB/km, including splices. This is again slightly better than the figure given in Table 7.2 for fiber type B and splices with 1 μm offset. The fibers

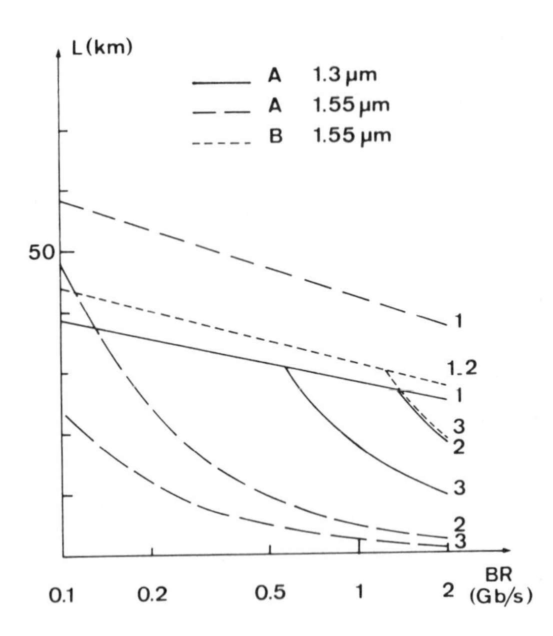

FIGURE 7.8 Repeater spacing as a function of bit rate for terrestrial systems (splice offset, 2 μm; 2.5 km between splices). Fibers A around 1300 nm and B around 1550 nm: 1, fibers operated at their exact respective zero chromatic dispersion wavelength λ_D and spectral width less than 5 nm, or 20 nm shifted wavelength and single-mode operation; 2, 20-nm shifted wavelength, and 2.5-nm rms spectral width; 3, 20-nm shifted wavelength, and 5 nm rms spectral width. Fiber A at 1550 nm: 1, single-mode operation; 2, 2.5-nm rms spectral width; 3, 5-nm rms spectral width.

were of triangular core index profile and surrounding index ring design and exhibited a much lower intrinsic attenuation than that assumed for step index core fiber type B. All together this shows that dispersion-shifted fibers could increase the repeater spacing by about 50% in terrestrial systems, with present high-performance splicing technology.

Evaluation of Undersea Systems. We consider a multimode microbending loss of 0.3 dB/km and the same sources as for terrestrial systems. Figure 7.10 shows the repeater spacing L as a function of the bit rate for the various combinations. It is seen that in all cases fiber B at 1550 nm provides the highest repeater spacing and

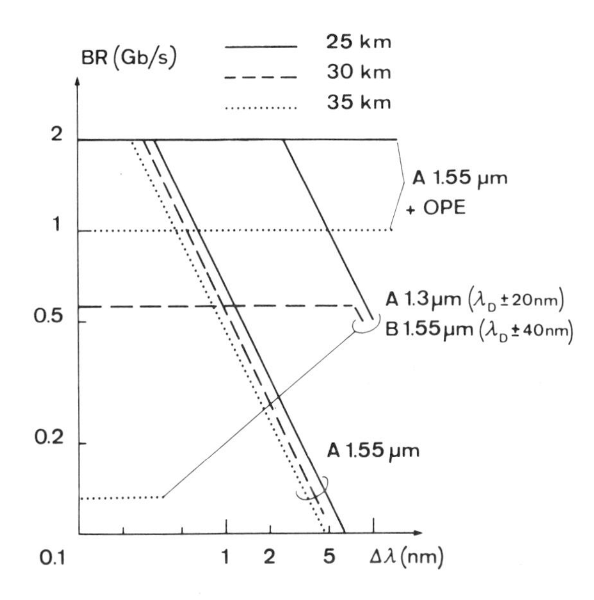

FIGURE 7.9 Maximum allowed bit rate as a function of rms source spectral width, for various terrestrial repeater spacings (splice offset 2 μm; splice spacing 2.5 km). Fiber A around 1300 nm is assumed to operate at a wavelength shifted by less than 20 nm from the zero-dispersion wavelength, while fiber B around 1550 nm has a tolerance of 40 nm. Horizontal lines are due to loss limitations, while diagonal lines are due to bandwidth limitations.

the smallest sensitivity to source spectral fluctuations. It thus appears to be the most interesting fiber for this application, and this is confirmed by Fig. 7.11, where we have plotted the maximum repeater spacing as a function of rms source spectral width. However, as reliability is a primary requirement for undersea systems, it may appear that the use of fiber A and an OPE (with a 3 dB insertion loss) is more attractive, because it is less sensitive to any laser degradation, while still allowing long repeater spacings.

Finally, although these combinations could offer repeater spacings of over 70 km, it is likely that the first generation of undersea optical systems will use the 1300-nm wavelength window and shorter spans between splices, because of available technology. Comparing Figs. 7.8–7.11 shows that it is reasonable to expect over 40-km repeater spacings up to 400 Mb/s at 1300 nm with fiber A. A better insensitivity to laser spectral characteristics and spectrum aging could be obtained using an OPE, which is a passive component.

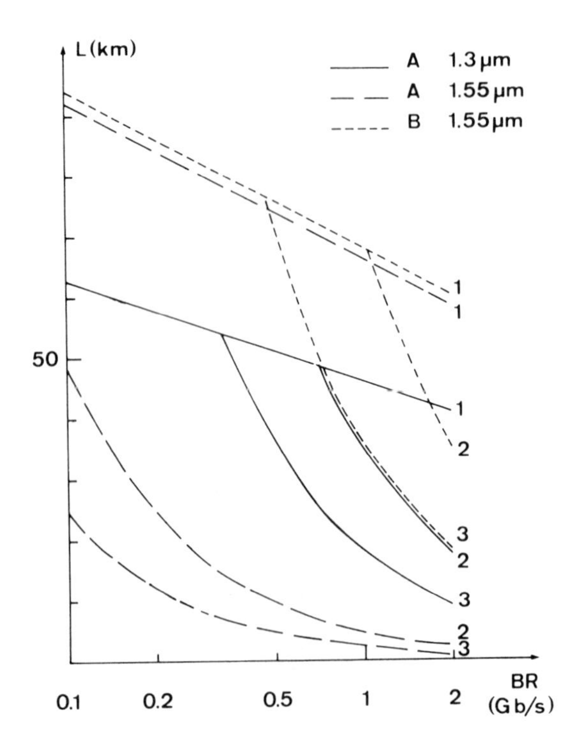

FIGURE 7.10 Repeater spacing as a function of bit rate for undersea systems (α_{mm} = 0.3 dB/km). Labels 1, 2, and 3 are the same as in Fig. 7.8.

Undersea transmission systems present an active application area for single-mode fiber technology, as this technology provides a more than 30% cost reduction as compared to coaxial cable technology. This is due to the very large repeater spacing achieved with single-mode fibers (50 to 150 km) as compared to coaxial cables (5 km). There is a large variety of applications ranging from unrepeatered transmission systems between a shore terminal and an offshore platform carrying a 3 Mb/s bit rate over 147 km at 1550 nm [29], to transoceanic systems such as TAT-8 (transatlantic), transmitting a bit rate of 280 Mb/s around 1300 nm with an average repeater spacing of about 60 km. Table 7.3 shows the design features for transoceanic undersea systems. For applications involving a total distance between the end terminals of less than 150 km (e.g., shore terminal to offshore platform, continental shore to islands, crossing a bay . . .), there is a tendency to try avoiding the submerged repeaters by using the

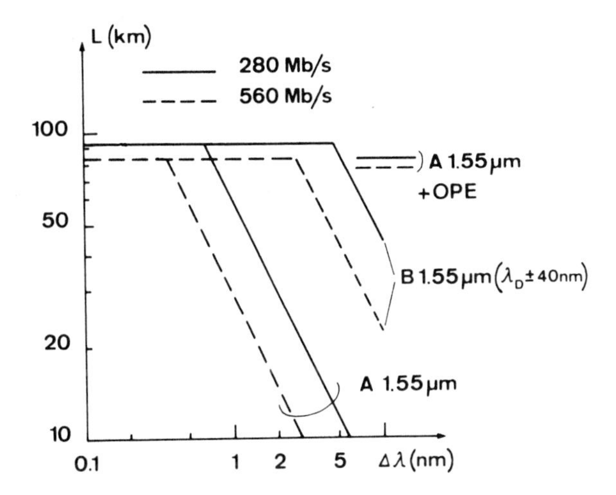

FIGURE 7.11 Maximum repeater spacing as a function of rms source spectral width for various bit rates (undersea link with α_{mm} = 0.3 dB/km). Fiber B around 1550 nm is assumed to operate at a wavelength shifted by less than 40 nm from the zero-dispersion wavelength.

1550-nm window: either with dispersion-unshifted fiber if the bit rate is low (BR < 10 Mb/s) or with dispersion-shifted fiber. Ultimately, the availability of DFB lasers will allow us to use dispersion-unshifted fibers even for higher bit rates.

For applications exceeding a total distance of 150 km, today's technologies do not allow us to avoid submerged repeaters (as of 1988). As the reliability of the lasers is of utmost importance because of the cost of recovering a faulty repeater for repair, only the 1300-nm window is usable today. However, as soon as 1550-nm DFB lasers are found to be reliable enough, it is likely that the 1550-nm window will be selected, because of the possible decrease of about 60% in the number of repeaters. In the meantime, dispersion-shifted fibers might be used at 1550 nm with normal lasers and still would provide a decrease of about 30% in the number of repeaters.

The TAT-8 is the first transoceanic optical fiber transmission systems to be installed (in 1988). A comparable system was installed between France and Corsica in 1987, operating at 140 Mb/s (instead of 280 Mb/s for TAT-8) [30]. The total link length is 392 km and nine submerged repeaters were used. The maximum sea depth is 2700 m. The type A fibers are placed with 1.1% slack in the helical grooves of a plastic rod and can move within their

TABLE 7.3 Design features for transoceanic undersea systems

	United Kingdom	United States	France	Japan
System length (km)	7500	8000	10,000	10,000
Water depth (m)	7500	7500	>6500	8000
Bit rate (Mb/s)	140–280	280	140–280	260–400
Wavelength (nm)				
1	1300	1300	1300	1300
2	1550	1550	1550	1550
Number of digital lines	1–5	<3	1–4	2–3
Supply current (A)	1.2	1	1	1
Reliability (MTBF, years)	10	8	15	10
Repeater span (km)	25–50	25–50	25–50	25–50
Repeater power/fiber (W)	3–5	4	5	3–5
Laser redundancy	—	≤3 standby	≤3 standby	≤1 standby
Optical loss (dB/km) (cabled + spliced)	<1	<1	<1	<1
Cable strength (kN)	90	80	80	75–100

Source: Ref. 24.

grooves to accommodate cable elongation (up to 100 kN tensile strength without fiber stress). A vault of steel wires provides the cable with the necessary longitudinal strength and protects the optical core from seawater pressure (see Fig. 7.7). Each repeater encloses four regenerators (2 × 2 systems), which are equipped with InGaAs-PIN receivers. Each regenerator includes one working and three standby transmitters using 1300-nm BH lasers to achieve the required system lifetime of 25 years. The working transmitter is selected by single-mode optomechanical switches.

Laboratory System Experiments. We now analyze some laboratory experiments, which are represented in Fig. 7.1. Remember that these laboratory experiments generally use uncabled fibers, and very few splices with optimization, and are thus not directly comparable to the predicted repeater spacings obtained here for cabled, spliced links with system margins.

The main characteristics of the experiments carried out in Refs. 1 and 2 are shown in Table 7.4. The experiment in Ref. 1 is typical of future trends in terrestrial systems: high bit rate and moderate repeater span. In addition to these characteristics it is worth noting that obtaining these performances required adjustment of the laser bias current to decrease its line width. The experiment in Ref. 2 is more representative of undersea links, and all the technologies involved in this experiment were designed and suited for transatlantic links (moderate bit rate, maximum length). In this last case, the same technologies made it possible to build up a link with an 84-km repeater span at a 420-Mb/s data rate. Note finally that the receiver sensitivity is better in both cases than the one we used for our calculations, and that the uncabled fiber loss is smaller than the one used in the theory. We turn next to the experiments in Refs. 4 and 7, which correspond roughly to the same philosophy as that used in the experiments noted above but at a 1550 nm wavelength. The characteristics of these experiments are shown in Table 7.5. The experiment in Ref. 7 uses a step-index fiber whose zero-dispersion wavelength is around 1310 nm and is thus similar to our fiber type A. Obtaining this high bit rate × length product despite the high chromatic dispersion of this fiber at the 1550 nm operating wavelength is possible because of the use of a very narrow spectral width source, namely, a DFB laser.

On the other hand, Ref. 4 uses a fiber with a zero-dispersion wavelength of about 1550 nm, which has typical parameters close to those of our fiber type B (see Table 7.1). This yields an extremely high bit rate × length product (103 Gb × km/s) without a special requirement on the source line width. Note, however, that the resulting performance is not much better than the similar experiment carried at 1300 nm with a 88.6-Gb × km/s product in Ref. 1 (see Table 7.4). Actually, the loss of this high-Δn fiber at 1550 nm is higher than that of fiber type A by about 0.2 dB/km, despite the smaller number of splices per unit length (1 for 26 km in Ref. 4, 1 for 6 km in Ref. 7), and if we compare these figures with our predictions for undersea systems (Table 7.2), it appears that whereas the loss of fiber type A is well approximated, the loss of fiber type B may be more difficult to decrease down to its theoretical

TABLE 7.4 Principal characteristics of experimental 1300-nm laboratory single-mode-fiber links

	Ref. 1	Ref. 2
Bit rate (Gb/s)	2	0.274
Repeater spacing (km)	44.3	84
Fiber type	Step	T-DIC
Δn	0.3 ± 0.05%	$\underset{\sim}{\sim}$0.7% (+0.5%, −0.2%)
Core diameter (μm)	10.7 ± 0.05	$\underset{\sim}{\sim}$7.5
λ_c (μm)	1.16 ± 0.03	1.23 ± 0.02
Splices	7, fusion	12, fusion + 3, epoxy
Total loss (dB)	25.3	38
Average loss (dB/km)	0.57	0.38
Source rms $\Delta\lambda$ (nm)	1.5	—
Launched power (dBm)	−1.2	—
Margin (dB)	2.9	—
Detector/amplifier	Ge-APD-Si bipolar transistor	InGaAs-PIN-FET
Laser	InGaAsP-BH	InGaAsP-BH
Coupling loss (dB)	4	4

limit. As discussed in Sec. 3.4.1, this requires very good control of the drawing conditions, and is improved by triangular core index profiles.

7.3.4 Analog Systems

One application of this type of system in found in Ref. 31, where a color video signal with a 53-dB unified weighted signal-to-noise ratio (CCIR) is transmitted using a pulse frequency modulation (PFM) method over 68 km of fiber at 1300 nm, and 102 km at 1520 nm. The fibers are the same as those used in Ref. 21, and the total link loss was 28.8 dB at 1300 nm, including five fusion splices and one connector (a 2-dB loss). The lasers provided −12 dBm of launched power at 1300 nm, and −5 dBm at 1520 nm, with a source

TABLE 7.5 Principal characteristics of experimental 1550-nm laboratory single-mode-fiber links

	Ref. 4	Ref. 7
Bit rate (Gb/s)	2	0.140
Repeater spacing (km)	51.5	223
Fiber type	Step	Step
Δn	0.01	0.0045
Core diameter (μm)	4.5	9
λ_c (μm)	? ($\simeq 1$)	1.2 ± 0.07
Splices	2, fusion	37, fusion
Total loss (dB)	27.7	49.3
Average loss (dB/km)	0.54	0.22
Source rms $\Delta\lambda$ (nm)	—	0.2
Launched power (dBm)	−2.2	+1.5
Margin (dB)	1.4	1.2
Detector/amplifier	p^+nn^- Ge-APD-Si bipolar transistor	Ge-APD
Laser	InGaAsP-BH	Ridge-waveguide DFB
Coupling loss (dB)	4.5	4

rms spectral width of about 2 nm. The detector consists of an InGaAs-PIN-FET combination, providing minimum detectable power of −43.8 dBm and −39 dBm, respectively, at 1300 nm and 1520 nm (for the video signal-to-noise ratio noted above). The PFM carrier frequency is 30 MHz with a peak deviation of 7 MHz, yielding a 10 MHz base bandwidth. It has been shown that there was no power penalty due to dispersion effects at 1300 nm (close to the fiber's zero-dispersion wavelength), whereas a 3-dB power penalty was incurred at 1520 nm. Note, however, that this penalty is far below what would be incurred at 140 Mb/s with the same sources.

Another application area concerns microwave delay lines, related to radar applications, interferential radioastronomy, and so on. An application to radar delay lines has been discussed in Ref. 32. The purpose of the delay line is to allow the subtraction of a 1-ns return radar pulse from the next 1-ns return radar pulse, both pulses being separated by about 1 ms, in order to determine the speed of the moving target. This therefore requires a 200-km delay line with an overall bandwidth of more than 1.5 GHz. Obviously, the quality of the output signal requires very linear ultra-low-noise repeaters, and such a repeater is described in Ref. 32 at 850-nm wavelength. It is concluded that only very few cascaded repeaters can be used to preserve the required signal-to-noise ratio, and thus the fibers should be operated in the long-wavelength region.

Finally, the effect of laser-mode partition noise seems to be much more dramatic in analog systems than in digital systems, and the product (frequency × fiber length × source rms width × chromatic dispersion) should be limited to about 0.01, which is about 10 times less than for digital systems [33].

7.4 COHERENT DETECTION TRANSMISSION SYSTEMS

The use of coherent optical detection is allowed by the spatial coherence of the LP_{01} mode transmitted by single-mode fibers. It is well known that this detection method may provide significant improvements in terms of minimum detectable power, and a detailed study of the various optical modulation/detection methods is given in Ref. 34, some results being summarized in Fig. 7.12. The best reported results for laboratory experiments [11,12] show that the improvement over direct detection may be very significant, although not as large as theoretically evaluated. At 400 Mb/s, the receiver sensitivity was −41 dBm for direct detection in Ref. 8 and −49 dBm for frequency-shift-keying (FSK) heterodyne detection in Ref. 11. At 2 Gb/s, the improvement reported is only from −37 dBm [9] to −39 dBm with phase-shift-keying (PSK) heterodyne detection [12]. Detailed studies of system design are given in Refs. 35 and 36. We discuss here the particular features and requirements of a PSK heterodyne differential detection system, as it has been shown in these studies to yield a good trade-off between system performance and technical difficulties.

7.4.1 System Analysis

Transmitter. A typical transmitter is shown in Fig. 7.13a. The single-longitudinal-mode laser diode is temperature stabilized, and its emission optical frequency is controlled by an automatic optical frequency control loop (AOFC) such as those described in Refs. 37–39, which makes it possible to obtain a stability of better than 1 MHz over long periods. Regarding spectral purity, natural laser line widths of about 1 MHz at 35% above threshold have been measured [40], and the use of optical feedback through reflection from an

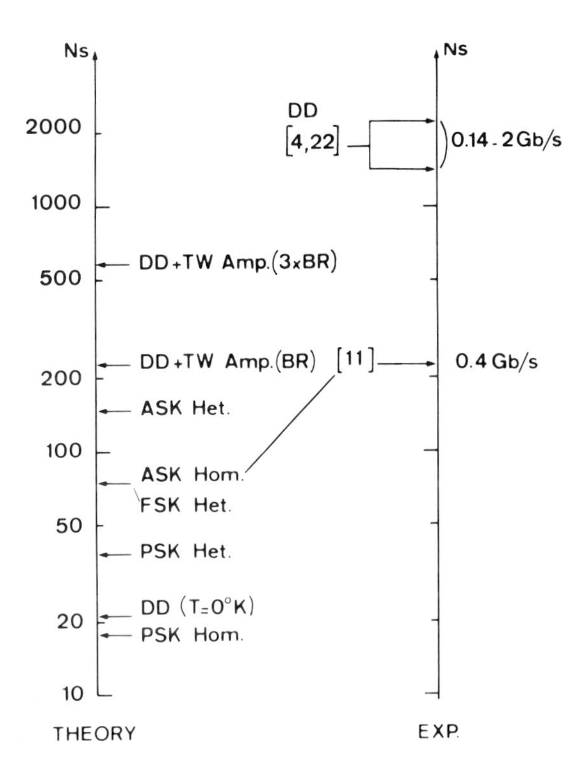

FIGURE 7.12 Theoretical and experimental minimum detectable number of signal photons per bit for a 10^{-9} bit error rate at 1550 nm. ASK, FSK, and PSK are amplitude, frequency, or phase shift keying, respectively, with coherent heterodyne or homodyne detection. DD is direct detection (intensity modulation) and has been evaluated theoretically for no thermal noise (T_e = 0 K), or from the experimental values from Refs. 4 and 22 and traveling-wave (TW) optical preamplifiers with a spectral filter having a width equal to the bit rate, or to 3 × BR. Experimental results are labeled through their respective references. Theoretical calculations have assumed a detector's quantum efficiency q = 1, and that only 50% of the incoming signal photons reached the detector for coherent systems because of the mixer losses.

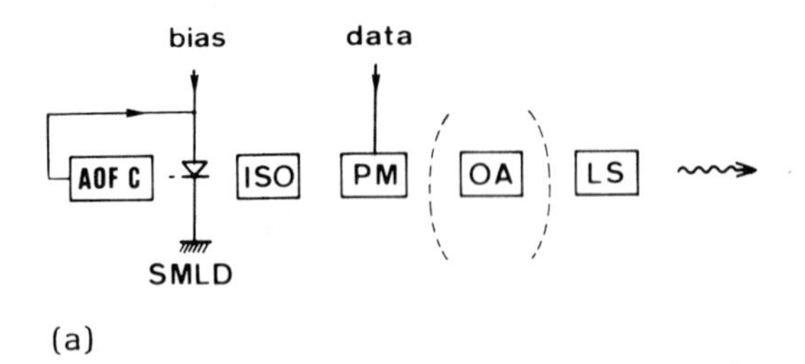

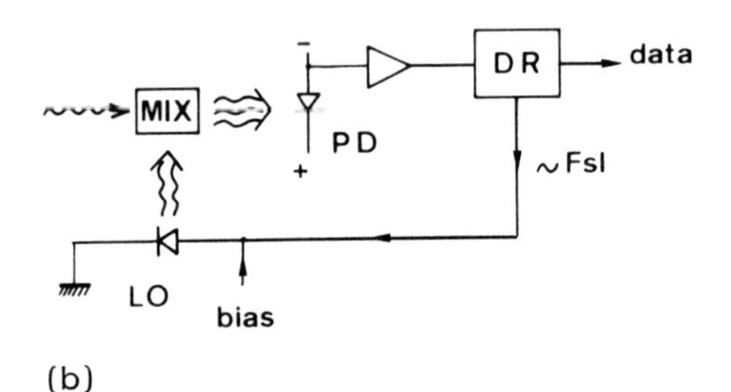

FIGURE 7.13 (a) Transmitter for PSK heterodyne systems. AOFC, automatic optical frequency control; SMLD, single-longitudinal-mode laser diode; ISO, optical isolator; PM, phase modulator; OA, optical amplifier (optional); LS, launching system. (b) Basic receiver for PSK heterodyne systems. MIX, optical mixer; LO, optical local oscillator; PD, photodetector; DR, data recovery.

external grating [39], a mirror [41], or a 1-m-long optical fiber [42], even made it possible to obtain a line width of less than 100 kHz. These figures are quite compatible with PSK differential detection systems for bit rates above 140 Mb/s as the line width should be less than about 0.002 times the bit rate (whereas for FSK, the line width can be as large as the bit rate) [36]. The isolator prevents spectral instabilities in the laser, which could be caused by light reflections from the following devices, and it can be realized with the techniques described in Sec. 6.4.2 or in Ref. 43. The phase modulator is a well-known device that is easy to manufacture using integrated optics techniques.

The optional optical amplifier could be used to maximize the power launched into the fiber. It has been shown in Ref. 35 that this amplifier should be of the traveling-wave type, operated in the saturated gain regime, and that its broadband noise power can be neglected. The power-handling capability of the fiber is determined by nonlinear optical effects (see Chap. 9), and a rough estimation shows that stimulated Raman scattering limits the launched power to about 3 W, whereas stimulated Brillouin scattering brings the limit to about 20 mW at 100 Mb/s and 200 mW at 1 Gb/s [36]. If this power capability can be fully exploited, it appears that the total allowed link loss becomes independent of the bit rate as the minimum detectable power also increases linearly with the bit rate, as will be shown below. If polarization-maintaining fibers are used, the launching system obviously has to maximize the coupling efficiency into the fiber and must also adapt the state of polarization of the light to one of the polarization eigenstates of the fiber.

Receiver. A basic receiver is shown in Fig. 7.13b, where we assumed that the light entering the mixer has a stable SOP, by the use of either polarization-maintaining fibers (see Sec. 2.4.3), or a polarization stabilization loop (Refs. 44 and 45 and Sec. 6.4.1). Note, however, as will be discussed below, that the particular case of differential PSK allows us to design a polarization diversity receiver which is therefore insensitive to the transmitted SOP fluctuations [46–48].

The optical mixer has to adapt the SOPs and the field geometry of the transmitted and local oscillator waves to each other in order to maximize the mixing efficiency. Practically, this last function can be achieved with either a bulk-optical beam splitter, or a SMF X-coupler, as described in Sec. 6.2.3. Usually, the coupler transmission coefficient will be chosen high (e.g., 80 to 90%, leading to a small cross-coupling coefficient of 20 to 10%), to fully exploit the weak signal power, this being allowed by a strong local oscillator power. Let us recall that such SMF X-couplers show very small insertion losses, and one can assume that the signal attenuation by the mixer is between 2 and 3 dB. The optical local oscillator is similar to that of the transmitter and should exhibit similar spectral characteristics.

The output signal photocurrent of the photodetector takes the form [35]

$$I(t) = MsP_1 + 2Ms(P_sP_1)^{\frac{1}{2}} \cos[2\pi F_{s1}(t)t + U_{s1}(t) + m(t)] \qquad (7.5)$$

where M is the multiplication factor of the photodetector; s its sensitivity; P_s and P_1 the signal and local oscillator powers, respectively ($P_s << P_1$), falling onto the detector; F_{s1} the intermediate frequency (optical carrier frequency – local oscillator optical frequency); U_{s1} the relative phase of the signal carrier with respect to the local oscillator; and m the phase modulation (m = 0 or π for antipodal PSK modulation).

Note that in usual heterodyne systems, the intermediate frequency F_{s1} should exceed four times the bit rate. Differential phase demodulation is performed as in microwave systems, by splitting the preamplified current into two parts, one being delayed by one bit duration (1/BR) before being recombined with the other part in a quadratic mixer whose output passes through a low-pass filter (Fig. 7.14). Assuming that the intermediate frequency $F_{s1}(t)$ fluctuates slowly compared to the bit rate BR, the output from the low-pass filter reads

$$s(t) = \frac{1}{8} G^2 M^2 P_s P_1 \cos\left[2\pi \frac{F'_{s1}(t)}{BR} t + 2\pi \frac{F_{s1}(t)}{BR} + U_{s1}(t) - U_{s1}\left(t - \frac{1}{BR}\right) + m(t) - m\left(t - \frac{1}{BR}\right) \right] \quad (7.6)$$

where G is the preamplifier gain and F'_{s1} is the derivative of F_{s1} with respect to t.

The desired information is all contained within the term $m(t) - m(t - 1/BR)$, which amounts to 0 or $\pm\pi$, depending on the modulation variation between the successive bits. However, this information is flawed by various noise sources, which are discussed below.

First, it is desirable to choose $F_{s1} = K(BR/2)$ (where K is an integer, greater than 8) to maximize the variation of s(t) when m jumps from 0 to π. This can be achieved by properly adjusting the temperature and bias current of the local oscillator. From this point of view, the value of F_{s1} should be maintained within $0.07 \times BR$ from its nominal value for about 0.5 dB optical power penalty (which leads to 10 MHz for BR = 140 Mb/s). This can easily be achieved by the use of the intermediate-frequency feedback loop shown in Figs. 7.13b and 7.14, based on the same principle as that described in Ref. 38, which lead to a stability of 1 MHz (for one oscillator, and thus a 2-MHz worst case for F_{s1}).

The term $F'_{s1}(t)t$ arises from the quantum phase noise of both the transmitter's and the receiver's optical oscillators and is related to their line widths. For a negligible contribution to the bit error rate, the full width at half maximum of both laser spectra should not exceed BR/140, that is, about 1 MHz for a bit rate of 140 Mb/s [36].

The other optical noise sources are the intensity fluctuation of the oscillators (which are likely to be very small in such well-controlled lasers), and the phase fluctuations due to the fluctuations of the fiber transmission delay (temperature fluctuations, acoustic waves, which influence the terms U_{s1}). The fiber turbulence term $[U_{s1}(t) - U_{s1}(t - 1/BR)]$ can be approximated by $U'_{s1}(t)/BR$ (U'_{s1} is the time derivative of U_{s1}), and it is limited by the same

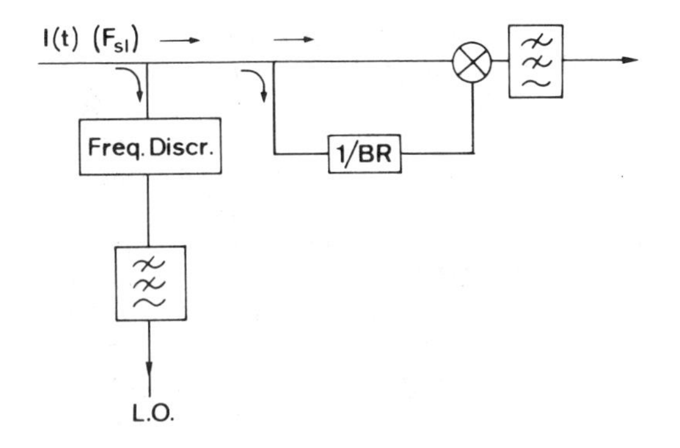

FIGURE 7.14 Details of the data recovery system for a differential PSK heterodyne system, with the demodulator and the intermediate-frequency feedback loop on the laser oscillator.

figures as the optical oscillators linewidths: $U'_{s1} < 2\pi \times BR/140$. This will be discussed below.

Finally, the electrical power signal-to-noise ratio is given approximately [35,36]

$$SNR_P = \frac{qN'_s/M^x}{1 + 2kT_e/R_1 s\ e\ P_1 M^{2+x}} \tag{7.7}$$

where q is the quantum efficiency of the photodetector, N'_s the number of signal photons per bit (emerging from the mixer and falling onto the detector), x the excess noise factor, k the Boltzmann constant, T_e the equivalent noise temperature of the load impedance R_1, and e the electron charge. The bit error rate with maximum a posteriori decoding is given by Ref. 34.

$$BER = \frac{1}{2}\ \mathrm{erfc}\ (SNR_P)^{\frac{1}{2}}$$

From Eqs. (7.7) and (7.8), it can be calculated that the improvement in sensitivity compared to direct intensity modulation detection can reach about 17 dB for a PSK heterodyne differential [34]. Theoretical results for other modulations schemes are also shown in Fig. 7.12, together with an experimental result obtained for FSK heterodyne detection [11], which shows a practical improvement of 9 dB over direct detection [8].

The attainment of this ultimate sensitivity improvement requires that the local oscillator power P_1 be sufficiently high to suppress the thermal noise effect [see Eq. (7.7)]. If a PIN photodetector ($M = 1$) is considered together with a 50-Ω load resistance, calculations based on the normal T_e values show that a power of several milliwatts would be necessary [35,36], which would hardly be acceptable by the photodetector. The required local oscillator power can be decreased by the use of APDs [$M > 1$, as P_1 decreases as M^{-2-x}, whereas N'_s increases as M^x only; see Eq. (7.7)] or PIN-FET combinations. However, the requirement for high-frequency operation ($F_{s1} \geq 4$ BR) in heterodyne systems may be incompatible with such combinations, and other possibilities could be envisaged with resonant circuits as load impedance (allowed by the small relative bandwidth $\simeq 2$ BR/F_{s1}).

Polarization Diversity Receivers. Up to now we assumed a stable SOP for the signal wave, which requires either polarization-maintaining fibers or active polarization feedback loops. However, it is possible to design a polarization diversity receiver, in the particular case of differential PSK demodulation, which is insensitive to the SOP of the transmitted wave [46–48]. This requires simply the use of a polarization beam splitter at the output of the fiber, which separates the wave into two orthogonal SOPs (Fig. 7.15). Each of the resulting waves has a randomly variable amplitude and phase (which reflect the random variations of the initial SOP), but also a well-defined SOP which makes it possible to use a standard receiver as described above in each arm. The two output signals from the two data recovery systems obey an equation similar to Eq. (7.6), each having a randomly variable value of P_s and a random phase U_{s1}. The random phase U_{s1} is eliminated as discussed above, because the SOP fluctuations remain much slower than the bit rate; and by adding the two signals we obtain a stable output because the sum of the two intensities P_s is constant (this assumes that the product of the various transmission and gain coefficients, from the polarization beam splitter up to the electronic adding circuit, is equal in both arms).

Because of the use of two basic (stable SOP) receivers here, there is an intrinsic power penalty of 3 dB, increased by the practical polarization beam splitter insertion loss (about 1 dB), compared to the case where the transmitted wave has a stable SOP. However, this may prove more efficient than active polarization feedback loops which have rather high insertion losses, or polarization-maintaining fibers, as it is not certain that they can be manufactured with a loss increase of less than 0.05 dB/km compared to standard circular core fibers (for a 100-km repeater span, this amounts to a 5-dB loss increase, comparable to the sensitivity

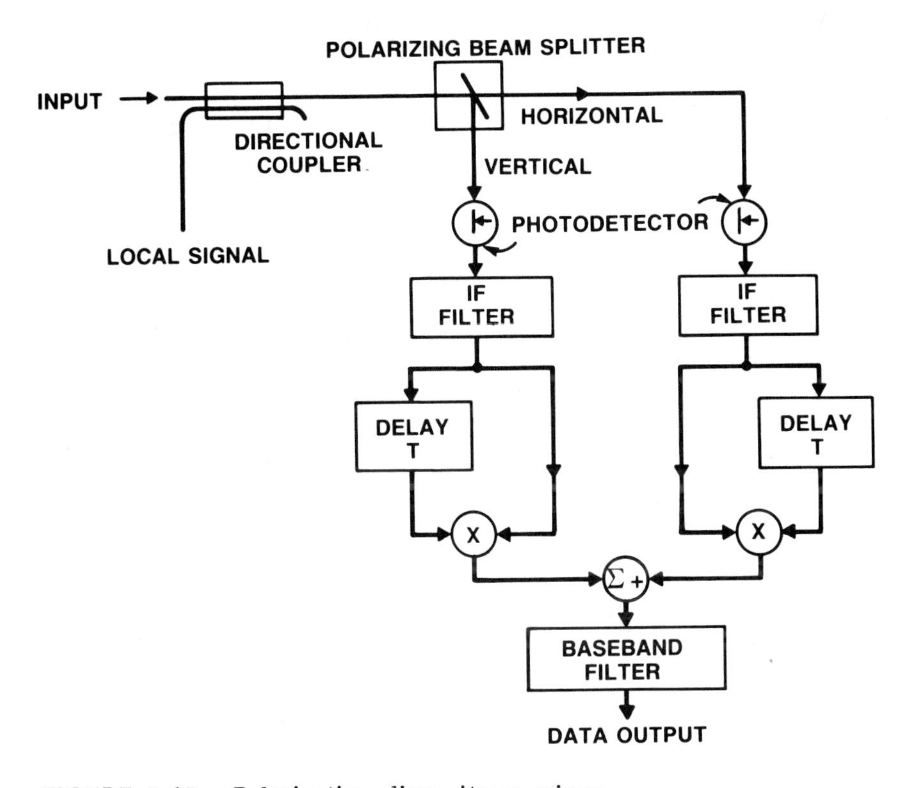

FIGURE 7.15 Polarization diversity receiver.

penalty observed here). A sea trial of an FSK heterodyne transmission system has shown the practical feasibility of polarization diversity reception [49].

Transmission Line. The transmission line itself is composed of a single-mode-fiber cable with joints, and may bring repeater spacing limitations through chromatic dispersion, birefringence properties, loss, or turbulence. Loss limitations are obviously similar to those discussed in Sec. 7.2.2. If we consider a minimum loss of about 0.4 dB/km, a theoretical receiver sensitivity improvement of 17 dB as compared to direct detection, and an additional loss of 2 to 7 dB in the receiver because of additional optical elements (mixer, polarization control, polarization diversity receiver), we find that practical coherent detection systems will increase the repeater spacing by 25 to 40 km compared to direct detection systems, with the same launched power. This is roughly comfirmed by Fig. 7.1. However, direct detection systems could use higher power sources with larger line widths and yield the same repeater spacing increase, because of Raman, rather than Brillouin scattering, launched power limitations. On the other hand, it has been shown in Sec. 4.4.2 that worst-case chromatic dispersion limitations are on the order of 40 Gb $\times$ km$^{\frac{1}{2}}$/s, but almost disappear if dispersion-free fibers or electronic group delay compensation is used.

Birefringence properties lead to two effects. The first is the output SOP stability, which was discussed in Sec. 2.4.3 for polarization-maintaining fibers, and immediately above for polarization diversity receivers. The second associated effect is due to birefringence dispersion, which was discussed in Sec. 4.4.2, allowing bit rates up to 10^4 Gb $\times$ km/s with low birefringence fibers (if chromatic dispersion effects are compensated) but only 100 Gb $\times$ km/s if linear polarization-maintaining fibers are used.

Finally, the last possible limitation arises from phase turbulences due to temperature- or acoustically incuded phase fluctuations in the fiber (see Chap. 8). The temperature-induced turbulence has been evaluated experimentally for a bare fiber [50], and if we consider that the corresponding optical frequency turbulence should not exceed BR/140 (as for the source line width), we obtain at 1550 nm (worst-case limitation)

$$L\ (\text{km}) \leq 7 \times 10^6\ \frac{\text{BR (Mb/s)}}{140} \tag{7.9}$$

This can clearly be neglected, even if the actual fiber conditioning (cable) increases its temperature sensitivity by a factor of 100 or so. For acoustically induced turbulence there is no experimental evaluation; however, we will get an order of magnitude estimate by using the typical fiber sensitivity to pressure, the sea noise level (see Sec. 8.2.1), and a discussion analogous to that in Ref. 50. The sea noise level spectral power density is about $(10^{-6}/f^2)$ Pa^2/Hz and the pressure sensitivity coefficient of a bare fiber of length L (km) at 1550 nm is about 1.8×10^{-2} L rad/Pa. We thus obtain the optical frequency fluctuation spectral power density as $(1.8 \times 10^{-5}\ L/2\pi)^2$ Hz^2/Hz, which has to be multiplied by the detection bandwidth (2 BR). The rms value has then to be compared to BR/140, yielding the limitation

$$L\ (\text{km}) \leq 2 \times 10^7 \left[\frac{\text{BR (Mb/s)}}{140}\right]^{\frac{1}{2}} \tag{7.10}$$

This again is of no practical importance, even if.the fiber coating increases its sensitivity by a factor of 1000 or so, or if the cable is in a more noisy environment than the assumed ocean bottom.

It thus appears that loss limitations are far more dramatic than all other limitations and do not provide very attractive repeater spacing increase over that of direct detection, for compensating the increased system complexity. However, the heterodyne detection principle itself functions in narrow optical band detection, as optical frequencies are translated into electrical frequencies, and thus the optical spectrum is filtered by the electronic filters associated with the receiver circuitry. This allows efficient rejection of the broadband amplified spontaneous emission noise generated by optical amplifiers, and thus permits their use rather than the use of regenerative repeaters, as at the same time bit-rate limitations do not require pulse reshaping over distances of less than 1000 to 10,000 km. Such ultra-long regenerative repeater spacings can be attained potentially by cascading nonregenerative repeaters (optical amplifiers) each 40 to 60 km [36]. If such practical arrangements can be attained in the future, heterodyne detection systems could offer the very attractive feature for undersea links of requiring simpler repeaters than direct detection links, as even very narrow optical filtering should not make it possible to cascade many optical amplifiers in direct detection systems. Experiments have been reported which tend to confirm this possibility [13,14], as shown in Fig. 7.1.

7.4.2 Wavelength-Division Multiplexing

Wavelength-division multiplexing is of interest for increasing the total capacity of one fiber. Although single-mode fibers still have a very high capacity at a single wavelength, it may happen that wavelength-division multiplexing will be required if large-scale video services are introduced in the telecommunications network.

Because of the very narrow line width and stable frequency of the transmitters, coherent detection allows very fine wavelength-division multiplexing, with channels separated by 10 to 100 GHz (0.1 to 1 nm at 1550-nm wavelength). The stable frequency separation between channels could be obtained by the use of feedback loops based on diffraction gratings as in Ref. 39, and the wavelength multiplexers and demultiplexers could be simple dispersive elements such as monochromators. Obviously, regenerative repeaters would use as many receivers and transmitters as different channels, but the very wide amplification bandwidth of the traveling-wave optical amplifiers ($\simeq$3000 GHz) would allow very simple nonregenerative repeaters, as a single amplifier would serve for all channels simultaneously.

An ultimate coherent transmission system could thus use, for example, 100 channels separated by 10 GHz, each channel carrying a 1-Gb/s data rate with regenerative repeater spacings of 1000 km or more. Clearly, in such a system the sensitivity of coherent detection receivers is no longer the dominant feature, but rather the possibility of using simple nonregenerative repeaters and many channels with simple multiplexers.

7.4.3 Laser Amplifiers

The advent of very low loss single-mode fibers has increased interest in laser amplifiers, which have received little attention in the past (e.g., Refs. 51 and 52), largely for two reasons:

1. Single-mode fibers can provide almost dispersion-free propagation, and the only reason for repeaters is energy attenuation by the fiber.
2. Single-mode fibers provide spatial filtering of the highly multimode noise generated by laser amplifiers (spontaneous emission), which is necessary for rendering these devices potentially useful.

Recent theoretical and experimental investigations have been carried out on such optical amplifiers [13,14,53–65] and we will briefly summarize the main characteristics of traveling-wave semiconductor laser amplifiers (TWLAs), which seem to be the most promising of all the structures studied, despite their more difficult technology. Later, we discuss the potential applications of such amplifiers to single-mode-fiber transmission systems.

Characteristics of TWLA. Basically, such an amplifier is composed of a laser diode whose facet reflectivity has been decreased to less than 0.1% in order to suppress the oscillator behavior due to the initial Fabry–Perot structure. This is technologically difficult and generally requires the use of very well controlled antireflection coatings and possibly of facets tilted with respect to the waveguide axis. Obviously, the laser diode should have two pigtails, one on each facet, to make possible in- and out-couplings.

Figure 7.16 shows the qualitative aspect of the signal optical frequency spectrum, before and after amplification by such an amplifier. The gain and amplified spontaneous emission (ASE) noise curves typically have a full width at half maximum of 3000 GHz (or 25 nm at 1550 nm). This broadband ASE noise imposes an optimum signal input power (or an optimum position of the amplifier in a link) for getting the maximum link gain, defined as:

$$\text{Link gain} = \text{TWLA net gain} - 10 \log \frac{P_0\ (\text{TWLA})}{P_0\ (\text{no-TWLA})} \qquad (7.11)$$

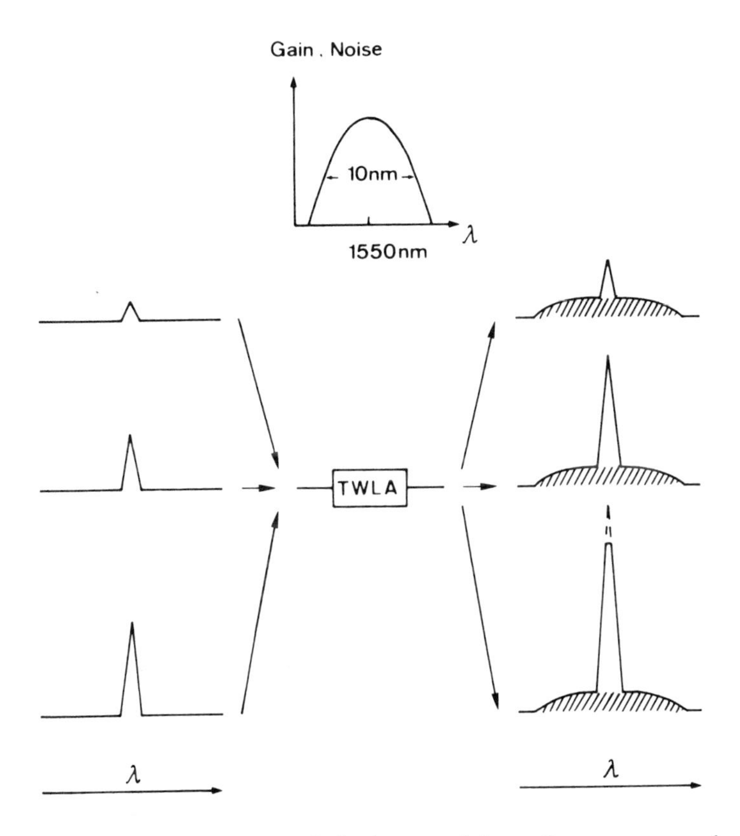

FIGURE 7.16 General features of traveling-wave semiconductor laser amplifiers. The gain and amplified spontaneous emission (ASE) noise curves have a full width at half maximum of typically 3000 GHz (or 25 nm at 1550 nm). The shaded area represents the ASE noise contribution to the output spectral power density of the wave.

where P_0 is the minimum detectable power for a given bit error rate, which is increased by the insertion of the TWLA because of the degradation of the optical power signal-to-noise ratio.

For the very weak input signal shown in the upper curve of Fig. 7.16, the link gain will be very small (if not negative), because the ASE noise brings a very important contribution to the output power, yielding a very poor optical signal-to-noise ratio which increases P_0 (TWLA). At the opposite extreme, the strong input signal shown in the lower curve also yields a poor link gain, because of saturation in the amplifier gain, thus decreasing its net gain. Finally, the middle curve illustrates an intermediate case where a useful link gain can be obtained.

From a practical point of view, however, the broadband ASE noise contribution should be decreased as much as possible by filtering out the ASE noise. Ideally, one should use a filter that blocks out all the noise outside the signal bandwidth (and this can be attained with coherent detection, which permits the use of electronic filters), but in the case of optical filters it seems difficult to achieve such a narrow filter spectral width. The use of a rather large filter width degrades the link gain and decreases the tolerance on the optimum input signal power. For example, typical estimations shows that with a TWLA net gain of 20 dB, a filtering bandwidth equal to the signal bandwidth yields a link gain of more than 19 dB for an input number of signal photons between 5×10^3 and 10^6, while a filtering bandwidth equal to 3000 times the signal bandwidth yields a maximum link gain of 19 dB, and the tolerance for the number of input signal photons (at a 1-dB link gain compression) is reduced to $10^5 \leq N_s \leq 10^6$ [56].

We can summarize the experimental data as follows [58,59]:

Maximum net gain obtained at 850 nm (including insertion losses): 20 dB.
Maximum input power for a 1-dB net gain compression: −15 dBm.
Optical signal-to-noise ratio (at 850 nm) for an optical filter bandwidth of 120 GHz (0.3 nm) and an input optical power of −32.6 dBm: 40 dB, for a net gain varying between 5 and 20 dB, in a 1-MHz detection bandwidth.
Less than 2 dB gain sensitivity to optical state of polarization.

Finally, we note that an experimental link gain of 14 dB, with an input power range for a 1-dB link gain compression of −17 dBm to −27 dBm, has been obtained with a resonant Fabry–Perot amplifier and an optical filter of 120 GHz (or 0.3 nm, around 850 nm) [56].

Applications of TWLA. Traveling-wave amplifiers should find promising applications either as front-end detection preamplifiers in direct detection systems or as on-line amplifiers (repeaters) in both direct and coherent detection systems. As for detection preamplifiers, experiments carried out at 850 nm [54] showed a 3-dB receiver sensitivity improvement, but with a resonant-type amplifier (whose ASE noise is smaller than that of TWLA). Evaluations carried out for TWLA and the direct detection performances of Refs. 4 and 22 show that the association of a TWLA and such detectors could provide an improvement in the receiver sensitivity, provided a very narrow optical filter is used (see Fig. 7.12). The application to on-line amplification, as nonregenerating repeaters seems to be much more interesting, as the characteristics discussed above show that substantial link gains can be envisaged. With the very efficient filtering of coherent detection, many amplifiers (about 150) can be cascaded for providing very long regenerating repeater spans of several thousands of kilometers. Although the performances are more limited in direct detection systems if realistic optical filters are envisaged, such amplifiers could provide interesting improvements in the overall link performances, especially at high data rates where the electronic circuitry of regenerating repeaters is quite complicated.

An experimental 1550-nm system using two optical amplifiers and direct detection showed a net gain of 26 dB [60]. Other experiments carried out at 1550 nm with coherent detection showed net gains of 8 dB for a single amplifier [65], 21 dB for two amplifiers [64], and 48 dB for four amplifiers [13]. This offered the possibility of reaching a 372-km transmission distance at 400 Mb/s.

ACKNOWLEDGMENT

Fruitful discussions with Ph. Dupuis, F. Favre, M. Joindot, A. Leclert, M. Monerie, and J. C. Simon concerning coherent detection systems and optical amplifiers are sincerely acknowledged.

REFERENCES

1. J. I. Yamada and T. Kimura, *IEEE J. Quantum Electron. QE-18*(4):718–727 (1982).
2. P. K. Runge, C. A. Brackett, R. F. Gleason, D. Kalish, P. D. Lazay, T. R. Meeker, J. C. Williams, and D. P. Jablonowski, in *OFC'82 Digest of Technical Papers*, Optical Society of America, Washington, D.C., Apr. 1982, postdeadline paper PD7.
3. J. J. Bernard, C. Brehm, A. Tardy, L. Fersing, Y. Gausson, and M. Jurczyszyn, *Proc. Opto 82*, Paris, May 1982, pp. 273–275.
4. J. Yamada, A. Kawana, H. Nagai, T. Kimura, and T. Miya, *Electron. Lett. 18*(2):98–100 (1982).
5. N. S. Bergano, R. E. Wagner, H. T. Shang, and P. F. Glodis, in *OFC'85 Digest of Technical Papers*, Optical Society of America, San Diego, CA, Feb. 1985, paper TuM2.
6. L. C. Blank, L. Bickers, and S. D. Walker, in *OFC'85 Digest of Technical Papers*, Optical Society of America, San Diego, CA, Feb. 1985, postdeadline paper PD 7.
7. L. Bickers, L. C. Blank, and S. D. Walker, *Electron. Lett. 21*(7):267–268 (1985).

8. V. J. Mazurczyk, N. S. Bergano, R. E. Wagner, K. L. Walker, N. A. Olsson, L. G. Cohen, R. A. Logan, and J. C. Campbell, *Proc. ECOC'84*, Stuttgart, Germany, Sept. 1984, postdeadline paper PD 7.
9. B. L. Kasper, R. A. Linke, K. L. Walker, L. G. Cohen, T. L. Koch, T. J. Bridges, E. G. Burkhardt, R. A. Logan, R. W. Dawson, and J. C. Campbell, *Proc. ECOC'84*, Stuttgart, Germany, Sept. 1984, postdeadline paper PD 6.
10. K. K. Korotky, G. Eisenstein, A. H. Gnauck, B. L. Kasper, J. J. Veselka, R. C. Alferness, L. L. Buhl, C. A. Burrus, T. C. D. Huo, L. W. Stulz, K. Ciemiecki Nelson, L. G. Cohen, R. W. Dawson, and J. C. Campbell, in *OFC'85 Digest of Technical Papers*, Optical Society of America, San Diego, CA, Feb. 1985, postdeadline paper PD 1.
11. K. Iwashita, T. Imai, T. Matsumoto, and G. Motosugi, *Electron Lett.* *22*(3):164–165 (1986).
12. A. H. Gnauck, R. A. Linke, B. L. Kasper, K. J. Pollock, K. C. Reichmann, R. Valenzuela, and R. C. Alferness, in *OFC'87 Digest of Technical Papers*, Optical Society of America, Reno, NV, Feb. 1987, postdeadline paper PD 10.
13. N. A. Olsson, M. G. Oberg, L. A. Koszi, and G. Przybylek, *Electron. Lett.* *24*(1):36–37 (1988).
14. M. G. Oberg, N. A. Olsson, L. A. Koszi, and G. J. Przybylek, *Electron. Lett.* *24*(1):38–39 (1988).
15. L. Jeunhomme, *IEEE J. Quantum Electron.* *QE-18*(4):727–732 (1982).
16. A. D. Pearson, in *IOOC'81 Digest of Technical Papers*, Optical Society of America, Washington, D.C., Apr. 1981, paper WA3.
17. C. Brehm and C. Le Sergent, CGE Research Laboratories, unpublished work.
18. W. T. Anderson and P. F. Glodis, *AT&T Bell Laboratories Tech. J.* *63*(3):425–430 (1984).
19. Y. Okano, K. Nakagawa, and T. Ito, *IEEE Trans. Commun.* *COM-28*:238–243 (1980).
20. S. Yamamoto, H. Sakaguchi, and N. Seki, *IEEE J. Quantum Electron.* *QE-18*(2):264–273 (1982).
21. D. J. Malyon and A. P. McDonna, *Electron. Lett.* *18*(11): 445–447 (1982).
22. D. R. Smith, R. C. Hooper, P. P. Smyth, and D. Wake, *Electron. Lett.* *18*(11).453–454 (1982).
23. J. J. Bernard and J. Guillon, *Proc. OFC'86*, Optical Society of America, Atlanta, GA, pp. 108–110 (1986).
24. I. Yamashita, Y. Negishi, M. Nunokawa, and H. Wakabayashi, *ITU Telecommun. J.* *49*(2):118–124 (1982).
25. K. Mochizuki, Y. Namihira, and H. Yamamoto, *Electron. Lett.* *19*:743–745 (1983).
26. N. Uesugi, Y. Murakami, C. Tanaka, Y. Ishida, Y. Mitsunaga, Y. Negishi, and N. Uchida, *Electron. Lett.* *19*:762–763 (1983).

27. J. Stone, *IEEE J. Lightwave Tech.* *LT-5*(5):712–733 (1987).
28. A. R. Hunwicks, P. A. Roscher, L. Bickers, and D. Stanley, *Electron. Lett.* *24*(9):563–537 (1988).
29. R. L. Lynch, L. J. Marra, and J. J. McNulty, *Proc. Suboptic 86*, Versailles, France, pp. 94–99 (1986).
30. J. H. Wilbrod, A. P. Leclerc, and F. Gobin, *Proc. ECOC'87*, Helsinki, Finland, pp. 21–24 (1987).
31. D. J. T. Heatley, *Electron. Lett.* *18*(9):369–371 (1982).
32. C. T. Chang, *IEEE J. Quantum Electron.* *QE-18*(4):741–745 (1982).
33. G. Grosskopf, L. Küller, and E. Patzak, *Electron. Lett.* *18*(12): 493–494 (1982).
34. Y. Yamamoto, *IEEE J. Quantum Electron.* *QE-16*(11):1251–1259 (1980).
35. F. Favre, L. Jeunhomme, I. Joindot, M. Monerie, and J. C. Simon, *IEEE J. Quantum Electron.* *QE-17*(6):897–906 (1981).
36. Y. Yamamoto and T. Kimura, *IEEE J. Quantum Electron.* *QE-17*(6):919–935 (1981).
37. T. Okoshi and K. Kikuchi, *Electron. Lett.* *16*(5):179–181 (1980).
38. F. Favre and D. Le Guen, *Electron. Lett.* *16*(18):709–710 (1980).
39. S. Saito and Y. Yamamoto, *Electron. Lett.* *17*(9):325–327 (1981).
40. F. Favre and D. Le Guen, in *IOOC'81 Digest of Technical Papers*, Optical Society of America, Washington, D.C., Apr. 1981, paper MJ3.
41. K. Kikuchi and T. Okoshi, *Electron. Lett.* *18*(1):10–12 (1982).
42. F. Favre, D. Le Guen, and J. C. Simon, *IEEE J. Quantum Electron.* *QE-18*(10):1712–1717 (1982).
43. T. Aoyama, K. Doi, H. Uchida, T. Hibiya, Y. Ohta, and K. Matsumi, *Proc. 7th Eur. Conf. Opt. Commun.*, Copenhagen, Sept. 8–11, 1981, paper 8-2.
44. Y. Kidoh, Y. Suematsu, and K. Furuya, *IEEE J. Quantum Electron.* *QE-17*(6):991–994 (1981).
45. R. C. Alferness, *IEEE J. Quantum Electron.* *QE-17*(6):965–969 (1981).
46. T. Okoshi, S. Ryu, and K. Kikuchi, *Proc. IOOC'83*, Tokyo, pp. 386–387 (1983).
47. B. Glance, *IEEE J. Lightwave Tech.* *LT-5*(2):274–276 (1987).
48. L. D. Tzeng, T. W. Cline, W. L. Emkey, and C. A. Jack, *Proc. OFC'88*, New Orleans, LA, pp. 32–33 (1988).
49. S. Ryu, S. Yamamoto, Y. Namihira, K. Mochizuki, H. Wakabayashi, *Electron. Lett.* *24*(7):399–400 (1988).
50. T. Musha, J. Kamimura, and M. Nakazawa, *Appl. Opt.* *21*(4): 694–698 (1982).
51. S. D. Personick, *Bell Syst. Tech. J.* *52*(1):117–133 (1973).

52. G. Zeidler and D. Schicketanz, *Siemens Forsch. Entwicklungs Ber.* *2*:227–234 (1973).
53. Y. Yamamoto, *IEEE J. Quantum Electron. QE-16*(9):1073–1081 (1980).
54. Y. Yamamoto and H. Tsuchiya, *Electron. Lett. 16*(7):233–235 (1980).
55. S. Kobayashi and T. Kimura, *Electron. Lett. 16*(7):230–232 (1980).
56. J. C. Simon, I. Joindot, J. Charil, and D. Hui Bon Hoa, in *IOOC'81 Digest of Technical Papers*, Optical Society of America, Washington, D.C., Apr. 1981, paper MH2.
57. T. Mukai and Y. Yamamoto, *IEEE J. Quantum Electron. QE-17*(6):1028–1034 (1981).
58. J. C. Simon, *Electron. Lett. 18*(11):438–439 (1982).
59. J. C. Simon, J. L. Favennec, G. Drillet, and J. Charil, *Proc. 8th Eur. Conf. Opt. Commun.*, Cannes, Sept. 1982, paper C29.
60. T. Mukai and T. Saitoh, *Electron. Lett. 23*:216–217 (1987).
61. G. Eisenstein, R. M. Jopson, R. A. Linke, C. A. Burrus, and U. Koren, *Electron. Lett. 21*:1076–1077 (1985).
62. J. C. Simon, B. Landousies, Y. Bossis, P. Doussiere, B. Fernier, and C. Padioleau, *Electron. Lett. 23*:332–334 (1987).
63. I. W. Marshall, M. J. O'Mahony, and P. D. Constantine, *Electron. Lett. 22*:253–254 (1986).
64. N. A. Olsson, *Electron. Lett. 21*:1085–1087 (1985).
65. R. C. Steel and I. W. Marshall, *Electron. Lett. 23*:296–297 (1987).

8

Sensor Applications

8.1 INTRODUCTION

The application of single-mode fibers in sensing external stimuli such as temperature, pressure, acoustic waves, current, magnetic and electric fields, and rotations is a very important and quickly developing activity. There obviously are possible ways to use single-mode-fiber sensors based on the same principles as those for some multimode fiber sensors (such as microbending losses induced by pressure transducers), but we will concentrate here on sensor designs that rely on two original properties of single-mode fibers compared to multimode fibers: the spatial coherence of the LP_{01} mode and the nondepolarizing properties of the SMF. These sensors thus exploit either phase or SOP modulation of the transmitted wave through external stimuli, which can be detected by coherent detection methods. A very detailed discussion of optical fiber sensors, particularly SMF sensors, can be found in Ref. 1.

8.2 TEMPERATURE, PRESSURE, AND ACOUSTIC SENSORS

8.2.1 Interferometric Sensors

General Principles. The basic arrangement consists in building a Mach–Zehnder interferometer with one SMF in each arm, only one of them being affected by the external stimulus (Fig. 8.1). A temperature or pressure variation will induce a modulation of the phase for the optical wave transmitted through the exposed SMF. As in a

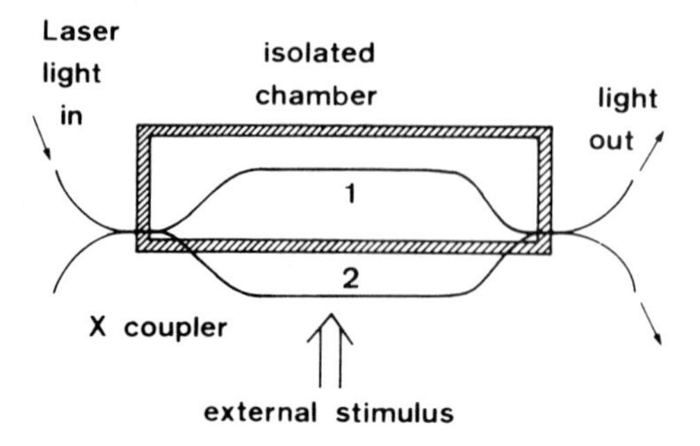

FIGURE 8.1 Basic interferometric sensor arrangement.

homodyne phase-modulation transmission system, this phase variation $d\phi$ will be detected using coherent detection methods. Such sensors have been used for acoustic wave detection [2], strain sensing [3], and pressure and temperature sensing [4].

Note that instead of a single-beam Mach–Zehnder interferometer, it is possible to use multiple-beam interferometers of the Fabry–Perot type by using a single fiber with high reflectivity coatings on its ends [5]. The transmitted intensity is a function of the phase delay in the cavity, as in any Fabry–Perot interferometer, and allows a high sensitivity; the sensitivity improvement over the Mach–Zehnder design is approximately $1.3F^{\frac{1}{2}}$, where F is the finesse of the fiber Fabry–Perot.

Sensitivity of Pure-Silica Fibers. The sources of the induced phase modulation are the variation of the normalized propagation constant b (through variations of core diameter, index difference and thus V; see Chap. 1), the variation of the refractive index (hereafter written n for simplicity), and the variation of the fiber length L. The first effect has been shown to be negligible [4], and approximating the propagation constant of the LP_{01} mode by kn, where k is the free-space propagation constant of the light, allows us to write

$$\frac{d\phi}{L} = kn \frac{dL}{L} + k\ dn \tag{8.1}$$

A detailed study of the sensitivity is found in Ref. 4 for pure-silica fibers (which obviously is an asymptotic case, as we at least need a doped silica region and a primary coating). For a pressure change dP, we have

$$\frac{dL}{L\,dP} = -\frac{1-2\nu}{Y} \tag{8.2}$$

$$\frac{dn}{dP} = \frac{1-2\nu}{Y}\frac{n^3}{2}(p_{11} + 2p_{12}) \tag{8.3}$$

where ν is the Poisson's ratio of silica ($\nu \simeq 0.16$), Y is the Young's modulus ($Y = 6.5 \times 10^{10}$ N/m^2), $n = 1.46$, $p_{11} = 0.12$, and $p_{12} = 0.27$. Using Eqs. (8.1)–(8.3) for a wavelength of 633 nm, we obtain

$$\frac{d\phi}{L\,dP} = (-15.16 + 10.66) \times 10^{-5} = -4.5 \times 10^{-5} \text{ rad/Pa}\cdot\text{m} \tag{8.4}$$

Compared to this theoretical figure for pure-silica fibers, a value of -6.8×10^{-5} rad/Pa·m has been measured in Ref. 4, the discrepancy possibly being due to the presence of doped regions and of plastic coatings, as will be seen below.

For the temperature sensitivity, the thermal expansion coefficient of silica (5×10^{-7}/°C) and its refractive index sensitivity of -10^{-5}/°C lead to a phase variation, at a wavelength of 633 nm and for a temperature change dT, of

$$\frac{d\phi}{L\,dT} = -92 \text{ rad/°C}\cdot\text{m} \tag{8.5}$$

Measured values of 83 rad/°C·m [4] and 99 rad/°C·m [6] agree more or less with this theoretical value and show again that the exact fiber structure may significantly influence the overall sensitivity.

Sensitivity Modifications. In the case of temperature a more detailed study [6] takes into account the effects of strains induced by the difference in thermal expansion coeffkient between the pure-silica regions, the doped-silica regions, and the plastic jackets. This shows that we generally have the combined effects of temperature and strains. The strain effect has been shown to become very important when the plastic jacket is thick, and a sensitivity of 261 rad/°C·m has been measured in a standard ITT single-mode fiber with its 0.25-mm-diameter silicone coating and its 0.7-mm-outside-diameter Hytrel jacket [6].

Similarly for pressure sensitivity, the fiber composition and jacket design may strongly affect the overall sensitivity essentially through a longitudinal strain contribution [7]. Several ways appear possible to reduce, or even cancel, the pressure sensitivity, by using the fact that the strain and the index contributions are of opposite signs [Eqs. (8.2)–(8.4)]: special fibers with a low Y value for the core (K_2O-SiO_2 glass) and a high Y value for the cladding; or conventional high-silica fibers with a 0.015-mm-thick nickel primary coating and a 0.1-mm-outside-diameter Hytrel jacket [7].

On the other hand, the pressure (or acoustic) sensitivity can be increased by surrounding the fiber first with a soft material coating (such as silicone, which does not affect the pressure transmission because of its high compressibility), and then with a thick jacket of a material having a high Young's modulus and a low bulk modulus (high compressibility) [8]. Such a structure is shown in Fig. 8.2a, and the corresponding acoustic sensitivity is displayed in Fig. 8.2b. Table 8.1 shows the measured acoustic sensitivity for various fiber jackets, corresponding to the structure of Fig. 8.2a. Finally, winding the bare fiber around a hallow cylinder has been found to yield a sensitivity increase by a factor of 500 for a Lucite cylinder and 2000 for a polyethylene cylinder [9], compared to the results of Eq. (8.4).

Detection. Before analyzing the detection scheme, let us recall that the field transfer matrix of an X-coupler relates the output fields to the input fields by

$$\begin{pmatrix} S_1 \\ S_2 \end{pmatrix} = \begin{pmatrix} \hat{t} & \hat{c} \\ -\hat{c}^* & \hat{t}^* \end{pmatrix} \begin{pmatrix} E_1 \\ E_2 \end{pmatrix} \tag{8.6}$$

where S_1 and S_2 are the output fields in each arm of the X-coupler, E_1 and E_2 are the input fields, and $\hat{t}$ and $\hat{c}$ are the field transmission and cross-coupling coefficients, respectively. For a lossless coupler, we have

$$|\hat{t}|^2 + |\hat{c}|^2 = 1 \tag{8.7}$$

If we assume that in the interferometer of Fig. 8.1 a power P is launched as indicated and that both couplers are identical, the signal photocurrents delivered by photodetectors placed in each output arm will be

$$\begin{aligned} I_1 &= (|\hat{t}|^4 + |\hat{c}|^4 - \hat{t}^2|\hat{c}|^2 e^{-i\phi} - \hat{t}^{*2}|\hat{c}|^2 e^{i\phi})\, sP \\ I_2 &= (2|\hat{t}|^2|\hat{c}|^2 + \hat{t}^2|\hat{c}|^2 e^{-i\phi} + \hat{t}^{*2}|\hat{c}|^2 e^{i\phi})\, sP \end{aligned} \tag{8.8}$$

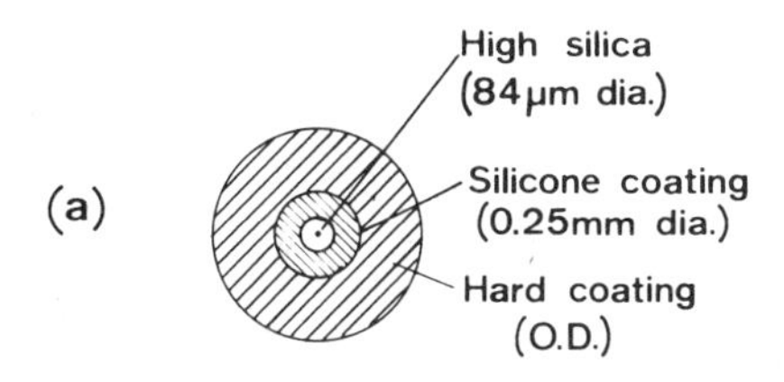

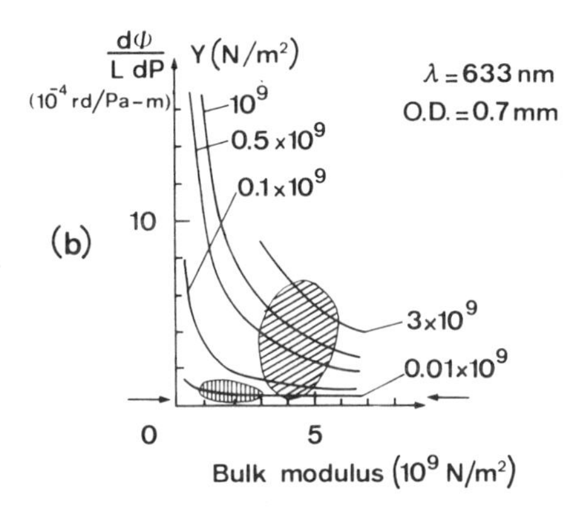

FIGURE 8.2 Pressure-sensitivity modification by fiber jackets. (a) Fiber structure. (b) Acoustic sensitivity as a function of bulk modulus of the hard coating for various Young's modulus of the hard coating. The arrows point to the asymptotic sensitivity of a bare fiber. The upper shaded area corresponds to plastics, the middle to UV curable coatings, and the lower to rubbers. (From Ref. 1.)

where ϕ is the phase difference between the waves transmitted through arms 1 and 2 and s is the sensitivity of both detectors. In our case, ϕ can be separated into a fixed part and the phase variation $d\phi$, which is of interest here, and we can also enter the argument of $\hat{t}^2|\hat{c}|^2$ into the phase term. For the sake of simplicity, we will also assume $|\hat{t}|^2 = |\hat{c}|^2 = 0.5$, and we can thus write

$$I_1 = 0.5sP[1 - \cos(\phi_0 + d\phi)]$$
$$I_2 = 0.5sP[1 + \cos(\phi_0 + d\phi)] \tag{8.9}$$

TABLE 8.1 Acoustic sensitivity measured for various coated SMFs at 633 nm compared to bare fiber theoretical sensitivity

Fiber coating	No	0.25-mm-diameter silicone +		
Fiber jacket	No	Hytrel (0.5 mm O.D.)	Teflon EFP (0.69 mm O.D.)	Polypropylene (0.62 mm O.D.)
Pressure sensitivity (10^{-4} rad/Pa·m) at 633 nm	0.45	2.9	5.8	5.8

Source: Ref. 1.

Although the use of a single photodetector could in principle yield the desired information ($d\phi$), it seems far superior to use both photodetectors and electronically perform the normalization operation $S = (I_2 - I_1)/(I_2 + I_1)$, thus yielding a signal S equal to $\cos(\phi_0 + d\phi)$ independently of the source power (this strictly requires $|\hat{t}| = |\hat{c}|$). As small variations $d\phi$ are to be detected, it is necessary for maximum sensitivity that $\phi_0 = 90°$ (quadrature condition). If slow temperature or pressure variations are to be detected, one possible way is to adjust the interferometer path-length difference for the nominal operating conditions and to find methods for rendering the fiber insensitive to the undesired phenomenon (temperature if pressure is to be measured), which could disturb the measurement.

Another possibility consists of applying a high-frequency phase modulation through a piezoelectric cylinder placed in one arm, around which the fiber is coiled, thus producing a phase-modulated signal $\cos[\phi_0 + d\phi + \phi_m \cos(\omega_m t)]$, which can be demodulated in a complex way to recover $\phi_0 + d\phi$ (Ref. 1, SHET detection).

In the case of acoustic wave detection, the signal frequency spectrum is quite different from that of slow temperature or pressure variations, and this can be exploited to design an active feedback loop minimizing the effects of these noise sources, by continuously maintaining the quadrature condition [10]: the very low frequency part of the signal $S = (I_2 - I_1)/(I_2 + I_1)$ is used for driving a piezoelectric phase modulator placed on one arm of the interferometer, in order to maintain the very low frequency signal equal to 0. It appears, then, that the output signal S is equal to $\sin(d\phi)$. With this method, a minimum detectable phase shift of about 2×10^{-7} rad has been obtained with a semiconductor laser and a detection bandwidth of 1 Hz [10]. With a 1-m-long bare fiber arranged in the basic interferometer structure of Fig. 8.1, and the sensitivity given by Eq. (8.4), this corresponds to a minimum detectable acoustic pressure of about 0.002 Pa, which can be reduced to about 1 μPa by winding the fiber around a polyethylene hollow cylinder [9]. These figures have to be compared to the value of about 0.001 Pa for the Knudson sea-noise level, and to about 20 μPa for state-of-the-art piezoelectric sensors.

The method described above for solving the signal fading problem in the interferometer has two main drawbacks: It works only when the signal frequency spectrum is well separated from the frequency spectrum of the perturbations; and it requires a piezoelectric cylinder, which is generally bulky and weighs more than all the other components of the sensor together. Other methods based on two-wavelengths interferometry or two-polarizations interferometry or using a 3×3 coupler as the output coupler have been suggested and experimented with [11].

Another problem arises from the fact that the coherent detection process used in these sensors requires that the states of polarization at the output of both fibers be the same. As in telecommunication applications of coherent detection systems, this can be solved with polarization maintaining fibers (see Chap. 2), or polarization feedback loops [12] (see Refs. 44 and 45 of Chap. 7), or a polarization diversity receiver as described in Sec. 7.4.2.

Finally, the laser light source characteristics also introduce performance limitations. First, the normalization scheme discussed above [$S = (I_2 - I_1)/(I_2 + I_1)$] totally rejects the source amplitude noise only if the X-couplers are prefectly balanced. Practically, however, the noise is reduced by a factor of 10 only, and typical semiconductor lasers have an intensity noise of about 10^{-6} at 1 kHz, with a $(\text{frequency})^{-\frac{1}{2}}$ variation [1]. This means that the minimum detectable phase shift is about 2×10^{-7} rad at 1 kHz. Next, the Mach–Zehnder interferometer always exhibits some path-length difference between its two arms, which should be maintained below the coherence length of the laser source. Typical single-longitudinal-mode lasers have coherence lengths of about 4 m (50 MHz line width), but this can be degraded by strong optical feedback, weak satellite modes, and so on. The laser spectral width should thus be controlled, possibly with the help of an isolator and the methods discussed in Sec. 7.4.2. Finally, any laser's emission frequency fluctuation is transformed into a phase noise by the unbalanced path length. It has been observed that a 1-m difference increases the phase noise by a factor of 3000 compared to a perfectly balanced interferometer. With a free-running single-longitudinal-mode laser, it would be necessary to match the path lengths to better than 1 mm for getting a minimum detectable phase shift of 1×10^{-6} rad. This can be improved by stabilizing the laser's emission frequency (Sec. 7.4.2), or by the phase modulator discussed above for maintaining the interferometer at quadrature, or by adopting the polarimetric sensor in which the path lengths are balanced (the remaining difference is given by length × normalized fiber birefringence, that is, about 10^{-4} relative path-length difference).

8.2.2 Polarimetric Sensors

General Principles. Instead of using two fibers for building the Mach–Zehnder interferometer, the polarimetric sensors use only one fiber with a strong linear birefringence (either built in or induced), called hereafter the bias birefringence. The two arms of the interferometer become the two linear eigenstates of polarization, and the external stimulus modulates the phase difference between these two eigenstates, which has thus to be detected. This can also be seen as a modulation of the output SOP when both eigenstates are excited, but the first representation is closer to the interferometric sensor conception. It is essential in these sensors that the bias birefringence be strong enough to avoid any coupling between both eigenstates.

In practice, both eigenstates should be equally excited (linear input SOP oriented at 45° from the birefringence axes, or circular input SOP), and the equivalent of the output X-coupler seen in Fig. 8.1 is constituted by a polarization beam splitter oriented at 45° from the fiber's birefringence axes, which thus provides two output signals. It is easy to see that these output signals are 180° out of phase, and the same detection methods as above can be used here. The quadrature condition obviously corresponds to a fiber's circular output SOP, which can be obtained by adjusting the fiber length or using a Babinet–Soleil compensator or an active feedback loop controlling a linear birefringence-inducing device (such as a piezoelectric cyclinder on which the fiber is tightly wound) instead of a phase modulator as above. Alternatively, a passive detection scheme using four photodetectors and insensitive to the quadrature condition has been reported [13].

Practical Designs. The first proposal for such sensors uses a low intrinsic birefringence fiber wound under tension on an acoustically compliant cylinder [14]. The high linear birefringence induced by tension coiling the fiber (see Sec. 2.3.1) serves as bias birefringence, and the modulation is induced by the deformations of the cylinder when subjected to external pressures or temperature variations. The signal emerging from the electronic circuitry similar to that described above is thus simply $\sin[d(\delta\beta L)]$, where $\delta\beta$ is the linear birefringence that can be obtained from Sec. 2.3.1 and L is the fiber length. As these sensors rely on a differential phase change, the sensitivity is reduced compared to the interferometric sensors, and a sensitivity reduction by a factor of 40 compared to the design of Ref. 8 has been reported [14].

Another application of this principle is found in Ref. 15, where a beam-forming sensor is proposed. The fiber has a high built-in linear birefringence $\delta\beta$ and it is twisted at N turns per meter, so that the bias birefringence is much greater than the twist-induced circular birefringence (see Chap. 2). Values of 9 mm for the beat length and 10 turns/m were used. The fiber is then fixed on the surface of an acoustically compliant hollow cylinder or embedded in it with its axis parallel to the cylinder axis. It is essential here that the coupling between the linearly polarized eigenstates due to the twist can be neglected, the only affect of the twist being to rotate the birefringence axes uniformly. The acoustic wave impinging on the cylinder causes it to vibrate, yielding lateral anisotropic forces on the outer surface of the fiber with a longitudinal periodicity linked with the acoustic wavelength and the angle of incidence. These longitudinally propagating lateral foces induce a linear birefringence (see Sec. 2.3.1) which combines with the bias birefringence, but with a relative axis orientation uniformly varying along the fiber because of the twist. With the same launching and detection schemes as above, the signal amplitude is given by [15]

$$S \sim \frac{\sin[(K_a \cos\theta + 4\pi N)L/2]}{K_a \cos\theta + 4\pi N} + \frac{\sin[(K_a \cos\theta - 4\pi N)L/2]}{K_a \cos\theta - 4\pi N} \qquad (8.10)$$

where K_a is the acoustic wave vector in the cylinder, θ the angle between the acoustic wave vector and the cylinder axis, and L the interaction length. Equation (8.10) shows that this sensor is frequency and direction selective, as the maximum signal occurs for $|K_a \cos\theta| = 4\pi N$. This selectivity is due basically to a phase-matching phenomenon between the acoustic wave and the rotating bias birefringence.

8.2.3 Sensor Multiplexing

Many of the applications of interferometric fiber optic sensors require multisensor systems. A particularly demanding application is the passive hydrophone array, where acoustic antennas involving up to 1000 sensors are contemplated. An all-fiber multisensor network has several advantages over conventional electrical multiplexing schemes. First, an all-optical telemetry system capable of supporting a large number of sensors is feasible thanks to the large transmission bandwidth of single-mode fibers. Second, the network operates passively without the need for electrically active components, thus eliminating the need for a bulky remote electrical power supply and retaining a low intrinsic susceptibility to electromagnetic interference.

Ideally, a multiplexed sensor array or network would be addressed by one single input—output fiber (or possibly one input and one output fiber). Each sensor encodes the optical carrier with information on the corresponding sensed physical parameter, and the total encoded optical output is conveyed to a demultiplexer which separates the information relevant to each sensor into the appropriate number of channels. Two requirements are thus to be fulfilled by the multiplexing approach: it shall be able to address each sensor individually and shall allow for the demodulation of the interferometric signals produced by each sensor. A review of various multiplexing techniques is presented in Ref. 16.

A number of addressing schemes are conceptually possible, such as time, frequency, wavelength, coherence, and polarization-division addressing. However, time-division multiplexing (TDM) has received the most attention and has achieved respectable performances in terms of phase detection sensitivity and sensor cross-talk levels. It can be implemented with a frequency-modulated laser diode [17] or a pulsed laser. In the case of the pulsed-laser source, an approach consists in launching the light pulse into a long fiber line, made up of a serial array of Mach–Zehnder interferometers tapped along a single-fiber bus [18]. This approach yielded an experimental sensitivity of 10 μrad/$\sqrt{\text{Hz}}$ and a cross-talk level of -30 dB for a two-sensor element array. Finally an approach combining a pulsed-laser source and two-state frequency modulation has been developed [19,20]. The single-fiber line comprises regularly spaced reflective splices (about 0.5% reflectivity), and each sensor element consists of the Fabry–Perot interferometer achieved by the fiber section between two successive splices. This reflective configuration is conceptually similar to the transmissive design reported in Ref. 18. The unbalanced nature of the reflectometric interferometer necessitates the use of a high-coherence laser. With 200 m of single-mode fiber between each reflecting splice, optical pulse pairs with pulse lengths of 1.5 μs and separations of 0.5 μs are generated with an optical frequency difference of 700 kHz between the two pulses of a given pair (the frequency shifting and the time gating is achieved through an acousto-optic Bragg cell). Hence the first pulse of a pair reflected by a splice combines with the second pulse of the same pair reflected by the preceding splice, as the round-trip travel time between two successive splices (2 $\times$ 200 m yields 2 μs) matches exactly the time difference between the two pulses (1.5 + 0.5 μs). The interference between these two reflected pulses produces an intermediate frequency carrier at 700 kHz which is phase-modulated by the pressure of the acoustic wave received by the corresponding fiber section. A sea trial of a six-hydrophone array based on this technique has been reported [21].

8.3 CURRENT AND MAGNETIC FIELD SENSORS

Optical fibers are very attractive for current sensing, especially on high-voltage lines, because of their intrinsic electrical isolation properties. This is achieved by means of several different principles based on fundamentally different physical phenomena, and we discuss them below.

8.3.1 Magneto-Optical Sensors

These sensors are based on the fact that a magnetic field H, parallel to the fiber axis, induces a nonreciprocal circular birefringence $\delta\beta_c$ given by (see Sec. 2.3.4)

$$\delta\beta_c = 2V_f H \tag{8.11}$$

where $V_f = 4.6\ 10^{-6}$ rad/A is the Verdet constant of silica (at 633 nm). Two main sensor configurations exploit this effect.

Polarimetric Sensors. In a fiber where no birefringence (either linear or circular) other than the magneto-optically induced one exists, a linear SOP launched at the input of the fiber will be rotated by an angle $V_f \times H \times L$, where L is the interaction length between the fiber and the longitudinal magnetic field [if the magnetic field is not uniform along the fiber, and H and L should be understood as the sum of elementary contributions H(z) dz]. Several sensors based directly on the measurement of the SOP rotation angle, either after one-way transmission [22,23] or after two-way propagation by polarization optical time-domain reflectometry ([24]; see Sec. 5.3.4), have been reported. However, the unavoidable residual linear birefringence (either built-in or bending induced) leads to large sensitivity variations with environmental conditions changes, because it affects the nominal SOP in the absence of a magnetic field; in the case of POTDR measurement, the method is limited to alternative currents.

Another configuration which is very close in its principles to the polarimetric acoustic sensors described in Sec. 8.2.2 has been proposed [25]. As the magnetic field induces a circular birefringence instead of a linear one for acoustic waves in Sec. 8.2.2, the bias birefringence is circular (instead of linear) and is obtained by twisting a fiber having a very low intrinsic linear birefringence. Thus the general principle of operation of this sensor is identical to that of acoustic polarimetric sensors if we replace the acoustic wave by the magnetic field and the linearly polarized eigenstates by circular ones. The only differences with the acoustic wave polarimetric sensor are that, whatever its orientation, the input linear SOP always excites the two circular eigenstates equally. The feedback loop for maintaining the quadrature condition (fiber's output SOP linear, at 45° from the polarization beam-splitter axes) uses a rotating motor to which the fiber is attached, instead of a linear birefringence modulator, for controlling the total circular birefringence by adjusting the fiber twist.

The output signal is slightly different from what would be obtained by simply using the same formulas as for the acoustic polarimetric sensor, because here we cannot neglect completely the residual linear birefringence [25]:

$$S = \sin\left[\left(1 - \frac{\delta\beta_\ell^{\,2}}{8\pi^2 g^2 N^2}\right) 2MV_f J\right] \tag{8.12}$$

where $\delta\beta_\ell$ is the residual linear birefringence, g = 0.14 (see Sec. 2.3.3), N is the fiber twist rate in turns per meter, M is the number of fiber turns around the current conductor, and J is the current. It can be seen from Chap. 2 that realistic fiber manufacturing conditions, bendings of radius 5 to 10 cm, and twist rates of 5 to 20 turns per meter lead to a scale factor of 0.99 or higher in Eq. (8.12). If we consider a minimum detectable phase shift of 5×10^{-7} rad reported in Ref. 10 for similar detection methods and the usual current frequency range (50 to 100 Hz), we obtain a minimum detectable current of about 50 mA for 1 turn or 1 mA for 50 turns.

Ring Interferometer. This arrangement for a current sensor is similar in many aspects to the Sagnac effect rotation sensor (see Sec. 8.4.1), as it exploits the fact that the Faraday effect is nonreciprocal, just like the Sagnac effect (this can be understood from the fact that a linear SOP launched in a fiber with no birefringence except that induced by the magnetic field and reflected back at the output will return with a rotation equal to twice the rotation incurred for one-way propagation). This type of sensor is very attracting, as it is insensitive to all other reciprocal effects, such as temperature- and pressure-induced birefringences and phase shifts, provided that both optical paths of the counter-rotating waves are strictly identical [26].

The only significant difference from the rotation sensor stems from the fact that the Faraday effect is polarization dependent, contrary to the Sagnac effect. The phase shift for a right circular polarization is opposite to the phase shift for a left circular polarization. This requires using circular polarization maintaining fibers, and launching the same circular polarization at both fiber ends. If the fiber were perfectly maintaining circular polarizations, the output electrical signal from the detector (using the techniques of detection and phase modulation described in Sec. 8.4.2) would be proportional to $\sin(2MV_fJ)$ for a ring interferometer comprising M fiber turns around a straight conductor carrying a current J. However, becuase of residual linear birefringence, a scale factor equal to $(1 - \delta\beta_\ell/2\pi gN)$ (with the same definitions as above) is applied to the phase shift. As this scale factor can range between 90 and 100%, it can lead to unacceptable measurement uncertainties. It can easily be suppressed by using a secondary electrical coil around the fiber, driven by the detection electronics in order to cancel the total phase shift. The measurement of the current traversing this secondary coil when the zero is achieved, together with knowledge of the number of turns in this coil, allows us to determine the current traversing the main conductor without knowing the scale factor nor the Verdet constant.

8.3.2 Indirect Techniques

These techniques basically use the two-fiber interferometer design shown in Fig. 8.1. For inducing the required phase shift, the influence of the current is transformed either into a pressure by coating the fiber with a magnetostrictive jacket [27–29] and placing it inside a current conductor solenoid, or into a heating effect by coating the fiber with aluminum and passing the current directly through this coating [30]. All the problems and measurement methods associated with the interferometric pressure or temperature sensors discussed in Sec. 8.2.1 are applicable here, and we do not discuss them further. For a 1-m-long magnetostrictive sensor using a nickel-coated fiber a sensitivity of about 30 rad/A was reported [30], yielding a minimum detectable current of about 10 nA and a good linearity over 4 decades. In the case of a fiber wound around a current conducting cylinder made of copper surrounded by a strip of metallic glass as magnetostrictive material, minimum detectable magnetic fields of 5×10^{-9} Oe per meter of fiber were achieved [29]. In the case of the heating sensor, the sensitivity varies as the inverse of frequency, and at 50 Hz it is about 10 rad/A^2 for a 1-m-long sensor, yielding a minimum detectable current of about 0.1 mA [30].

8.4 ROTATION SENSORS: THE FIBER OPTIC GYROSCOPE

The fiber optic gyroscope, often abbreviated FOG, holds a very specific position in the fiber sensor field. It allows one to detect rotation with respect to inertial space, because of an enhanced Sagnac effect induced in a multiturn single-mode fiber coil. The effect produces a difference between the transit times of the co-rotating and the counter-rotating waves. This is detected in a passive two-wave ring interferometer where it generates a phase difference proportional to the rotation rate.

First demonstrated in 1976 [31], the subject has since justified an important R&D effort in many teams around the world [32]. Several of the basic single-mode fiber optic components (couplers, in-line polarizers, polarization controllers), polarization-preserving fiber and superluminescent diodes, were first developed for this application.

8.4.1 Basic Reciprocal Configuration

The Sagnac effect depends only on the geometrical parameters of the fiber coil, without any influence of its material and guidance properties [33]. The phase difference is

$$\Delta\phi_s = \frac{2\pi LD}{\lambda c}\,\Omega_{/\!/} \tag{8.13}$$

where L is the coil length, D is the coil diameter, λ is the source wavelength in vacuo, and c is the light velocity in vacuo. The rotation rate $\Omega_{/\!/}$ is the component that is parallel to the equivalent area vector (i.e., usually the "axis") of the coil, because the Sagnac effect is the result of a line integral along the closed path of the ring interferometer.

As is any optical interferometer, the FOG sensitivity is limited by photon noise. Detection of phase shifts as small as 10^{-6} to 10^{-7} rad should then be possible with several microwatts of returning light power, but when compared to the total phase accumulated along the coil (about 10^9 rad for 100 m of silica fiber at a 840 nm wavelength), this means detecting one part in 10^{15} to 10^{16}. To get this incredible number happens to be feasible because of the fundamental principle of reciprocity of light propagation in a linear medium. This requires only the use of a truly single-mode waveguide filter (single spatial mode and single polarization state) at the common input–output port of the ring interferometer [34]. This ensures that, at rest, both counterpropagating waves follow exactly the same path in opposite directions. Then they suffer exactly the same propagation delay through the coil and their phase difference is nulled out. This yields a so-called reciprocal configuration (Fig. 8.3) which drastically improves the stability of the interferometer. Good alignments are obviously still needed, but only to optimize a high throughout power in the total system, which is a much less stringent requirement than ensuring a phase stability of 10^{-7} rad, that is, a mechanical stability of 10^{-5} nm.

This interferometer response is then a raised cosine law of the rotation-induced Sagnac phase shift $\Delta\phi_S$. Biasing is obtained with a reciprocal phase modulation generated asymmetrically at one end of the fiber coil [34,35] (Fig. 8.4). Both counterpropagating waves suffer the same phase modulation ϕ_m, but one experiences the modulation at the beginning of the coil whereas the other experiences it at the end. They interfere with modulation $\Delta\phi_m$ of their phase difference:

$$\Delta\phi_m(t) = \phi_m(t) - \phi_m(t - \tau) \qquad (8.14)$$

where τ is the transit time through the coil. With respect to this phase modulation, the interferometer behaves like a delay line filter with a sinusoidal transfer function (Fig. 8.5). It can be shown that when the modulation uses the "eigen"frequency, $f_e = 1/(2\tau)$, its efficiency is maximum and there is perfect rejection of direct current and the even harmonics, which makes this method perfect, even with defects (nonlinearity and intensity modulation, in particular) of the modulator [36]. The biased response of the FOG is obtained by demodulation with a lock-in amplifier of the f_e component of the modulated interference signal (Fig. 8.6).

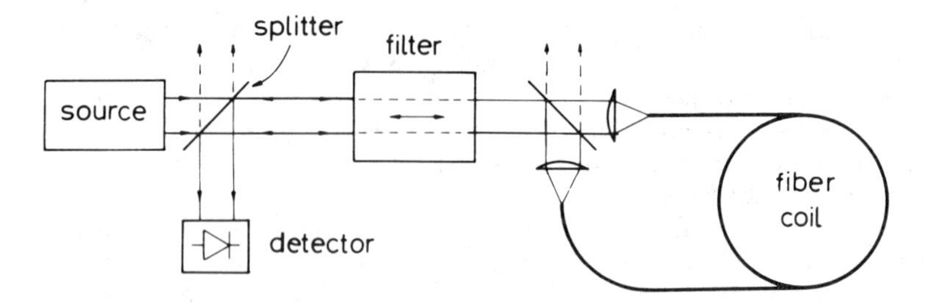

FIGURE 8.3 Reciprocal configuration using a truly single-mode filter at the common input–output port of the ring interferometer.

The use of a filter at the input–output port and of a modulator operated at the eigenfrequency of the coil yields a biased sine response of the FOG interferometer with very good stability of the zero signal at rest of the FOG (Fig. 8.7).

8.4.2 Low Noise and Good Long-Term Bias Stability

The concept of reciprocal configuration has been widely accepted, improvements are still needed to get the best possible performances. In particular, it is very important to use a broadband source. As a matter of fact, the FOG can be analyzed as an interferometer with two primary waves which are perfectly in phase at rest because of reciprocity, but there are also a lot of secondary parasitic waves generated by back-reflection, backscattering, and crossed-polarization coupling. If the source is coherent, the two primary waves serve as a local oscillator for an efficient coherent detection of these various low-power parasitic waves. The use of a broadband source

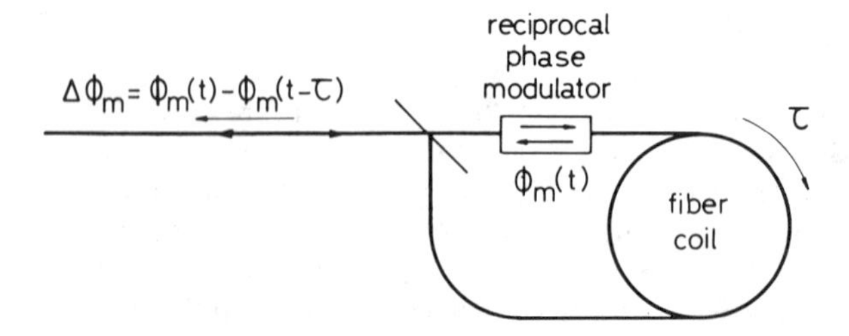

FIGURE 8.4 Principle of FOG biasing using a reciprocal phase modulator at one end of the coil and the coil as a delay line.

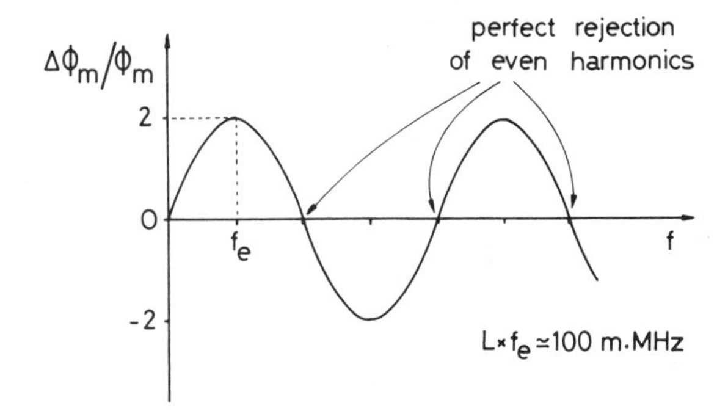

FIGURE 8.5 Transfer function of the ring interferometer considered as a phase modulation delay line filter.

avoids this problem because the path difference between the primary waves and the parasitic waves is much longer than the coherence length of the source, which eliminates these spurious interferences. Notice that this does not limit the dynamic range of the system because practical FOGs are working only on the first fringes of the Sagnac interferometer. This was demonstrated as a very good solution [37] to reduce the noise induced by coherent Rayleigh backscattering [38]. In particular, superluminescent diodes (SLDs) [32] are commonly used as the optimal source of FOGs, because of their broad spectrum and their high coupling efficiency into a single-mode fiber. This applies also to back-reflections at the various interfaces, that cannot be completely eliminated, even when using tilted interfaces [39].

Furthermore, such a broadband source is also useful to ensure good reciprocity. As a matter of fact, the spatial filtering obtained with a single-mode fiber can be considered as perfect, but the practical rejection of polarizers does not provide enough filtering of the birefringence nonreciprocities. It was shown [40] that the bias error depends on the amplitude rejection ratio of the polarizer [i.e., 60 dB of rejection can yield a bias error as high as 10^{-3} rad] if a coherent source is used. It was thought obvious, that the use of a polarization-preserving fiber would reduce these birefringence nonreciprocities. This was demonstrated experimentally [41], but the main reason for improvement was due to the intrinsic birefringence of the polarization-preserving fiber because it yields depolarization with a broadband source. An analysis of the problem shows that the birefringence of the input lead is very important [42]. A model of the behavior of the fiber coil can also be given [43]. It shows that the problem of polarization coupling is actually

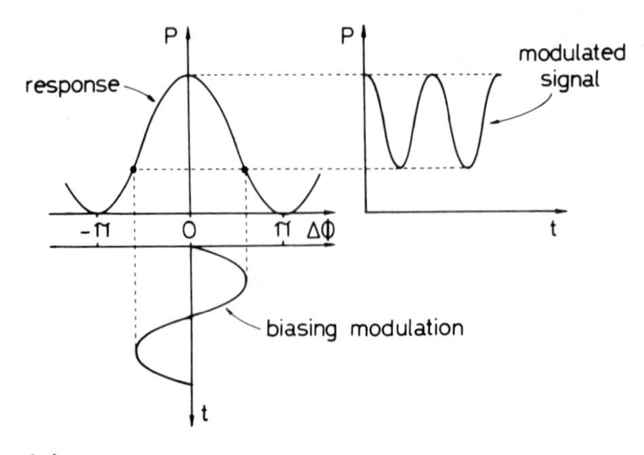

(a)

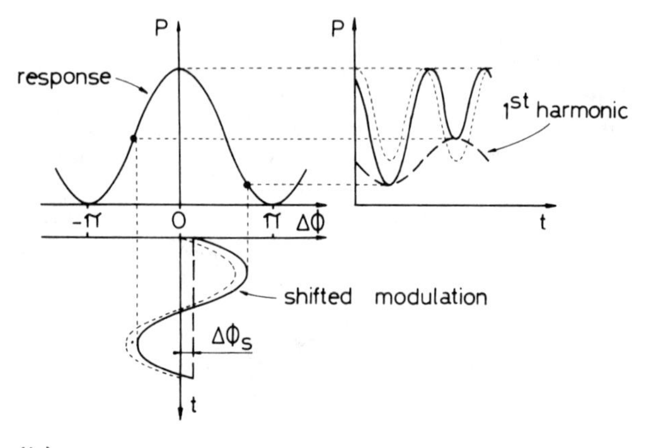

(b)

FIGURE 8.6 Modulated interference signal: (a) Even harmonics of the modulation frequency when there is no nonreciprocal phase difference, (b) first harmonic component, which provides a biased signal at the modulation frequency.

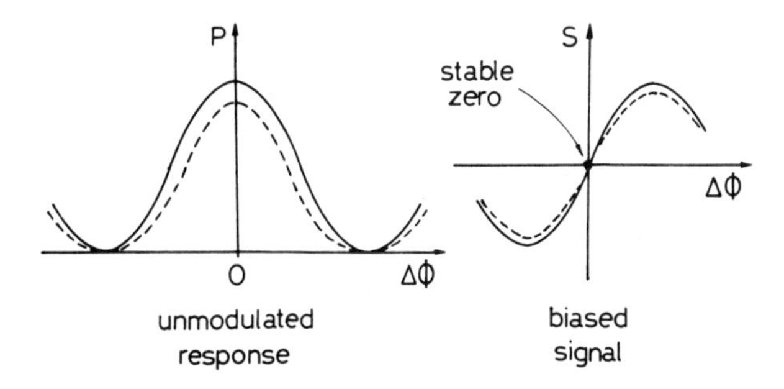

FIGURE 8.7 Baseband unmodulated ring interferometer response and demodulated biased signal.

very similar to the problem of backscattering and back-reflection—a randomly distributed effect is not very harmful, but symmetrical coupling points must be avoided.

In practice, residual birefringence nonreciprocities are the limit of the bias stability because of a lack of perfect polarization filtering but it is important to notice that a powerful test method [44,45] has been developed to control the birefringence and polarization characteristics of the components and the connections, which can yield localized polarization couplings. This is based on "white light" interferometry, and it should become a test method for any high-birefringence fiber system, as widely used as OTDR for ordinary fiber systems.

In addition, broadband sources happen to be the ideal solution to eliminate the influence of the nonlinear Kerr effect [46,47], which could have been a drastic limit to high performance. Finally, some additional design features are useful to eliminate residual parasitic effects: symmetrical winding to limit the effect of temperature transients, vibration and acoustic noise [48,49], untwisted winding of polarization-preserving fiber to avoid the residual influence of the environmental magnetic field [50].

Assuming that these various operating conditions are respected, FOGs with a phase noise level smaller than 10^{-6} rad/$\sqrt{\text{Hz}}$ and a phase bias error smaller than 10^{-6} rad have been demonstrated experimentally [32]. Brassboards are typically using a coil of a few hundred meters with a 50 to 100 mm diameter, which yields noise and bias stability in the 0.1°/h range. Obviously, good mechanical ruggedness and low loss are helpful and most systems are now using in-line all-fiber components or a multifunction integrated optic circuit. The most commonly used wavelength is the 800 to 850 nm window because the intrinsic loss of the fiber is not a problem over only a few hundred meters, but there is also a tendency to move toward the single-mode telecommunication standards of 1300 and 1550 nm because of radiation hardening.

8.4.3 High-Scale Factor Accuracy

High sensitivity is important with any measuring instruments, but gyroscopes also have drastic requirements involving the stability and the linearity of the scale factor over a very wide dynamic range. As a matter of fact, a gyroscope measures a rotation rate which is integrated to obtain the change in angular orientation. All the past errors accumulate and deteriorate the precision of the final measurement. One needs to detect very small rotation rates but also to measure very precisely the value of this rate when the system is rotating quickly. A scale factor accuracy of 100 ppm, which would be very good in most instruments, is only considered as medium accuracy for a gyroscope.

So far, we have only been concerned by the zero bias of the FOG. Such a system is intrinsically analog, which often yields problems of practical dynamical range and accuracy. Such a difficulty has been overcome by using digitalizing closed-loop processing. A phase nulling scheme with acousto-optic frequency shifters was first proposed [51,52] but this destroys the intrinsic reciprocity of the ring interferometer [53], which degrades the bias stability.

However, this can be avoided with an additional phase ramp modulation, often called serrodyne modulation in radar processing where it is commonly used. This requires the use of wide bandwidth phase modulators, which is actually the fundamental technical advantage of integrated optics. In particular, so-called digital-phase ramp processing [43,53] (Fig. 8.8) brings a quasi-perfect solution to this problem. It uses a double feedback control loop to provide an incremental signal of rotation with a self-calibration of the efficiency of the integrated optic phase modulator on 2π rad, the periodicity of the interference fringes. Furthermore, this processing scheme can be used with a so-called Y-tap configuration (Fig. 8.9) [53], which provides optimal simplicity while respecting all the requirements of the reciprocal configuration. This combines high scale factor accuracy and linearity with low long-term bias drift.

Such a phase-nulling scheme yields a precise measurement of the Sagnac phase difference, but this one remains proportional to the length and diameter of the coil, and also to the optical frequency (inverse of the wavelength). A good scale factor accuracy then requires temperature modeling of the coil, but also good control of the mean wavelength of the source. This last point presents a difficult engineering problem because it is necessary to implement

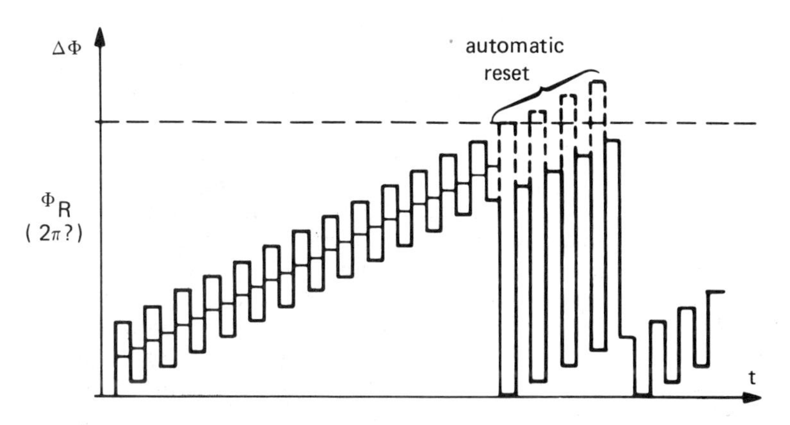

FIGURE 8.8 Digital phase ramp modulation with automatic 2π reset.

such control in a compact device, but some solutions have been proposed to reach the 100 to 10 ppm range. They use either an external control spectrometer or interferometer [54,55] or the wavelength dispersion in the fiber coil [45,56].

8.4.4 Future Prospect

After a decade of research, the FOG is now actively entering its development phase. Efforts are mostly directed toward a compact

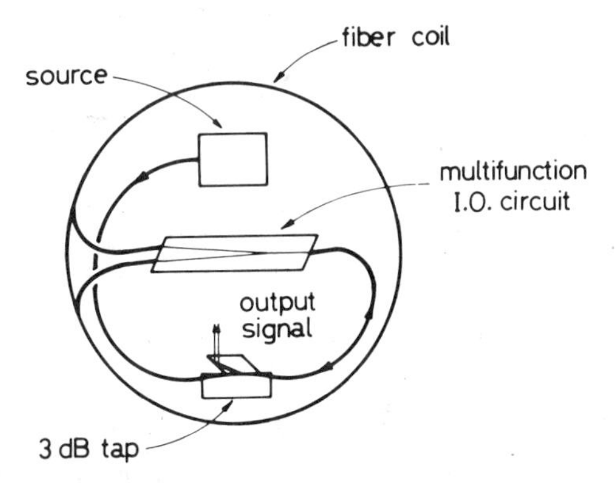

FIGURE 8.9 "Y-tap" configuration of the FOG, using a multifunction integrated optic circuit composed of a Y junction with a polarizer on the common trunk and phase modulators on both branches. Notice the tilted parallelogram shape to avoid back-reflections.

medium-accuracy tactical gyroscope. A reasonable medium-term engineering goal is a 50 cm^3 FOG with a dynamic range higher than 1000°/s, a bias stability better than 1°/h, and a scale factor accuracy better than 100 ppm. Such a device will find many applications in inertial guidance and stabilization, but it will also remain a fascinating subject of applied physics. Was it reasonable, at the beginning, to expect to detect a path change smaller than 10^{-5} nm after several hundred meters of propagation, one part in 10^{16}?

Notice that besides this two-wave interferometer FOG, there is also research interest in a passive resonant fiber gyroscope [32]. However, this alternative faces very difficult technical problems (as a very narrow source, for example) and although there has been progress in component and system understanding, it seems that it will not compete successfully against the two-wave FOG, which offers a much higher simplicity.

ACKNOWLEDGMENT

Section 8.4 on the fiber optic gyro has been revised by Dr. Hervé C. Lefevre. His contribution to this chapter is gratefully acknowledged.

REFERENCES

1. T. G. Giallorenzi, J. A. Bucaro, A. Dandridge, G. H. Sigel, Jr., J. H. Cole, S. C. Rashleigh, and R. G. Priest, *IEEE Trans. Microwave Theory Tech. MTT-30*(4):472–511 (1982).
2. J. A. Bucaro, H. D. Dardy, and E. F. Carome, *Appl. Opt. 16*(17):1761–1762 (1977).
3. C. D. Butter and G. B. Hocker, *Appl. Opt. 17*(18):2867–2869 (1978).
4. G. B. Hocker, *Appl. Opt. 18*(9):1445–1448 (1979).
5. T. Yoshino and T. Ose, in *CLEO'82 Digest of Technical Papers*, Optical Society of Amerifa, Washington, D.C., Apr. 1982, paper WL6.
6. N. Lagakos, J. A. Bucaro, and J. Jarzymski, *Appl. Opt. 20*(13):2305–2308 (1981).
7. N. Lagakos and J. A. Bucaro, *Appl. Opt. 20*(15):2716–2720 (1981).
8. N. Lagakos and J. A. Bucaro, *IEEE Trans. Microwave Theory Tech. MTT-30*(4):529–535 (1982).
9. G. W. McMahon and P. G. Cielo, *Appl. Opt. 18*(22):3720–3722 (1979).
10. A. Dandridge and A. B. Tveten, *Appl. Opt. 20*(14):2337–2339 (1981).

11. S. K. Sheem, T. G. Giallorenzi, and K, Koo, *Appl. Opt. 21*(4): 689–693 (1982).
12. A. D. Kersey, A. Dandridge, and A. B. Tveten, *Proc. OFS'88*, Optical Society of America, New Orleans, LA, pp. 44–47 (1988).
13. S. C. Rashleigh, *Appl. Opt. 20*(9):1498–1499 (1981).
14. S. C. Rashleigh, *Opt. Lett. 5*(9):392–394 (1980).
15. S. C. Rashleigh and H. F. Taylor, *Electron. Lett. 17*(3):138–139 (1981).
16. A. D. Kersey, A. Dandridge, and A. B. Tveten, *Fiber Optic and Laser Sensors V*, Proc. SPIE 838, pp. 184–193 (1987).
17. F. Farahi, A. S. Gerges, J. D. C. Jones, and D. A. Jackson, *Electron. Lett. 24*(1):54–55 (1988).
18. A. D. Kersey and A. Dandridge, *Proc. OFS'88*, Optical Society of America, New Orleans, LA, pp. 80–83 (1988).
19. J. P. Dakin, C. A. Wade, and M. L. Henning, *Electron. Lett. 20*(1):53–54 (1984).
20. S. W. Thornton and M. L. Henning, *Proc. OFS'84*, VDE-Verlag, Stuttgart, Germany, pp. 369–373 (1984).
21. M. L. Henning and C. Lamb, *Proc. OFS'88*, Optical Society of America, New Orleans, LA, pp. 84–91 (1988).
22. A. M. Smith, *Appl. Opt. 17*(1):52–56 (1978).
23. A. Papp and H. Harms, *Appl. Opt. 19*(22):3729–3745 (1980).
24. J. N. Ross, *Electron. Lett. 17*(17):596–597 (1981).
25. S. C. Rashleigh and R. Ulrich, *Appl. Phys. Lett. 34*(11):768–770 (1979).
26. J. L. Lesne and B. Dessus, *Conférence Horizons de l'Optique*, Pont-à-Mousson, France, Apr. 1980.
27. A. Dandridge, A. B. Treten, G. H. Sigel Jr., E. J. West, and T. G. Giallorenzi, *Electron. Lett. 16*:408–409 (1980).
28. N. Hartman, D. Vahey, R. Kidd, and M. Browning, *Electron. Lett. 18*(5):224–226 (1982).
29. K. P. Koo and G. H. Sigel, Jr., *Opt. Lett. 7*(7):334–336 (1982).
30. A. Dandridge, A. B. Tveten, and T. G. Giallorenzi, *Electron. Lett. 17*(15):523–525 (1981).
31. V. Vali and R. W. Shorthill, *Appl. Opt. 15*:1099–1100 (1976).
32. Fiber Optic Gyros: 10th Anniversary Conference, *Proc. SPIE 719* (1986).
33. H. J. Arditty and H. C. Lefèvre, *Opt. Lett. 6*:401–403 (1981).
34. R. Ulrich, *Opt. Lett. 5*:173–175 (1980).
35. J. M. Martin and J. T. Winkler, *Proc. SPIE 139*:98–102 (1978).
36. R. A. Bergh, H. C. Lefèvre, and H. J. Shaw, *Opt. Lett. 6*: 502–504 (1981).
37. K. Böhm, P. Russer, E. Weidel, and R. Ulrich, *Electron. Lett. 17*:352–355 (1981).
38. C. C. Cutler, S. A. Newton, and H. J. Shaw, *Opt. Lett. 5*: 488–490 (1980).

39. H. J. Arditty, H. J. Shaw, M. Chodorow, and R. Kompfner, *Proc. SPIE 157*:138–148 (1978).
40. E. C. Kintner, *Opt. Lett. 6*:154–156 (1981).
41. W. K. Burns, R. P. Moeller, C. A. Villaruel, and M. Abebe, *Opt. Lett. 8*:540–542 (1983).
42. R. J. Frederiks and R. Ulrich, *Electron. Lett. 20*:330–332 (1984).
43. H. C. Lefèvre, J. P. Bettini, S. Vatoux, and M. Papuchon, *Agard Conference Proc. 383*:9A1–9A13 (1985).
44. K. Takada, K. Chida, and J. Noda, *Appl. Opt. 26*:2979–2987 (1987).
45. H. C. Lefèvre, *Proc. SPIE 838*:86–97 (1987).
46. S. Ezekiel, J. L. Davis, and R. W. Hellwarth, *Fiber-Optic Rotation Sensors*, pp. 332–336, Springer Verlag, Berlin (1982).
47. R. A. Bergh, B. Culshaw, C. C. Cutler, H. C. Lefèvre, and H. J. Shaw, *Opt. Lett. 7*:563–573 (1982).
48. D. M. Shupe, *Appl. Opt. 19*:654–655 (1980).
49. N. J. Frigo, *Proc. SPIE 412*:268–271 (1983).
50. K. Hotate and K. Tabe, *Appl. Opt. 25*:1086–1092 (1986).
51. J. L. Davis and S. Ezekiel, *Proc. SPIE 157*:131–136 (1978).
52. R. F. Cahill and E. Udd, *Opt. Lett. 4*:93–95 (1979).
53. H. C. Lefèvre, S. Vatoux, M. Papuchon, and C. Puech, *Proc. SPIE 719*:101–112 (1986).
54. H. Chou and S. Ezekiel, *Opt. Lett. 10*:612–614 (1985).
55. R. F. Schuma and K. M. Killian, *Proc. SPIE 719*:192–193 (1986).
56. E. Udd and R. F. Cahill, *Proc. SPIE 719*:17–23 (1986).

9

Nonlinear Optical Effects

9.1 INTRODUCTION

Since early studies on nonlinear optical effects generated in optical fibers (e.g., Refs. 1–3), there has been a continuously growing interest in the use of optical fibers as an interaction medium for stimulated Raman scattering (SRS), stimulated Brillouin scattering (SBS), self-phase modulation (SPM), and four-wave mixing (FWM). One reason for this is that despite the small cross section of glasses for most of these effects (although liquid core fibers have been used), they can be efficiently generated due to the possible power confinement on long lengths in low-loss fibers. In addition to that, experiments based on the impregnation of silica single-mode fibers with molecular hydrogen yielded a very large wavelength shift for SRS, and seem to open the way for large possibilities in the application of SMF to obtain various nonlinear effects with very efficient cross sections [4]. Moreover, single-mode fibers are of special interest because of their small core diameter, which requires lower power levels, as all these effects depend directly on the light intensity.

Being interested here in single-mode fibers, we will not discuss the four-wave mixing process, as it requires at least two propagating modes for fulfilling the associated phase-matching conditions, or working around the zero chromatic dispersion wavelength [5]. A general review of these effects can be found in Ref. 6. Finally, as all the effects discussed here are usually studied with high-power sources, it is worth noting that the damage power limit for pure silica is about 10^{10} W/cm^2, or 5 kW for an 8-μm mode diameter.

9.2 STIMULATED RAMAN SCATTERING

When an optical wave is incident in a medium, the incoming photons scatter, producing a phonon by exciting molecular vibrations, and a "Stokes" photon with a smaller frequency (longer wavelength), so that as usual the total energy is conserved. In glasses, the *lack of wave-vector selection rules* (no phase-matching condition) contrary to what happens in crystals, yields a broad Raman band instead of a very narrow band corresponding to a well-defined frequency shift. Reciprocally, if the temperature is relatively high, some molecules of the medium are still vibrating and the corresponding phonons may interact with the incident optical wave for producing a frequency-up-shifted "anti-Stokes" wave. At typical temperatures of 300 K, the anti-Stokes scattering intensity is much weaker than the Stokes intensity, and they would be of equal intensities only at infinite temperature (because there would be many excited molecules by the thermal effects). Now, a weak signal at the Stokes wavelength injected together with a strong pump will be amplified with a gain (per unit length) proportional to the pump intensity (whereas an anti-Stokes signal would be absorbed). For large pump intensities and/or long interaction lengths (both being easily obtained in fibers), the spontaneously scattered Raman light is amplified and may reach intensities comparable to those of the pump (which is depleted accordingly). There is *no threshold power*; however, this regime is usually called the stimulated Raman scattering, and leads to the definition of a critical power corresponding to equal pump and Stokes output intensities.

It thus appears that SRS in fibers can find useful applications as broadband coherent light generators or light amplifiers. On the other hand, it sets a limitation on the maximum power that can be handled by a fiber in high-data-rate systems, because the Stokes wavelength usually has a different velocity from the pump (because of chromatic dispersion), leading thus to a pulse broadening.

A theoretical study of the behavior of SRS in optical fibers can be found in Ref. 7. We will concentrate on the Stokes waves, as the anti-Stokes components are always relatively weak.

9.2.1 Basic Data

The wavenumber $(1/\lambda)$ decrease of the Stokes wave relatively to the pump wave has been determined for various vitreous bulk samples [8]. For SiO_2, an average shift of 440 cm^{-1} was found; 420 cm^{-1} for GeO_2 with a relative cross section of 9 compared to silica; 810 cm^{-1} for B_2O_3 with a relative cross section of about 5; and 640 cm^{-1} and 1390 cm^{-1} for P_2O_5 with relative cross sections of about 6 and 3.5, respectively. Results obtained on high-silica-content glass fibers [9] show a splitting of the SiO_2 peak into 439-cm^{-1} and 496-cm^{-1} shifts, the latter being twice as intense as the former. The same study shows that P_2O_5-doped silica fibers exhibit a peak at 1330 cm^{-1} with a relative intensity of 0.7 compared with the 496-cm^{-1} silica peak. Note finally that the pure silica shift has a full width at half maximum of about 200 cm^{-1} (or 25 nm at the 1000-nm pump wavelength).

The required minimum power (power threshold) for practically obtaining the first Stokes has been estimated to be [10]

$$P_t = \frac{16\pi w_0^2 \alpha}{G_r[1 - \exp(-\alpha L)]} \tag{9.1}$$

where w_0 is the mode field radius as defined in Sec. 1.2.2, α the fiber loss per unit length, G_r the Raman gain, and L the interaction length. For pure silica, the Raman gain G_r at a pump wavelength λ (expressed in nanometers) is

$$G_r = \frac{9.75 \times 10^{-11}}{\lambda} \quad \text{m/W} \tag{9.2}$$

As will be seen below, practical measurements indicate that the Raman gain in normal GeO_2-doped fibers is about five times higher than in pure silica fibers, as expected qualitatively from the indications above about the respective cross sections of doped and undoped silica.

The use of more unusual dopants may make it possible to vary the wavenumber shift and/or the gain, for various applications. It has been reported that after a high-silica-content single-mode fiber has been pressurized in H_2 at 5×10^7 Pa ($\sim$500 atm) and room temperature, it retained interstitial molecular hydrogen at an equivalent pressure of about 1.5×10^6 Pa ($\sim$15 atm) [4]. This led to a large Stokes shift of 4136 cm^{-1} with a narrow line width of about 20 cm^{-1} (FWHM; or approximately 1.5 nm at 883-nm Stokes wavelength and a 647-nm pump wavelength), and a gain of about $(3.2 \times 10^{-11}/\lambda)$ (m/W with λ in nanometers). This is an improvement by a factor of more than 2×10^4 in gain over H_2 in a conventional gas cell [4].

In some cases, Eq. (9.1) has to be corrected. If short pulses are used for the pump wave, the chromatic dispersion of the fiber will make the pump pulse and the Stokes pulse separate after a length which can be estimated as [11]

$$L_e = \frac{\Delta t}{1.4|\tau(\lambda_p) - \tau(\lambda_s)|} \tag{9.3}$$

where Δt is the full width at half maximum of the pump pulse and $\tau(\lambda_p)$ and $\tau(\lambda_s)$ are the group delay times per unit length of the pump and the Stokes waves, respectively, which can be calculated from Chap. 4. The interaction length that enters Eq. (9.1) should be taken as the smallest of the physical fiber length or L_e.

Second, if the source line width is larger than the Raman line width, the gain G_r is divided by the ratio of the source line width to the Raman line width. For high-silica-content fibers this should not happen practically, because of the large Raman line width.

Finally, Eq. (9.1) holds if the fiber maintains the state of polarization of the Stokes wave identical to that of the pump wave over the interaction length. If this is not the case, P_t should be multiplied by some factor greater than 1, which peaks at 2 for completely scrambled polarizations [12]. Ref. 13 presents design rules for optimizing optical fibers with enhanced SRS effects.

9.2.2 Applications

CW Raman Fiber Laser. Before beginning our discussion, we should note that the term laser as used here does not mean that there are feedback mirrors at both ends of the fiber: it simply refers to the stimulated character of the Raman emission. A CW Raman laser has been obtained with a 35-km-long single-mode fiber, pumped by a CW YAG laser emitting at 1064 nm [14]. The fiber has an absolute index difference of 0.86%, a SiO_2-GeO_2 core diameter of 7.4 μm, and a loss of 1.13 dB/km at 1064 nm. The first Stokes (about 1120 nm) appears for a launched pump power of 600 mW. Using the expressions above for a polarization scrambled interaction (justified by the great fiber length) and a pure-silica core yields a predicted threshold of 3.2 W. This shows that the Raman gain for GeO_2-doped silica is about five times higher than that for pure silica. At 1.25 W pump power, the first Stokes output power reaches 0.08 mW, identical to the pump output power; the Stokes output power increases linearly with the pump power up to 4.8 W pump power, where the second Stokes begins to build up. At 4 W pump power, the pump is totally converted to the first Stokes with an output power of 0.4 mW.

Tunable Raman Oscillator. By placing mirrors at each fiber end, the Stokes emission takes place at a wavelength such that the Stokes pulse round-trip time is a multiple integer of the pump pulse spacing. Using a dispersive fiber allows us to select the emission wavelength, and tunability is obtained by translating one mirror [15]. At the same time, measuring the emission wavelength as a function of the mirror's position allows us to measure the fiber chromatic dispersion.

Such an oscillator has been obtained [15] with a mode-locked YAG laser emitting 200-ps pulses with a repetition rate of 100 MHz and an average power of 1.1 W at 1064 nm. The fiber is 600 m long, with an absolute index difference of 0.7%, a pure-SiO_2 core diameter of 5.3 μm, and a loss of 3 dB/km at 1064 nm. Because of dispersion, the interaction length as determined by Eq. (9.3) is about only 100 m. The power threshold is about 450 mW, and at 700 mW about 60% of the pump is converted to the first Stokes, at the wavelength of maximum emission. Tunability is obtained between 1100 nm and 1125 nm with a line width of about 1.5 nm.

Pulsed Raman Laser. A study of the threshold power as a function of fiber length has been carried out in Ref. 11, with a Q-switched, mode-locked YAG laser emitting 500-ps pulses at 1064 nm. The fiber has an absolute index difference of 0.3%, a SiO_2-GeO_2 core diameter of 8.7 μm, and a loss of 1.65 dB/km at 1064 nm.

As predicted by Eq. (9.1), the threshold power varies almost as 1/L up to L = 200 m, where it becomes stable. With the fiber characteristics from which the chromatic dispersion is deduced (Sec. 4.3.1), and the Stokes shift of about 50 nm at a 1064-nm pump wavelength, Eq. (9.3) indicates an interaction length of about 240 m, which is in reasonable agreement with the experimental figure. When the fiber length is increased above 200 m, the first Stokes threshold is obtained for 8 W of peak pump power, whereas Eq. (9.1) indicates 54 W for pure SiO_2 and a polarized wave. This seems again to indicate a higher Raman gain for GeO_2-doped silica, as expected from Ref. 8. The second Stokes (1180 nm) appears at 20 W of peak pump power and the third Stokes (1240 nm) at 30 W. A very broadband Raman laser has been obtained in Refs. 16 and 17 with a Q-switched, mode-locked YAG laser emitting 300-ps pulses at 1064 nm.

In Ref. 16 the 176-m-long fiber has an absolute index difference of 0.4%, a pure SiO_2 core diameter of 6 μm, and a loss of 4 dB/km at 1064 nm. The first Stokes threshold is 70 W of peak pump power, in very good agreement with the predictions of Eqs. (9.1)–(9.3) for pure silica and a polarized wave. With 1 kW of peak pump power, five Stokes (1120 nm, 1180 nm, 1240 nm, 1300 nm, 1440 nm) are obtained. The frequency spacing between successive Stokes is slightly modulated by the spectral loss of the fiber and by self phase modulation (SPM), which broadens the higher Stokes orders. The component at 1440 nm shows such a broadening that the tall of this peak presents a residual output peak power density of about 7 mW/nm at 1600 nm.

In Ref. 17 the 234-m-long fiber has an absolute index difference of 0.35%, a SiO_2-GeO_2 core diameter of 9.6 μm, and a loss of

1 dB/km at 1064 nm. The threshold for the first Stokes order is obtained at 27 W, which again indicates that the presence of GeO_2 increases the Raman gain [Eqs. (9.1)–(9.3) indicate 84 W for pure silica and a polarized wave]. With 400 W of peak pump power, the higher-order Stokes components yield an output peak power density of about 100 mW/nm at 1700 nm, and the anti-Stokes components yield 10 mW/nm at 800 nm. With 400 W of pump power, the first Stokes appears for a fiber length of 2 m and is maximum for 10 m; the second Stokes appears for 5 m and peaks at 20 m; the third Stokes appears for 10 m and peaks at 40 m.

Fiber Raman Amplifier. Various experiments have shown that a CW or pulsed (Q-switched) Nd:Yag laser with high power could be used as a pump source for amplifying weak optical pulses around 1300 or 1550 nm, through SRS in a fiber [6]. This could be used to boost emitted pulses before transmission through a long fiber line, or as an optical preamplifier before detection of the weak transmitted pulses in fiber communication systems.

9.3 STIMULATED BRILLOUIN SCATTERING

Brillouin scattering is similar to Raman scattering, but the corresponding phonons are "acoustical" instead of "optical," as they are related to lattice vibrations. The Brillouin wavenumber (or frequency) shift is thus much smaller. Also, it should be remembered that SBS always requires phase matching, contrary to SRS for the Stokes components. In bulk materials, simple considerations concerning phase-matching relationships (conservation of energy and of total wave vector) show that the Brillouin frequency shift (acoustic phonon frequency) ν_b can be written

$$\nu_b = \frac{2n}{\lambda} V_a \left(\frac{1 - \cos\theta}{2}\right)^{\frac{1}{2}} \tag{9.4}$$

where V_a is the acoustic velocity, n the material refractive index, λ the light wavelength in vacuum, and θ the angle between the scattered wavevector and the initial pump wavevector. In single-mode fibers, we can consider only $\theta = 0$ or $\theta = \pi$, the latter corresponding to a nonzero frequency shift, contrary to the former. Thus interesting phase matching can be obtained only with counterpropagating waves. Whereas the SRS Stokes waves are usually emitted by the output end of the fiber, SBS Stokes waves are emitted by the fiber end where the pump is launched. A complete review of SBS in single-mode fibers can be found in Ref. 18.

9.3.1 Basic Data

The Brillouin–Stokes frequency shift of silica is normally $-17/\lambda$ (GHz if λ is in micrometers) [19], but can be reduced by finite geometry effects of the fiber structure [20]. At the same time, the Brillouin linewidth at a wavelength λ (expressed in micrometers) is [21]

$$F_b = \frac{29.6}{\lambda^2} \quad \text{MHz} \tag{9.5}$$

The Brillouin gain G_b is theoretically 5.8×10^{-11} m/W, but has been experimentally found to be 4.3×10^{-11} [21]. Note that it is almost independent of the wavelength and three orders of magnitude stronger than the SRS gain.

The pump power threshold for the first Stokes–Brillouin component is given by [10]

$$P_t = \frac{21\pi w_0^2 \alpha}{G_b[1 - \exp(-\alpha L)]} \times \begin{cases} 1 \\ F_p/F_b \end{cases} \tag{9.6}$$

The first line in Eq. (9.6) holds if the pump line width is smaller than the Brillouin line width, whereas the second line holds if the pump spectral width F_p (in terms of frequency) is larger. Note that because of the backward nature of the Brillouin scattering, this power should be understood as a CW (or an average if periodic pulses are used) power. Polarization scrambling can again increase this value by as much as 100%.

9.3.2 Applications

SBS has been used for obtaining CW lasers with various cavity mirrors configurations. A first application with fiber ring resonators is described in Ref. 22, and a detailed study of this type of laser can be found in Ref. 23.

In Ref. 22 the pump is an argon-ion laser emitting at 514.5 nm with a line width of 25 MHz, and the fiber has an absolute index difference of 0.5%, a core diameter of 2.4 μm, a length of 9.5 m, and a loss of about 100 dB/km. The mirrors have a reflectivity of 50% and 4%, respectively. With no feedback, the threshold is about 500 mW pump power, which is in good agreement with the prediction of Eqs. (9.5) and (9.6), and when the mirrors are used the threshold decreases to about 250 mW. With 750 mW of pump power, the Brillouin–Stokes power increased to 9 mW. Another possible configuration consists in using a Fabry–Perot cavity with the pump laser's output mirror as one of the cavity mirrors and a high-reflectivity mirror for the other [24]. This made it possible to obtain 10 cascaded Brillouin–Stokes orders and weaker anti-Stokes orders, with 3 W of pump power and a 20-m-long fiber similar to that of Ref. 22.

9.4 SELF PHASE MODULATION

This effect arises from the dependence of the medium refractive index upon the field intensity (optical Kerr effect), resulting in a different transmission phase for the peak of a light pulse as compared to the trailing and leading edges of the pulse. This results in a modification of the pulse spectrum, occurring only for amplitude-modulated coherent sources such as mode-locked lasers or intensity-modulated single-longitudinal-mode lasers. Thus, contrary to the previous effects, SPM does not exist with constant-amplitude waves, and in addition to that, no power threshold exists for it.

9.4.1 Basic Data

For a carrier wave with optical pulsation ω and time-dependent power profile $P(t)$, the light spectrum emerging from the fiber is [25]

$$F(\Omega) = \frac{1}{2\pi}\int_{-\infty}^{+\infty} [P(t)]^{\frac{1}{2}} \exp[i\phi(t)] \exp[-i(\Omega - \omega)t]\, dt \qquad (9.7)$$

where $\phi(t)$ is given by

$$\phi(t) = 2\pi \frac{1 - e^{-\alpha L}}{\alpha L} Z_0 \frac{\delta n}{n} \frac{P(t)}{\pi w_0^2} \qquad (9.8)$$

where L is the fiber length, α the attenuation per unit length, λ the light wavelength, Z_0 the vacuum impedance, δn the optical Kerr effect constant, n the refractive index, and w_0 the mode field radius. The frequency deviation at time t is deduced from Eq. (9.8) by taking the derivative of $-\phi(t)$ with respect to t and dividing by 2π. Experimentally, a value of 1.27×10^{-22} $(m/V)^2$ has been found for δn [25].

9.4.2 Applications

In Ref. 25 a mode-locked argon laser emitting at 514.5 nm is used in conjunction with a 99-m-long fiber with a cutoff wavelength of 542 nm, a core diameter of 3.4 μm, and a loss of 17 dB/km. With a pulse input optical spectral width of 4.6 GHz (at 1/e) and 5 kW of peak pump power, the output optical spectral width is increased to 50 GHz, with an unchanged pulse shape.

In Ref. 26 SPM is used for frequency shifting. Equation (9.8) shows that with an asymmetric pulse, the output spectrum will be frequency-shifted with respect to the input spectrum. This has been achieved with a YAG laser emitting a 25-ps pulse at 1064 nm, whose shape is controlled with a diffraction grating and a transmittance mask of appropriate shape in front of it. With triangular pulses and several kilowatts of peak power, frequency shifts typically of 20 GHz are obtained after 10 m of fiber.

Soliton Propagation. If the chromatic dispersion of the fiber (as defined in Chap. 4) is positive, there exist critical pulse shapes and powers for which the pulse shape remains unchanged along the fiber (first-order soliton) or oscillates (higher-order solitons) [27]. This is due to the fact that, as seen in Eq. (9.8), SPM downshifts the optical frequency (upshifts the wavelength) in the leading half of the pulse, and if the chromatic dispersion is positive, this results in an increased group delay for the leading edge with respect to the central part, and thus a possible pulse narrowing. As the symmetric effect holds for the trailing half of the pulse, and as the frequency shift depends on the pulse shape and peak power, it can be understood that critical conditions may lead to soliton propagation.

For obtaining the Nth-order soliton at a wavelength λ with a chromatic dispersion C, the field pulse envelope launched into the fiber should satisfy [28]

$$E(t) \simeq \left(\frac{C}{\delta n}\,\frac{\lambda^3}{2\pi^2 c}\right)^{\frac{1}{2}} \frac{N}{t_0}\,\mathrm{sech}\left(\frac{t}{t_0}\right) \tag{9.9}$$

This has been checked experimentally [28] with a color center laser emitting 7-ps pulses at 1550 nm, and a 700-m-long fiber with a core diameter of 9.3 μm and a chromatic dispersion of 16 ps/nm × km. At low power levels, the transmitted pulse is broadened as expected from the chromatic dispersion, but when the input power increases, it narrows down to 2 ps for 5 W of input power. The first-order soliton (output pulse width identical to the input pulse width) is obtained for 1.2 W of input power, whereas the second-order soliton appears for 5 W, the third-order for 11.4 W, and the fourth order for 22.5 W, in reasonable agreement with Eq. (9.9). Another experiment combining Raman amplification and soliton propagation yielded unaffected pulse transmission over 10 km of fiber [29].

9.5 POWER LIMITATIONS

If we look back at all the effects noted above, we observe that SRS and SBS can limit the launched power in telecommunication applications of single-mode fibers. Above the power threshold for SRS, the apparent loss of the carrier wave increases because part of its power is converted to the first Stokes wave. As noticed in Ref. 14, above the threshold, the carrier transmitted power continues to increase up to a maximum value and the corresponding input power has been regarded as the maximum power that the fiber can accept. However, owing to the large wavelength shift associated with SRS in the region 1300 to 1500 nm (tpyically 80 nm), any presence of the first Stokes order even at low power levels will lead to an unacceptable pulse broadening even with first-order dispersion-free fibers. We will thus retain hereafter the threshold for first SRS Stokes as the maximum power the fiber can tolerate with respect to this effect.

On the other hand, SBS yields a much smaller frequency shift (12 GHz, corresponding to about 0.08 nm in the region 1300 to 1500 nm), which should be disturbing almost only for coherent detection systems, but we will consider that the power threshold for the first-order Stokes SBS is always the maximum tolerable power, although with direct detection a power twice as high could probably be accepted. Note additionally that the effect of pulse broadening expected from the Stokes SBS wave will be very weak, because only the part reflected by the input fiber end (4%), or backscattered in the forward direction, will actually be detected (SBS being a backward phenomenon).

At last, SPM will not be considered hereafter as a limiting factor, as most of the fibers and wavelengths envisaged for telecommunication applications correspond to a zero or a positive chromatic dispersion, thus yielding a pulse narrowing. Only higher-order solitons could be disturbing, but it can be checked that the required power levels are higher than those corresponding to SRS and SBS for bit rates less than 2 Gb/s.

9.5.1 Multi-Longitudinal-Mode Laser

The comparison of Eqs. (9.5) and (9.6) for SBS with (9.1) and (9.2) for SRS shows that, owing to the very large spectral width of such sources (typically, several nanometers of several 100 GHz in the region 1300 to 1500 nm), SRS is by far the most limiting factor in this case. For computing the power threshold we thus use Eq. (9.1), but as it appears that GeO_2-doped fibers exhibit a gain about five times higher than that of pure silica, we use for G_r

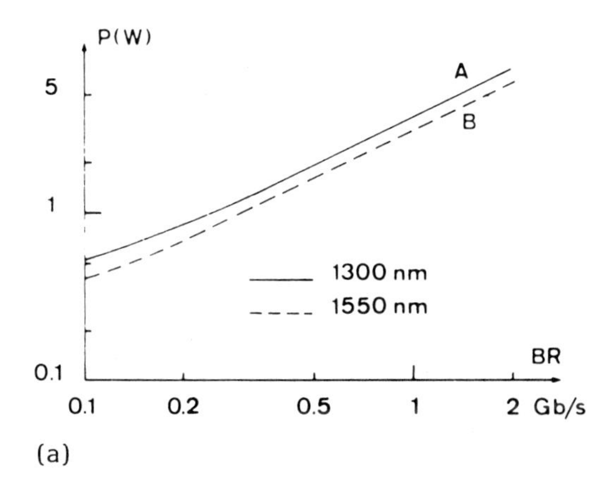

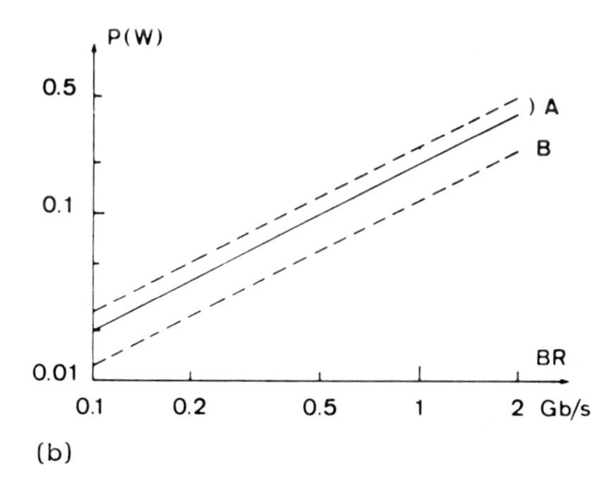

FIGURE 9.1 Maximum launching power in fiber types A and B as a function of bit rate BR (see Table 7.1 for fiber characteristics). (a) Multi-longitudinal-mode laser. (b) Single-longitudinal-mode laser and coherent systems.

$$G_r \simeq \frac{4.9 \times 10^{-10}}{\lambda \text{ (nm)}} \quad \text{m/W} \tag{9.10}$$

As these sources require dispersion-free fibers, we consider fiber type A of Sec. 7.2.2 at 1300 nm and fiber type B at 1550 nm for computing w_0. The loss α is taken as the minimum loss exhibited by these fibers among various system configurations (see Chap. 7). Finally, L_e is computed from Eq. (9.3) with $\Delta t = 1/(1.7 \times \text{bit rate})$, and from the second-order dispersion of these fibers (see Chap. 4). Figure 9.1 shows the corresponding power limitations as a function of the bit rate, and it is seen that the limiting powers exceed by far the figures that can be obtained with semiconductor lasers, but fall within the capability of 1320-nm YAG lasers.

9.5.2 Single-Longitudinal-Mode Laser

Comparing Eqs. (9.5) and (9.6) with Eqs. (9.1) and (9.10) shows that SBS becomes dominant if the source line width is less than about 1.5 GHz for CW operation, and even more for higher bit rates. As single-longitudinal-mode lasers have line widths of 100 MHz or even less, we have to deal here with SBS limitations. The maximum power is obtained from Eqs. (9.5) and (9.6), with F_p = bit rate, for fibers A at 1300 nm and 1550 nm and fiber B at 1550 nm (see Sec. 7.2.2). The results are shown in Fig. 9.1b and agree well with experimental results [30]. It is seen that for relatively low bit rates, these power limitations could be comparable to semiconductor laser outputs, especially if power optical amplifiers are used.

One may thus raise the question whether a 1320-nm YAG laser with a relatively large line width (thus yielding SRS limitations) would be more interesting than a 1550-nm SLM semiconductor laser if both are assumed to deliver power corresponding to their respective limitations. As seen from Fig. 9,1, the power allowed by SRS exceeds that allowed by SBS by about 10 dB, which corresponds to 40 km of fiber with a loss difference of 0.25 dB/km between 1300 nm and 1550 nm (see Chap. 7). It thus does not appear in these ultimate conditions that 1320-nm YAG lasers would provide any link-length increase.

It has also been shown that the modulation format could significantly influence the power limitations of SBS [18], since PSK raises the SBS threshold power by as much as 20 dB in some cases [18].

REFERENCES

1. R. H. Stolen, E. P. Ippen, and A. R. Tynes, *Appl. Phys. Lett.* *20*(2):62–64 (1972).

2. R. H. Stolen, *IEEE J. Quantum Electron.* *QE-11*(3):100–103 (1975).
3. J. Stone, *Appl. Phys. Lett.* *26*(4):163–165 (1975).
4. J. Stone, A. R. Chraplyvy, and C. A. Burrus, in *CLEO'82 Digest of Technical Papers*, Optical Society of America, Washington, D.C., Apr. 1982, post-deadline paper ThB5.
5. C. Lin, W. A. Reed, A. D. Pearson, and H. T. Shang, *Opt. Lett.* *6*:493–495 (1981).
6. C. Lin, *IEEE J. Lightwave Tech.* *LT-4*(8):1103–1115 (1986).
7. A. Saissy, J. Botineau, A. Azema, and F. Gires, *Appl. Opt.* *19*(10):1639–1646 (1980).
8. F. L. Galeener, J. C. Mikkelsen, Jr., R. H. Geils, and W. J. Mosby, *Appl. Phys. Lett.* *32*(1):34–36 (1978).
9. V. V. Grigoryants, B. L. Davydov, M. E. Zhabotinski, V. F. Zolin, G. A. Ivanov, V. I. Smirnov, and Yu. K. Chamorovski, *Opt. Quantum Electron.* *9*:351–352 (1977).
10. R. G. Smith, *Appl. Opt.* *11*:2489–2494 (1972).
11. Y. Ohmori, Y. Sasaki, and T. Edahiro, *Electron Lett.* *17*(17):593–594 (1981).
12. R. H. Stolen, *IEEE J. Quantum Electron.* *QE-15*(10):1157–1160 (1979).
13. C. Lin, *J. Opt. Commun.* *4*(1):2–9 (1983).
14. Y. Sasaki, Y. Ohmori, M. Kawachi, and T. Edahiro, *Electron. Lett.* *17*(8):315–316 (1981).
15. C. Lin, R. H. Stolen, and L. G. Cohen, *Appl. Phys. Lett.* *31*(2):97–99 (1977).
16. L. G. Cohen and C. Lin, *IEEE J. Quantum Electron.* *QE-14*(11):855–859 (1978).
17. K. I. Kitayama, Y. Kato, S. Seikai, and M. Tateda, *Appl. Opt.* *20*(14):2428–2432 (1981).
18. D. Cotter, *J. Opt. Commun.* *4*(1):10–19 (1983).
19. P. Labudde, P. Anliker, and H. P. Weber, *Opt. Commun.* *32*(3):385–390 (1980).
20. N. L. Rowell, P. J. Thomas, H. M. Van Driel, and G. I. Stegeman, *Appl. Phys. Lett.* *34*(2):139–141 (1979).
21. E. P. Ippen and R. H. Stolen, *Appl. Phys. Lett.* *21*(11):539–541 (1972).
22. K. O. Hill, B. S. Kawasaki, and D. C. Johnson, *Appl. Phys. Lett.* *28*(10):608–609 (1976).
23. D. R. Ponikvar and S. Ezekiel, *Opt. Lett.* *6*(8):398–400 (1981).
24. K. O. Hill, D. C. Johnson, and B. S. Kawasaki, *Appl. Phys. Lett.* *29*(3):185–187 (1976).
25. R. H. Stolen and C. Lin, *Phys. Rev. A* *17*(4):1448–1453 (1978).
26. M. Vampouille and J. Marty, *Opt. Quantum Electron.* *13*:393–400 (1981).
27. A. Hasegawa and F. Tappert, *Appl. Phys. Lett.* *23*(3):142–144 (1973).

28. L. F. Mollenauer, R. H. Stolen, and J. P. Gordon, *Phys. Rev. Lett.* *45*(13):1095–1098 (1980).
29. L. F. Mollenauer, R. H. Stolen, and M. N. Islam, *Opt. Lett.* *10*:229–231 (1985).
30. N. Uesugi, M. Ikeda, and Y. Sasaki, *Electron. Lett.* *17*(11): 379–380 (1981).

List of Symbols

Only notation that appears in more than one chapter is listed here.

a	Core radius
a'	Inner cladding radius
a_e	Core radius of the equivalent step-index fiber
A	Fiber outside radius
b	Normalized propagation constant of the LP_{01} mode
b_e	Normalized propagation constant of the LP_{01} mode in the ESI
B	Normalized birefringence
c	Light velocity in vacuum
$C(\lambda)$	Chromatic dispersion
$C'(\lambda)$	Second-order chromatic dispersion $[dC(\lambda)/d\lambda]$
d	Fiber core axes lateral offset at a joint
DIC	Depressed inner cladding

E	Electrical field of the optical wave
ESI	Equivalent step-index fiber
H	Magnetic field
HE_{11}	Fundamental mode (or LP_{01})
i	$(-1)^{\frac{1}{2}}$
$J_n(\cdot)$	Bessel function of order n
k	Free-space propagation constant of the light
$K_n(\cdot)$	Modified Bessel function of order n
L	Fiber length
L_b	Birefringence beat length
LP_{01}	Fundamental mode (or HE_{11})
LP_{11}	Second-order mode
n	Average refractive index
n_1	Maximum core refractive index
n_2	Inner cladding refractive index
n_3	Outer cladding refractive index
N	Number of turns per meter in a twisted fiber
N_j	Group refractive index of medium j
p	Cross-polarization
r	Radial distance from fiber axis
rms	Root mean square
R	Riber radius of curvature
S	Rms pulse width
SOP	State of polarization

t	Time
T	Birefringence dispersion
T-DIC	Thick depressed inner cladding
V	Normalized frequency
V_c	Normalized frequency at cutoff of the LP_{11} mode
V_e	Normalized frequency in the ESF
w_0	Mode field radius of the LP_{01} mode
W	Source rms spectral width in terms of pulsation
W-type	Narrow depressed inner cladding
z	Abscissa along fiber axis
Z_0	Vacuum impedance
α	Fiber loss per unit length
α_c	Curvature loss per unit length
α_{mm}	Microbending loss of the reference multimode fiber
α_s	Rayleigh scattering loss per unit length
α_{sm}	Microbending loss of single-mode fiber
α_t	Total loss per unit length of a cabled and spliced fiber link
β	Average propagation constant of the LP_{01} modes (two polarizations)
β_e	Average propagation constant in the ESI
δ	$\Delta'n/\Delta n$
$\delta\beta$	Birefringence
$\delta\beta_\ell$	Linear birefringence

$\delta\beta_c$ Circular birefringence

Δ Core–inner cladding relative index difference

Δ' Inner–outer cladding relative index difference

Δ_e Relative index difference of the ESI

Δn Core–inner cladding index difference

$\Delta' n$ Inner–outer cladding index difference

λ Light wavelength in vacuum

λ_c Second-order mode cutoff wavelength

λ_D Zero chromatic dispersion wavelength

ν Poisson's ratio

τ Group delay time of the LP_{01} mode, per unit length

φ Azimuth in cylindrical coordinates

ϕ Phase of the transmitted field

ω Optical pulsation

Ω Optical amplitude-modulation pulsation

Index

h may be kept